Bridge Hydraulics

Bridge Hydraulics

Les Hamill

School of Civil and Structural Engineering
University of Plymouth

CRC Press is an imprint of the
Taylor & Francis Group, an **informa** business
A SPON PRESS BOOK

CRC Press
Taylor & Francis Group
6000 Broken Sound Parkway NW, Suite 300
Boca Raton, FL 33487-2742

First issued in paperback 2019

Typeset in Sabon by
J&L Composition Ltd, Filey, North Yorkshire

ISBN-13: 978-0-419-20570-8 (hbk)
ISBN-13: 978-0-367-44763-2 (pbk)

British Library Cataloguing in Publication Data
A catalogue record for this book is available from the British Library

Library of Congress Cataloguing in Publication Data
A catalog record for this book has been requested

Visit the Taylor & Francis Web site at
http://www.taylorandfrancis.com

and the CRC Press Web site at
http://www.crcpress.com

Contents

Preface

This is intended as a useful handbook on the subject of bridge hydraulics. It includes references to articles published in 1997, just prior to its completion, so compared with similar books it is relatively up to date. It explores how to undertake the hydraulic analysis or design of a bridge, either single or multispan, with either rectangular or arched waterways. It describes how to calculate the afflux (backwater), how to improve the hydraulic performance of a bridge, and how to evaluate and combat scour. The intention is to provide a good introduction to the fundamentals for anyone not familiar with this specalised branch of engineering, with enough detailed information to appeal to those who are.

This book is, in a way, the result of a mistake. Near my home town many years ago a rather old, untidy, steel truss bridge was replaced by a very elegant masonry structure. The result was that flooding upstream got worse. This raised the question: how is the size of the opening in a river bridge determined? Initial enquiries revealed that estimating the magnitude of the design flood was relatively straightforward; it was converting this into the dimensions of a bridge opening that was difficult. An expert on the subject candidly and charmingly admitted that there was much that he (and practically everyone else) did not know or understand, so if anyone cared to fill in a few gaps . . . Hence my research interest and the book. Another reason is that bridges are interesting: many people stand on a bridge watching the floodwater pass underneath. Hopefully some of this interest is captured in the following pages.

Some engineers may question why a book on bridge hydraulics is needed when it is possible to find computer software that will do all the analysis and design for you. Such people frequently believe, because computers are capable of giving answers to 20 decimal places, that everything that comes out of them is correct and accurate. This is not true. Ignoring the fact that the input data may be inaccurate, there may be mistakes in the computer program. A sobering thought is that someone once said that if a piece of software is worth using then it must have an error in it somewhere!

Many years ago the author was invited to use the research facilities of a large, prestigious company. Part of the work involved digitising some complex shapes to determine their area. To provide a check on the accuracy obtained a square was also digitised each time. It was found that the calculated areas of the squares were in error by a very considerable margin. When this was pointed out to the company they held a hasty conference and came to the conclusion that a square was too simple for the complex software to be able to handle! They subsequently modified the software.

No-one would deny that computers have a fundamental role to play in modern engineering, but sometimes the basic science is insufficiently understood or too complex to be represented accurately by the software. Sometimes, as in the case of scour, there is not enough reliable field data to verify the base equations or computer models under all conditions. Nature does not realise that it must always act in strict accordance with human rules! For all of these reasons, there are times when physical modelling is strongly recommended, particularly when important or unusual projects are involved. Similarly, common sense, experience and engineering judgement are needed. The smart engineer will make a few check calculations without using the computer, just to ensure that the answer is of the correct order of magnitude and makes sense. Similarly, smart engineers will ensure that they understand the basic principles involved, because it may not be possible to obtain the optimum design otherwise. In this respect little has changed over the years, as the saga of Mr Nagler and Mr Goodrich illustrates.

Following Nagler's paper of 1918 on the 'Obstruction of bridge piers to the flow of water' there is a nice account of how this article was used by a Mr Goodrich to calculate the backwater from a proposed development in the USA. The value turned out to be 3 inches (75 mm), which was unacceptable, so the city attorneys applied for a restraining order to prevent further construction. However, following additional field measurements and a review of the computations Mr Goodrich obtained a negative backwater. At the final hearing another, well-experienced hydraulic engineer showed that about 1 inch (25 mm) of backwater could be expected. Mr Goodrich wrote that

> The explanation to the Court of the disappearance of the other 2 inches of backwater is not anticipated with any great pleasure, but it will be easier than to tell how the water is piled up higher below the bridge than above it.

He also considered that he was lucky to have discovered the limitations of the Nagler equation before the final hearing so that a much more embarrassing situation had been avoided. Later in the discussion Nagler pointed out that he had given several cautions regarding the general applicability of his work and stated that

> Engineers are too prone to select empirical formulas and coefficients from handbooks and apply them to entirely irrelevant cases, never

> inquiring as to the natural limitations on the applicability which intelligent use would place on them. Intelligent extension of experimental formulas and coefficients to practical problems is the highest type of engineering, but the blind application of formulas smacks of student days.

The modern parallel to the Goodrich and Nagler saga is over-reliance on computers, which has resulted in some notable 'failures'. A multistorey carpark that developed significant cracks shortly after opening springs to mind. This type of situation has been termed computer-aided disaster (CAD). More than once a design has had to be hastily modified at the last minute simply because an updated version of the software arrived and this yielded a significantly different answer from the same input data. Engineers have been encountered using hydraulic software to analyse and design bridges without having any idea of what it was doing or what it was based on.

Because there have been relatively few in-depth investigations of bridge hydraulics, the equations and research referred to in this book will also have been incorporated to a greater or lesser extent into the commercially available software. Therefore this book and the software may be considered complementary, and a possible use for the book may be to help explain the fundamentals and to provide a means of checking the output from the software. However, bearing in mind Nagler's comments, it is still up to the engineer to use it wisely.

Metric units have been used throughout. Where necessary, charts, tables and equations have been converted from English units.

Acknowledgements

The author wishes to thank all those who have contributed in any way to the preparation of this book. This includes everyone who helped supply information or photographs. Every effort has been made to obtain copyright permissions and to include acknowledgements where necessary. Any omissions notified will be rectified at the earliest opportunity.

Thanks are also due to the many people who have contributed to the author's own research over the years, including the staff of the former South West Water such as Bob Hutchings and Alan Rafelt, the County Bridge Engineers of Devon, Cornwall and Somerset, staff at the Environment Agency such as Tim Wood, and former colleagues such as Graham McInally .

Principal notation

A numerical subscript attached to a symbol usually indicates the location of the cross-section, or part of a cross-section, or the reach of a river according to context.

The bridge waterway opening may be referred to as the opening or the waterway.

The river is always referred to as the river or the channel.

a	Cross-sectional area of flow in a (part full) bridge waterway opening (m^2).
a_{MT}	Net area of bridge openings between the bed and the midtide level in an estuary (m^2).
a_W	Total cross-sectional area of a waterway opening when flowing full (m^2).
A	Total cross-sectional area of flow in a river channel (m^2).
A_C	Net area of flow between the channel bed and the critical depth line (m_2).
A_{MT}	Gross area of the channel between the bed and the midtide level in an estuary (m^2).
A_N	Cross-sectional area of flow between the channel bed and normal depth line (m^2).
A_P	Cross-sectional area of the submerged part of the piers (m^2).
b	Net width (i.e. excluding pier width) of bridge opening at bed level at 90° to flow (m).
b_P	Width of an individual bridge pier measured at 90° to the flow direction (m).
b_S	Width between abutments of a skewed bridge, measured along the highway centreline (m).
b_T	Top width of free water surface in a bridge opening (m).
B	Width of river channel (m).
B_R	Regime (Lacey) width of an alluvial channel measured at 90° to the banks (m).
B_T	Top width of water surface between the river banks (m).
C, C_d, C_D	Coefficient of discharge (dimensionless).

C'	USGS method base coefficient of discharge (dimensionless).
C_C	Coefficient of contraction (dimensionless).
C_F, C_S	Coefficients for free and submerged flow over a highway embankment (dimensionless).
d_S	Total scour depth as a result of contraction, piers, abutments and degradation (m).
d_{SA}	Depth of scour at abutment (m).
d_{SC}	Depth of scour in a contraction or bridge opening (m).
d_{SL}	Depth of local scour at piers and abutments (m).
d_{SP}	Depth of scour at a pier (m).
D	Diameter of the (uniform) material comprising a river bed, riprap etc. (m).
D_b	USBPR method differential ratio to calculate the fall in water level across embankments.
D_M	Effective mean diameter (m) = $1.25D_{50}$ in Chapter 8.
D_{50}	Median diameter at which 50% of material by weight is smaller than the size denoted (m).
e	Eccentricity (numerical ratio of abutment lengths, or conveyances or discharges).
E	Total energy (m).
E_S	Specific energy, i.e. energy calculated above bed level = $Y + V^2/2g$ (m).
E_{SC}	Critical specific energy (m) i.e. specific energy when the flow is at critical depth.
f_S, f	(Lacey's) silt factor for a sediment of diameter D.
F	Froude number.
F_M, F_A	Mean/average Froude number calculated from mean/average depth on floodplain (Chapter 8).
F_N	Froude number with normal depth flow (= F_4, dimensionless).
g	The acceleration due to gravity (9.81 m/s^2).
h	Height of water surface above the centre of curvature of an arch (m).
h_C	Average depth of flow along a constricted tidal estuary at the midtide level (m).
h_F	Head loss due to friction (m).
H	Elevation of water surface above a datum level (m).
H_b^*	USBPR method bridge afflux (m) without adjustment for piers, skew, or eccentricity.
H_D^*	USBPR method afflux at a dual bridge (m).
H_1	Elevation above datum of water surface (with bridge) at section 1.
H_{1A}	Elevation above datum of water surface (no bridge) at section 1 with an abnormal stage (m).
H_1^*	Maximum afflux at cross-section 1 with normal depth = $Y_1 - Y_N$ (m).
H_{1A}^*	Maximum afflux at cross-section 1 with an abnormal stage (m).

H_3^*	Distance of the water surface below the normal depth line at section 3 (m).
J	Proportion of bridge waterway blocked by piers or piles, or blockage ratio (HR method).
k	USGS method adjustment factors (various subscripts) to base coefficient of discharge.
k^*	USBPR method total backwater coefficient (dimensionless).
k_C^*	USBPR method total critical depth backwater coefficient (dimensionless).
K	Total conveyance of river channel (m^3/s).
K_b	Conveyance of the part of the approach channel equivalent to the bridge opening (m^3/s).
K, K_A, K_N	Yarnell, d'Aubuisson and Nagler coefficients for flow past piers (Chapter 5).
$K_{A\Phi}$	Abutment scour adjustment factor for angle of attack (equation 8.18).
K_{1A}, K_{2A}	Abutment scour adjustment factors for shape and angle of attack (equation 8.17).
K_{1P}, K_{2P}, K_{3P}	Pier scour adjustment factors for nose shape, angle of attack and bed form (equation 8.14).
K_σ	Scour adjustment factor to allow for grading of bed material.
L	Length of bridge waterway in the direction of flow (m), or reach length with subscripts.
L_A	Length of bridge abutments or embankments normal to the flow (m).
L_C	Length of a constricted tidal estuary (m).
L_E	Length of the bridge road embankment when overtopped, as for a weir (m).
L_S	Length of spur dyke in the direction of flow (m).
M	Bridge opening ratio = q/Q or a/A or b/B or K_b/K (dimensionless).
M_L	Limiting opening ratio (dimensionless) at which the flow is at critical depth.
n	Manning's roughness coefficient ($s/m^{1/3}$).
P	Wetted perimeter of a channel (m).
q	Quantity of flow that can pass through the bridge opening unimpeded (m^3/s).
q	Discharge per metre width in Chapter 7 (m^3/s per m or m^2/s).
q_C	Discharge per metre width at the critical depth in Chapter 7 (m^2/s).
Q	Total discharge (m^3/s).
Q_b	Discharge in the part of the approach channel corresponding to the bridge opening (m^3/s).
Q_F	Nominal discharge capacity of a waterway running full in Chapter 7 (m^3/s).

Q_{MT}	Maximum tidal discharge in an estuary at midtide level (m^3/s).
Q_{MAX}	Maximum total discharge in a tidal estuary, including any river flow (m^3/s).
Q_{100}	Discharge corresponding to the 1 in 100 year flood (m^3/s).
r	Radius of curvature of an arch, or radius of entrance rounding to waterway (m).
R	Hydraulic radius of channel (= A/P m).
R_S	Regime scoured depth of flow (m) corresponding to channel width B_R.
s_S	Specific gravity (relative density) of sediment, bed material, stone or riprap (dimensionless).
S_F	Longitudinal slope of total energy line (dimensionless).
S_O	Longitudinal slope of river bed (dimensionless).
$S_C{}^*$	USBPR method afflux scour correction factor (dimensionless).
T	USGS method, height of water surface above bottom chord of bridge deck (m).
T	Tidal period between successive high or low water levels (hours/seconds).
U^*	Shear velocity = $(gYS_F)^{1/2}$ (m/s).
U_S^*	Scour-critical shear velocity at which bed movement occurs.
V	Mean flow velocity (m/s).
V_C	Critical velocity (m/s), velocity when $F = 1.0$
V_E	Threshold velocity at which erosion of bed material starts (m/s).
V_{MAX}	Local maximum velocity in the bridge openings at midtide level in an estuary (m/s).
$V_{MEANMAX}$	Mean maximum velocity in an estuary or tidal inlet with normal spring tides (m/s)
V_{MT}	Average maximum velocity in the bridge openings at midtide level in an estuary (m/s).
V_N	Mean velocity when flow in a river channel is at normal depth (m).
VOL	Tidal volume of an estuary calculated between low and high tidal levels (m^3).
V_S	Scour-critical velocity needed to move bed material and start live-bed scour (m/s).
V_{SC}	Neill's competent velocity at which the flow can just move the bed material (m/s).
V_u	Mean upstream approach velocity at either section 1 or 2 (m/s).
V_V	Velocity in the voids of riprap (m/s).
V_{2A}	Average velocity at section 2, in the opening, at the abnormal stage that would exist without the bridge (m/s).
w	Width of a chamfer on the entrance to a waterway (m).
w	Median fall velocity of a particle in water in Chapter 8 (m/s).
W	Channel width near a minimum energy opening measured along a curved orthogonal (m).

W_{SP} Width of pier scour hole (m).

X Length approach embankment/abutments (m) for calculation of eccentricity.

Y Depth of flow measured from the bed (m).

Y^* Depth of flow required for a waterway opening to become permanently drowned (m).

Y_C Critical depth (m), corresponding to critical flow ($F = 1.0$) at minimum specific energy.

Y_d Downstream depth measured above mean bed level on the channel centreline (m).

Y_M, Y_A Mean depth, average depth (m). Numerical subscript indicates location of cross-section.

Y_{MT} Average depth at midtide level in an estuary (m).

Y_N Normal depth (m), e.g. as with uniform flow and predicted by the Manning equation.

Y_S Water depth above the springings of an arch (m), i.e. above where the arch starts.

Y_S Scoured regime depth of flow (m) in a natural channel constriction (Chapter 8).

Y_{SN} Normal regime scoured depth of flow in a bridge opening (m).

Y_{SMAX} Maximum regime scoured depth of flow in a bridge opening (m).

Y_u Upstream mean depth, the larger of the depths at sections 1 and 2 (m).

Y_1 Depth at section 1 (including the afflux) upstream of the bridge (m).

Y_{1A} Depth at section 1 without the bridge when abnormal stage exists (m).

Z Vertical height of bridge opening (to the top of an arch) from mean bed level (m).

ΔA_S Increase in cross-sectional area of flow caused by scour (m^2).

Δd Degradation depth or long-term reduction in bed level (m).

Δh Difference in elevation of water surface between sections 1 and 3 (m).

ΔH Differential head (m) across the bridge $= Y_u + a_u V_u^2/2g - Y_d$.

Δk^* USBPR method incremental coefficients (various) for afflux (dimensionless).

$\Delta Y_u/Z$ Reduction in proportional depth by entrance rounding (Chapter 7).

$\Delta Q/Q_F$ Increase in proportional discharge by entrance rounding (Chapter 7).

a Velocity head coefficient (dimensionless).

ε Proportion of energy recovered between two cross-sections (dimensionless).

ν Kinematic viscosity (m^2/s).

ρ Mass density (kg/m^3). Subscript s indicates bed sediment or stone riprap.

ϕ Angle of skew, angle of bridge embankments or piers to the approach flow.

Φ Angle of bridge approach/abutments relative to flow (equation 8.17).

1 Putting things into perspective

1.1 Why study bridge hydraulics?

Many people, indeed many engineers, who are not familiar with the subject imagine that constructing a bridge across a river is entirely a problem in structural engineering. They assume that the bridge opening can be made so large that it will completely span the river at such a height that floodwater will never rise anywhere near the deck. If this was always true there would be little need to study bridge hydraulics, but in reality things are rarely this simple. Economics often dictate the length of span and therefore how many piers have to be located in the river. Similarly economics, the geography of the site or the nature of the crossing (such as a railway line with a fixed vertical profile as in Fig. 1.1) may impose some restriction on the maximum permissible elevation of the deck. Consequently flood levels may rise to deck height or above. What initially appeared to be an elementary problem turns out to be quite complicated.

So why study bridge hydraulics? Four answers quickly spring to mind.

- Nobody can be allowed to build a new bridge that has piers and/or abutments in a river without first being able to prove by calculation or modelling that the resulting backwater will not cause, or significantly exacerbate, flooding of land and property upstream. This is becoming increasingly important as the demand for building land leads to construction on river floodplains that, by definition, are already prone to flooding.
- At locations where there is an existing bridge and significant flooding, an analysis may be required to determine how much of the flooding is caused by the bridge and how much by other factors – such as simply too much floodwater to be carried within the river channel. If the analysis shows the bridge to be at fault, then this may be sufficient justification to construct a new structure.
- If it is known that a bridge provides a significant obstacle to flow and is responsible for much of the flooding that occurs, with a knowledge of bridge hydraulics it may be possible to design improvement works that will help to alleviate the problem.

Fig. 1.1 Exwick rail bridge; part of the flood relief channel at Exeter, Devon. Note the rounded soffit and abutments, and elongated piers. A physical hydraulic model study of the scheme was undertaken, and this is to be recommended whenever novel designs or complex flows are involved.

- In addition to the nature and geometry of the river channel, the shape, spacing and orientation of the bridge piers and abutments will affect the flow through a bridge and the likelihood of scouring of the bed. Well designed bridges are not immune to this problem, while bridges that are badly designed hydraulically are even more likely to fail and collapse. Section 1.3 illustrates that this can have tragic consequences. How can a good design be obtained without a knowledge of bridge hydraulics?

This book is concerned with the hydraulic analysis and design of bridge waterway openings. It covers both single and multispan bridges with either rectangular or arched waterways. It is based on a century or more of study and research by investigators all over the world, most notably in North America. It can be used to estimate how a bridge, existing or proposed, will alter water levels at a site at any particular discharge. Conversely, it can be used to calculate the discharge from the observed water levels. It can be employed to analyse and design modern bridges with relatively long rectangular spans, or to investigate the hydraulic performance of the old masonry arch structures that are still numerous in many parts of the world. Other topics covered include how to improve the flow through a bridge, and how to evaluate and alleviate the problem of scouring of the foundations.

It is a tribute to past generations of engineers that bridges that were built for horses and carts, sometimes many centuries ago, are still standing and able to carry modern traffic. However, although the mode of transport may have changed, some of the problems faced by bridge engineers are timeless.

1.2 Early developments in bridge hydraulics

Possibly the earliest permanent river bridge of any significance was built somewhere between 810 and 700 BC across the River Euphrates at Babylon (Overman, 1975). The bridge was around 120 m long with stone piers 10 m wide and 22 m in length. A timber deck spanned between the piers, although parts of this were removed at night to stop thieves crossing from one side of the city to the other! The piers were built in the shape of boats, pointing upstream to reduce the resistance to flow. This early example of applied bridge hydraulics was probably a logical extension of the practice of using boats lashed together to form floating bridges.

A form of floating bridge was used by Xerxes in 480 BC to span the 1400 m wide Hellespont, which joins the Black and Aegean Seas, so that his army could invade Greece. When a violent storm destroyed one of the two bridges, Xerxes ordered that the Hellespont should receive 300 lashes and have a pair of fetters thrown into it. He also ordered that the men responsible for building it should have their heads cut off, which is a rather severe form of professional liability by modern standards!

The Romans probably represent the finest early bridge engineers, the quality of their road network being apparent even today. They recognised that beam bridges had a limited span, and so employed semicircular arches of up to 30 m. They knew how to build cofferdams and drive piles. A good example of their work is the bridge in Rome now called the Ponte Molle, which was built in 109 BC. Although the bridge was restored in 1808, four of its original arches, the largest spanning 25.7 m, are still standing. The bridge was used during the Second World War to carry tanks across the Tiber.

The Romans did not develop the segmental or elliptical arch, which makes larger spans practical. These came much later. They were restricted to the semicircular arch, which has the advantage that there is no lateral thrust to the piers. They did, however, realise that if the piers had cutwaters or starlings that were pointed at both ends (instead of boat shape) then there were fewer eddies downstream of the structure so its hydraulic efficiency was improved.

Roman engineering was so good it is something of a surprise to see engravings of the first stone bridge to be built over the Thames at London between 1176 and 1209. The most striking features are the houses, shops and chapel built on the bridge, combined with the narrowness of the openings (Fig. 1.2). The bridge had 20 arches with spans that varied between 3.0 m and 9.6 m. Most of the piers were about 5.5 m wide and were surrounded by starlings to a little above low water level, so that the total width of the openings was only about 59.7 m between the starlings and 137.2 m above them, while the river was 282.2 m wide (Ruddock, 1979). This is an opening ratio of 0.21 and 0.49 between and above the starlings respectively. Typically, the piers of Roman bridges were only a quarter of the width of the river, giving an opening ratio of 0.75. A ratio of 1.0, of course, represents an unconstricted channel. Thus London Bridge represented quite an obstacle to the flow of the river, and this gave rise to the famous 'fall' or difference in water level across the structure. The fall terrified the London boatmen, understandably so since it was measured at up to 1.45 m around the year 1736. However, the fall had its uses. In the eighteenth century water-wheels were placed in four of the openings and used to drive pumps that supplied the city with water. For this reason an Act of Parliament forbade anyone to reduce the fall, even though the rush of water through the openings was so great that there was a constant danger of the piers being damaged or undermined and the bridge collapsing. By 1800 this problem had become so acute that a Select Committee of the House of Commons recommended replacement, and this new bridge was officially opened in 1831. Despite its many faults, the old bridge had lasted over 600 years.

Notwithstanding its longevity, old London Bridge illustrates nicely that narrow openings and wide piers result in high waterway velocities, a large fall across the structure, scour, and damage. It also shows that British bridge building had progressed little over the centuries. On the other hand, the French had established the Corps des Ingénieurs des Ponts et Chaussées

Fig. 1.2 London Bridge was built between 1176 and 1209. The openings are only about half the width of the river, and vary in shape and size. Houses and shops line the bridge. (Copyright © The Musuem of London)

by 1720 and the Ecole des Ponts et Chaussées by 1747. When it was proposed in 1734 that a new bridge should be built across the Thames at Westminster, there were few engineers in Britain that were qualified for such a task. Therefore it is no great surprise that it was a Frenchman who calculated the fall through both old London Bridge and the proposed Westminster Bridge. This was probably one of the first times that such a calculation had been made but, although apparently accurate, it was of limited value because few laymen or engineers of the day could understand it. In his excellent book, Ruddock (1979) suggested that the method used must have been similar to that described by Robertson in 1758. Adopting modern notation and terminology, this was as follows.

If the depth is constant, the continuity equation becomes

$$BV_1 = bV_2 \tag{1.1}$$

where B is the width of the river channel, b is the total width of the bridge openings, and V_1 and V_2 are the corresponding flow velocities. Recognising that the live stream will contract between the piers, a coefficient of contraction with a value of 0.84 was introduced (based on observations by Isaac Newton of flow through sharp-edged orifices). Thus

$$V_2 = \frac{BV_1}{0.84b} \tag{1.2}$$

If the respective heights of fall from rest to cause velocities V_1 and V_2 are $h_1 = V_1^2/2g$ and $h_2 = V_2^2/2g$, then the fall of the water surface from upstream to underneath the arches $(h_1 - h_2)$ is simply the difference in the velocity heads:

$$\text{Fall} = \frac{V_2^2}{2g} - \frac{V_1^2}{2g} \tag{1.3}$$

Substituting for V_2 using equation 1.2 gives:

$$\text{Fall} = \frac{V_1^2}{2g}\left[\left(\frac{B}{0.84b}\right)^2 - 1\right] \tag{1.4}$$

For London Bridge in 1746 with V_1 measured at 0.965 m/s, $B = 282.2$ m and $b = 59.7$ m, the fall estimated using this equation was 1.46 m, almost the same as the largest fall measured ten years earlier. In contrast, the fall calculated for the Westminster Bridge, which had an opening ratio of 0.82, was 25 mm, while that actually measured by Labelye reached a maximum of about 13 mm.

Old London Bridge was built because in 1176 a chantry priest called Peter de Colechurch decided that there should be a bridge across the Thames. For many centuries there was a tradition of bridge building in the church. Monasteries were active in business and commerce, so good communications were important to them. However, their primary interest was not in the science of bridge construction, which perhaps explains why there was so little advance, or even a regression, from Roman times. The situation was better in France, where the Frères du Pont (Brotherhood of the Bridge) were at least as good as the Romans. Between 1178 and 1188 they constructed what was at the time the longest stone arch river bridge, the Pont d'Avignon across the Rhone. The largest of the 21 spans was 32.5 m. However, the arches were pointed like church windows, so they were wasteful and expensive. This style is typical of the period, and can be seen in Bishop Skirlaw's fifteenth-century bridge across the Tees at Yarm (Fig. 1.3).

To be economical and efficient the arches of a bridge should be as long and as flat as possible. Consequently the use of segmental arches by Taddeo Gaddi in 1345 for the Ponte Vecchio in Florence, Italy, was a major step forward. With segmental arches the abutments have to withstand a lateral thrust, the thrust being balanced at the piers. In 1371 a single segmental arch with a span of 71 m was built over the Trezzo River near Milan. This span was the longest ever built, a record that was to stand for 400 years. The ability to bridge large distances with a single span, or a few large spans, made things much easier from the point of view of bridge hydraulics, although the structural problems increased and many bridges collapsed.

Fig. 1.3 The original fifteenth century part of this bridge across the tidal River Tees at Yarm has characteristic pointed arches (nearest the camera). It was constructed with a road width of about 3.7 m, but this was increased in 1806, the 'join' just being visible. The combination of a spate in the river with a high tide has frequently resulted in the town being flooded to a depth of several metres, notably in 1771 and 1881.

Advances in bridge hydraulics came rather slowly, and it was not until the completion of the Pont Notre Dame in 1507 that a new feature was introduced. This was due to the ingenuity of a French priest, Giovanni Giocondo, who by increasing the span of an arch towards its edges was able to spring the arch from nearer the pointed ends of the piers. This resulted in a wider bridge for a given pier size. The same technique has also been used to widen existing bridges. However, the 'rounding' of the edge of the arch also gave better hydraulic characteristics. This feature became known as *cornes de vâche* because it looks as though the central piers have cow's horns. Chamfers on the voussoirs were used by Thomas Telford on the Over Bridge at Gloucester (1826–1830) and at Morpeth to bridge the River Wansbeck (1831). In the former case it was to help pass the frequent large floods that occur in the Severn, while at Morpeth the low street levels and restricted width between the existing river walls made a hydraulically efficient bridge desirable (Fig. 1.4). Possibly Telford also liked the architectural effect of this feature.

In 1720 the Corps des Ingénieurs des Ponts et Chaussées was given the responsibility of developing France's roads, which led to the creation of a special college, the Ecole des Ponts et Chaussées, in 1747. This effectively created the first civilian civil engineers and established France as a leader in the field of bridge building. Chief Engineer of the Corps was Jean Rodolphe Perronet, who in 1774 used five elliptical arches of 36 m span for the Pont de Neuilly over the Seine in Paris. This enabled even longer flatter spans to be constructed, and with this bridge the piers were only one-ninth of the arch width, so the opening ratio must have been about 0.89.

The next significant hydraulic development appears to have been on the Sarsfield Bridge (formerly Wellesley Bridge) built by Alexander Nimmo

Fig. 1.4 Thomas Telford's 1831 bridge over the River Wansbeck at Morpeth, Northumberland. Note the *cornes de vâche* at the arch springings.

between 1824 and 1835 at Limerick in Ireland. Nimmo curved the whole soffit of the arch, making the opening higher on the outside edges than in the centre. This made it more streamlined and enabled it to pass more easily the spring tides that rise part of the way up the arch. Modern equivalents are shown in Figs 1.1 and 7.1.

In 1826 Thomas Telford's Menai Strait Bridge was opened. This was the world's first great suspension bridge, having a span of 174 m. In a way it marks a turning point, because after this civil engineering had advanced sufficiently to provide an effective alternative to multispan masonry bridges with their attendant hydraulic problems, although it could be argued that this distinction goes to the world's first cast iron bridge built by Abraham Darby in 1777. This spanned 30 m across the Severn at Coalbrookdale in England. Of course stone bridges continued to be built until perhaps the 1930s in Britain. Today over 1000 masonry arch bridges remain in the county of Devon (George, 1982) with a further 400 or so in Cornwall. Many of these are quite old, with around 310 bridges in the two counties being either scheduled as ancient monuments or listed as buildings of architectural or historical importance. These bridges are protected and have to be looked after. The total number of preserved arch bridges in Britain must be quite large.

When the first major reinforced concrete bridge with two 69 m spans was built in 1904 in Germany, the days of masonry arch bridges became numbered. Reinforced concrete or prestressed concrete river bridges with rectangular openings have become the norm, although to most people they lack the charm and individuality of the old arch bridges – so much so that in the more environmentally aware Britain of the 1990s public opinion has resulted in the construction of one or two new masonry arch bridges (Fig. 1.5), even though this was not necessarily the cheapest option. However, the twentieth century remains dominated by reinforced concrete.

The end of the era in which the construction of arch bridges was common more or less coincided with the start of serious research into bridge hydraulics. In 1840, d'Aubuisson published a work in Paris containing an equation for flow through a bridge. In America, Nagler (1918) and Lane (1921) published significant contributions to the subject, to be followed by Yarnell (1934) and many others. They established the subject that is developed in this book, and it is easy to forget the difficulties that they faced. At the start of the twentieth century the subject of fluid mechanics or hydraulics was not as advanced as it is now. Additionally, the instruments available were simple and, by modern standards, crude. However, many of the principles of applied bridge hydraulics that they pioneered are equally valid today and can still be found in modern practice.

While many old bridges are still in regular use, it is wrong to think that all bridges built by our recent ancestors stood for centuries. For instance, the iron bridge constructed at Yarm in 1805 to replace Bishop Skirlaw's crashed into the river in 1806 before it had even been opened to traffic.

Fig. 1.5 This new three-span masonry bridge across the River Esk at Egton Bridge, North Yorkshire, was opened in 1994, replacing an unattractive 1930s steel structure. The bridge is an exact replica of the original 1758 structure, two spans of which were destroyed by flood in 1930.

Consequently the old stone bridge was widened, and is still in use (Fig. 1.3). Some were washed away during construction; others collapsed. Only the best examples of the bridge builders' work usually survives but, regardless of age, the possibility of failure is ever present.

1.3 Hydraulic causes of bridge failure

Smith (1976, 1977) studied 143 bridge failures that occurred throughout the world between 1847 and 1975. He grouped the causes of failure into nine categories, as shown in Table 1.1. Almost half of the failures were due to floods. One flood can wash away the foundations of a large number of bridges at the same time, particularly small structures. Perhaps the table also shows that engineers pay more attention to structural design than they do to hydraulic considerations? Or perhaps, since the excess flow may bypass the bridge as discussed in Chapters 6 and 8, the most severe consequences of the design flood being exceeded are scour and foundation failure? Either way the table illustrates the importance of appropriate hydraulic design if a bridge is to stand for centuries. The design process is considered in Section 1.4.

White *et al.*(1992) commented on bridge failures in the USA and observed that the most common cause was floods and the other actions of water. More specifically, two factors were identified: one was scour, and the second was debris piled against the structure. Sometimes these two factors can

Table 1.1 Causes of bridge failure, 1847–1975 (after Smith, 1976, 1977). Except in the remarks column, secondary causes are omitted: each failure is listed once

Cause of failure	*Number of failures*	*Comments*
Flood and foundation movement	70	66 scour; two earth slips; one floating debris; one foundation movement
Unsuitable or defective permanent material or workmanship	22	19 by brittle fracture of plates or anchor bars
Overload or accident	14	10 ship or barge impact
Inadequate or unsuitable temporary works or erection procedure	12	Inadequacy in permanent design a supplementary cause in one instance
Earthquake	11	
Inadequate design in permanent material	5	
Wind	4	
Fatigue	4	Three cast iron; one hastened by corrosion
Corrosion	1	
Total number of failures	143	

Reproduced with permission, Institution of Civil Engineers

be interrelated. Of the two, damage caused by scouring of the bottom material around the foundations tends to be the most prevalent. A study conducted in 1973 showed that of 383 bridge failures caused by catastrophic floods, 25% involved pier damage and 72% involved abutment damage. A more extensive study in 1978 showed that scour at bridge piers was just as significant as scour at the abutments. In 1985, in Pennsylvania, Virginia and West Virginia, 73 bridges were destroyed by flooding, including scour. During the floods in spring 1987, 17 bridges in New York and New England were damaged or destroyed by scour.

Scour in its widest sense may also include lateral erosion of the riverbanks in the vicinity of a bridge (Fig. 1.6). This may result in the flow approaching the bridge at a skewed angle instead of perpendicularly, greatly increasing the potential for failure of the piers, abutments and highway embankments. In one survey of 224 bridges in the USA it was discovered that 106 sites (47%) had hydraulic problems as a result of lateral stream erosion (Brice, 1984). Accumulation of debris was a problem at 26 sites (12%), with many more experiencing some form of scour.

Between 1964 and 1984 around 46 bridges were seriously damaged by

Fig. 1.6 Lateral erosion downstream of the railway viaduct at Fenny Bridges, Devon, in July 1968. The line of the original river bank is shown by the dashes. Changes of this type can significantly alter the angle at which the flow hits the piers and thus affect scour depths. (Reproduced by permission of the Environment Agency)

flood action in Devon and Cornwall, and a further 13 were actually destroyed (Hamill and O'Leary, 1985). This is a significant number, but with over 5000 bridges in the two counties it represents a relatively small proportion of the total. Thus bridge failures are thankfully rare, but not uncommon. There are around 60 counties in Britain, so the figures for Devon and Cornwall can be used to gain some idea of the national picture. It should also be remembered that Britain's climate is relatively moderate compared with some parts of the world. When typhoon Fran hit Kyushu in Japan in 1976, giving 1.95 m of rainfall, a record at that time, the result was 233 bridges washed away, 133 fatalities and 337 injured (Holford, 1977).

It is instructive to see what lessons can be learnt from these failures. The evidence suggests that there were three principal factors involved, either singly or in combination: an inadequately sized opening, scour, and the accumulation of flood debris.

1.3.1 *Inadequate waterway openings*

There is no universally applicable definition of what constitutes an inadequate opening, but if a bridge is seriously damaged or destroyed by flood then possibly the bridge waterway was not large enough. Regular flooding upstream may also indicate that the opening should be larger.

Most of the damage and destruction experienced in Devon and Cornwall has been the result of a small number of extreme storms. For instance, in July 1968 10 bridges were destroyed and a further 19 damaged (Criswell, 1968; George, 1982). The waterway openings of the bridges were too narrow so that the resulting water velocities could easily remove the sedimentary bed materials and erode the foundations. In general, the smaller and older bridges were the worst affected. In August 1952 the catastrophic flood at Lynmouth damaged or destroyed 28 bridges, the openings being too small for the flood flows (Fig. 1.7). The two rivers, the East and West Lyn, that flow into Lynmouth have gradients of 1 in 25 and 1 in 8 as they approach the town. With an estimated combined flow of $700\,m^3/s$ from a catchment of only $101\,km^2$, what were normally small streams turned into raging torrents carrying large boulders and other flood debris. No wonder the bridge openings were insufficient. The force of the floodwater must have been enormous. A 7.5 tonne boulder was found in the basement of a hotel. The tremendous battering received by the bridges combined with water levels that rose above the top of the waterways not surprisingly led to damage and destruction. When five of the larger waterways were replaced,

Fig. 1.7 Lynmouth, Devon, after the flood of August 1952. A reminder never to underestimate the power of water. Note the boulders piled against the Lyndale Hotel to a depth of around 9 m, and the water staining on the walls. Midway between the camera and the hotel are many human figures and a bulldozer, which gives some sense of scale. (Reproduced by permission of the *Western Morning News*, Plymouth)

each had the width and height of span increased by a factor of 2 or 3 (George, 1982).

The magnitude of the Lynmouth flood was something of a shock for British hydrologists. With 34 lives lost, 93 houses destroyed or later demolished as unsafe and 300–400 people rendered homeless, the event illustrated the unpredictability and power of nature. For bridge engineers it raised the question as to what magnitude and frequency of flood event a waterway should be designed for. This was not a new question. In 1811 Telford confessed that he could not ensure the safety of bridges in the Highlands of Scotland from the maximum spates he was beginning to think were possible (Ruddock, 1979). If he did it would be uneconomic. In 1812 he observed that '... it is only by degrees that the proper dimensions for a bridge can be ascertained', and 'The bridges now built will serve for a scale of measurement, each for its own river.'

These words are still true almost 200 years later. With climatic change possibly occurring, the prediction of floods is a risky business. It appears that large floods may happen more frequently than first thought. In 1982 George considered Devon floods and the waterways of bridges, and suggested the guidelines in Table 1.2. In the discussion arising from this paper, Canadian guidelines were quoted as ranging between 1 in 200 years for long freeway bridges to 1 in 25 years for short local bridges. George also recommended that due consideration should be given to the maximum historical flood, and to the possibility of unusual types of flood occurring. For the purpose of assessing the waterway area required to cope with the design flood, the Lacey relationship for alluvial channels (equation 1.5) was said to be surprisingly good as a check on opening width (Neill, 1973), although local factors must also be considered.

As far as the height of the opening is concerned, it has been suggested that the soffit should be 0.6–1.0 m above the design flood level, or on minor watercourses 600 mm above the highest known stage (Brandon, 1989; Richardson *et al.*, 1993; Highways Agency, 1994). This is easier than it sounds given the difficulty of calculating the design water level and the fact that the water surface may not be flat but consist of waves superimposed on a transverse slope, particularly if the bridge and approach embankments are skewed (Fig. 2.7 and Section 3.7). If there is no restriction on height

Table 1.2 Suggested return intervals for bridge waterways based on Devon floods

Category	*Flood interval (years)*
Immediately downstream of a community	1 in 1000 years
Motorways and trunk roads	1 in 150 years
Other roads in rural areas	1 in 30 years
Submersible bridges (pipe bridges)	1 in 5 years

After George (1982)

then, given that the frequency of major floods may be underestimated, the higher the better. There is then less risk of debris becoming stuck, flooding upstream and damage to the bridge (see Appendix A).

Of course, a bridge does not have to be damaged or destroyed before it is considered to have an inadequate waterway opening. If the opening is so small that it acts like a retarding dam and causes the inundation of a large residential or commercial area upstream then the bridge can hardly be considered to be adequate, even if it survives the flood unscathed (Fig. 1.8). Additionally, if the quantity of water stored by a bridge and embankment exceeds 25 000 m^3 above the natural level of any part of the adjacent land, then in Britain reservoir safety legislation applies (HMSO, 1975), and the bridge and embankment have to be designed as a dam and inspected by suitably qualified engineers throughout their life, increasing costs significantly. Although 25 000 m^3 sounds like a large volume, it is only a depth of 2.0 m on floodplains 60 m wide and 209 m long, which can be achieved easily. To avoid this the waterway must be made large enough to pass the design discharge with floodplain storage kept below the critical value or, alternatively, the approach embankments should be designed to allow excess flow to pass over them (see Chapter 6).

Just how much backwater can be tolerated from a bridge depends upon factors such as the frequency of flooding and the local land use: different values will be adopted for city centres and open farmland. However, there is another consideration, which is the relationship between the backwater and the velocity of flow through the waterway opening. Simple hydraulics

Fig. 1.8 This 2.06 m span bridge at Polgooth, Cornwall, is reported to have severely obstructed the stream and thus contributed to severe flooding in the village in November 1997. The bridge itself exhibits separation of the upstream arch ring; undermining of the foundations was suspected. It also suffered a partial collapse downstream, and is to be replaced by a reinforced concrete deck-type bridge. This example serves as a reminder that small bridges must also be designed with care. (Reproduced by permission of Cornwall County Council)

dictates that the larger the head difference across the bridge the greater the waterway velocity. At sites where scour cannot be tolerated then a backwater of only 0.3 m may represent an upper limit (Bradley, 1978). If the riverbed is stable or consists of rock then a larger value may be permissible.

1.3.2 *Scour effects*

Because the openings of a bridge are usually less than the full width of the river, the water accelerates as it approaches and passes through the waterways. Consequently the velocity is higher than it would otherwise be, and this can cause scour and undermining of the foundations of the bridge. The narrower the openings the larger the velocity, and the finer the material the more easily it can be transported (Fig. 1.14 and Table 1.3).

Scour can be very destructive. As mentioned earlier it was a problem with old London Bridge. In 1763 John Smeaton was engaged to stop the bridge being undermined while alterations were being undertaken (Ruddock, 1979). His favourite solution was to use rock or rubble bed armouring. He held a low opinion of the common alternative of 'paving' the bed between the piers or abutments. Although Smeaton was perhaps the best river engineer of his time, he too had problems with scour. In 1777 Smeaton designed Hexham Bridge, and while it was being constructed a flood scoured the gravel from under the upstream ends of all four piers. Smeaton revised his design to include sheet piling protection and completed the bridge. Then in March 1782 a combination of rain and snow-melt sent a large flood down the River Tyne, and this undermined one pier and brought down six arches within 30 minutes. Thus the man who had successfully built a lighthouse on the notorious Eddystone Reef 23 km (14 miles) off Plymouth had great difficulty in building a bridge across a river! Nor was Smeaton alone; many engineers before and after have had the same problem. Of course, the materials and plant available then were very limited by today's standards, so constructing bridge foundations in loose sediments where there was deep or fast-flowing water was no easy matter. Consequently the potential vulnerability of old bridges, many of which survive, should be remembered. It was mainly the older, smaller bridges that were destroyed in the East Devon Flood of 1968 described above. Many railway bridges are vulnerable, having been built during the railway boom of the nineteeenth century. Modern construction techniques make serious problems with scour less likely, but it should never be forgotten. Indeed, it has a rude method of reminding us of its existence.

In October 1988 severe flooding of the River Towy near Llandeilo in Wales caused one of the piers of Glanrhyd railway bridge to be undermined as a result of scour (Fowler, 1990). The pier collapsed, and subsequently the whole bridge failed. An early morning passenger train drove off the abutment. Four people were killed. In February 1989 scour caused part of the 127 year old viaduct over the River Ness in Scotland to collapse only half an hour after a train had passed over it (Fig. 1.9). Only a few months earlier the viaduct had passed a routine maintenance check. It was due for a more comprehensive examination later in the year, including an underwater inspection (Montague and Fowler, 1989). However, such inspections are not foolproof. Scour can occur quickly, and scour holes created during a flood can be refilled as the spate recedes. By the time an inspection can take place the bed may not be obviously different from that which existed prior to the event, so afterwards it is easy to underestimate the depth of scour that has occurred.

In the USA, a failure that caused loss of life was the collapse of the New York State Thruway Bridge over Schoharie Creek in April 1987; ten persons died. In Tennessee, in April 1989, the route 51 road bridge over the Hatchie River collapsed. This structure fell as a result of a series of floods in January, February and April, which is thought to have caused scouring around two of the piers, initiating the failure. Five vehicles drove off the bridge into 10 m of water, killing eight people (Fowler and Russell, 1989; White *et al.*, 1992). The bridge was 1280 m long with 143 spans, three of which were the main river spans. The river normally has a width of 30–50 m, but when flowing overbank the floodplain is as wide as the bridge. Factors that may have contributed to the disaster include the river changing course and/or the collection of debris on the piers (see Section 1.3.3). It was recognised that because of the fine riverbed material scour

Fig. 1.9 Part of the viaduct over the River Ness, which collapsed in February 1989. Scour was suspected to be the cause. (Reproduced by permission of John Paul Photography, Inverness)

could be a problem, and it is reported that the bridge had been inspected from above during the February floods with no settlement or misalignment being recorded. Paradoxically, the critical April flood was apparently the smallest of the three.

Scour is a potential problem with all river crossings, but sometimes it may occur where it is not expected. A change in the thalweg or main river course contributed to both the Glanrhyd and Hatchie bridge disasters. When there is a shift of course the stream will not hit the piers head on but from the side, greatly increasing the obstacle to flow and the potential scour depth (Figs 1.6 and 1.15b). Thus the bridge engineer must assess the likelihood of a change of course and the possible consequences. Another duty is to ensure that regular and comprehensive inspections are conducted, both above and below water (Richardson *et al.*, 1993). Of course, even if it is known that there is a high risk of scour, it is not possible to undertake an underwater inspection during a flood. Until recently, the only clue to impending foundation failure might be defects in the superstructure, such as a dip in the parapet. Now various sensors are available that are capable of monitoring scour as it happens and, if necessary, of initiating an alarm (Table 8.1 and Fig. 8.3).

The magnitude of the threat posed by scour is illustrated by a study conducted after the Glanrhyd and Inverness failures. This suggested that over 1000 rail bridges in the UK could be at serious risk and in need of urgent investigation and foundation work (Watson, 1990). The true number was subsequently found to be smaller, but this illustrates nicely the scale of the problem and the degree of uncertainty involved. Virtually every river bridge in the UK has been under scrutiny, and new guides have been developed (Highways Agency, 1994; Railtrack, 1995). Bridges in the USA have also been under scrutiny (e.g. Richardson *et al.*, 1993; Bryan *et al.*, 1995)

So why does a bridge stand for 127 years before failing? Why does a bridge fail during a flood smaller than experienced only a few months earlier? Luck may determine when a bridge first encounters an extreme event, the collection of debris is unpredictable, while some effects (e.g. scour and ageing) can be cumulative. There is also a philosophical answer in the form of an old saying that no one sets foot in the same river twice. In other words, no two floods are the same.

1.3.3 Debris

The accumulation of flood debris is one of the more unpredictable problems (Figs 1.10 and 1.11). Anything from leaves to whole trees, and garden sheds (Fig. 6.1) to Dutch barns have been seen floating downriver during flood. When debris becomes wedged across the waterway opening or caught on a pier it reduces the flow area and increases the water velocity, which in turn can increase scour. When combined with the increased

Fig. 1.10 Even the performance of a well-designed bridge can be adversely affected by unexpected flood debris; Ham Footbridge, River Sid, Sidmouth in July 1968. (Reproduced by permission of Hayes Studio, Sidmouth)

Fig. 1.11 Debris can be a particular problem in forested areas, as illustrated here by Whetcombe Barton Bridge on the River Teign, Devon, in 1960. The approximate flood level was at the bottom rail of the parapet. (Reproduced by permission of the Environment Agency)

backwater that inevitably occurs, it can also create a horizontal pressure that may lead to failure of the superstructure (see Appendix A).

Bridge piers, particularly those in the centre of the main channel, can collect large quantities of debris, which significantly reduces the hydraulic performance. This should be taken into consideration at the design stage. Single-span bridges with large central waterways may be trouble free provided nothing jams across the entire opening. For small single-span bridges the blockage can be as much as 90% of the opening (Highways Agency, 1994). The hazard is largest where the catchment above the bridge contains a large forested area with shallow roots on steep slopes. In the analysis and design of bridge waterways, debris is always one of the unquantifiable complications. Even if the hydraulic analysis is carried out diligently, flooding may be much worse than predicted as a result of debris.

The debris does not always have to be caught on the bridge to cause problems: debris that obstructs the river channel either upstream or downstream of the structure may be sufficient to change the water levels at the site significantly and thus change the hydraulic performance of the bridge. Alternatively, the blockage may cause the river to change course.

When analysing the damage that occurred to bridges in Devon and Cornwall between 1964 and 1984 it was apparent that the accumulation of debris was either a contributory factor or suspected of being so in most cases. In September 1989 a fisherman warned police 'of a gaping hole' beneath the main river pier of the Girvan viaduct in Scotland (Pease, 1989). Nearly one-fifth of the bridge pier had disappeared. One possibility is that debris trapped on the pier altered the flow of the river, causing the scour. A void 2 m deep by 4 m long was reported at the upstream end of the pier. The structure was built in the 1880s.

1.4 The hydraulic design of bridges

A good hydraulic analysis is an essential part of designing a successful bridge, as illustrated by Table 1.1. Some general guidelines and suggestions are given below, but these are not exhaustive, and must be modified to suit a particular project: the design of a crossing of national importance over one of the world's great rivers is not the same as constructing an access road to a few isolated properties. Nevertheless, a generalised design process is shown in Fig. 1.12, which may be a useful reminder of some of the steps involved.

One of the principal features of a successful bridge is an adequately sized opening, which necessitates the determination of a suitable design flood. Table 1.2 may be of use in selecting a suitable flood frequency, but the calculation of the flood magnitude is covered in great detail elsewhere and will not be repeated here (NERC, 1975).

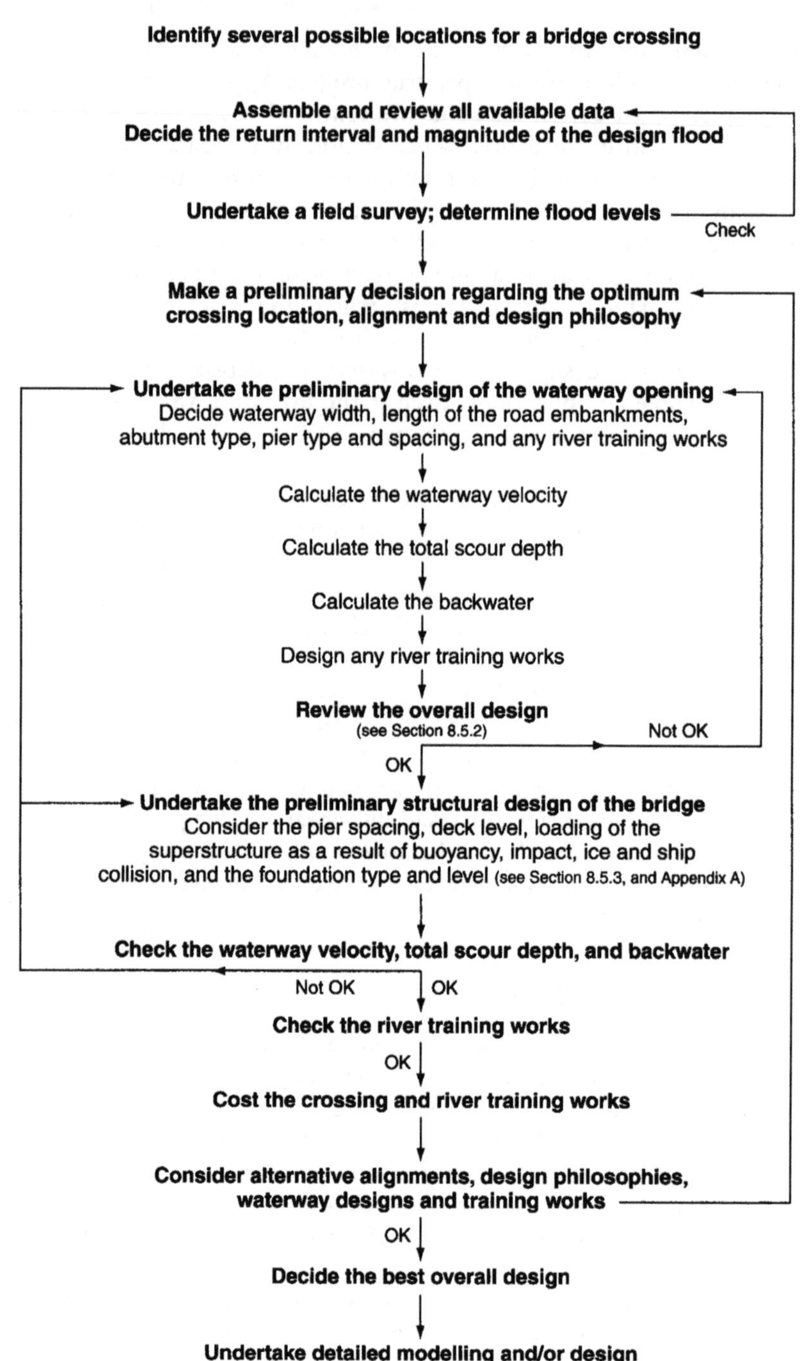

Fig. 1.12 Summary of some of the steps involved in the hydraulic design of a bridge.

1.4.1 Location

Usually the alignment of a highway or railway will be selected so as to minimise the overall cost, a large part of which may arise from the construction of the river crossings. Therefore it is sensible to seek an alignment that will minimise the cost of the bridges without significantly adding to the total length of the road. The optimum location and type of crossing is often the one that is most economical in terms of both initial construction and long-term maintainance (a cheap bridge that regularly suffers expensive flood damage is not the best option). It is normally cheapest and easiest to construct a crossing where the river channel is well defined and stable, such as where rock outcrops at or near the surface or there are deposits of stiff clay. In such conditions expensive river training works will probably not be needed (see Chapters 7 and 8). Therefore the alignment of the road and the location of the crossing should be decided with one eye on the local geology.

Usually it is cheaper to construct an embankment than a bridge or viaduct, so the approach road embankments should be as long as possible without making the bridge waterway so narrow that backwater or scour becomes a significant problem. Of course, if the waterway is made too narrow any economic advantage is lost because expensive river training works will be needed to compensate. The same concept, of an optimum opening width, applies to semi-stable or unstable channels that either migrate slowly or which are capable of a sudden and permanent change of course during a single flood. If such channels cannot be avoided then the best available site has to be found. The significant problem of lateral erosion and channel migration must always be remembered (Figs 1.6 and 1.13).

Stable river channels

Generally a bridge will not span the entire floodplain width, so the best site may be the one that offers the highest bridge opening ratio (Section 3.2): that is, where the flow in the main channel is largest compared with that over the floodplains. This helps to optimise the hydraulic efficiency and minimise the backwater. One of the key design decisions may be deciding the width of the waterway opening relative to the length of the highway embankments.

Semi-stable and unstable river channels

If the river channel is gradually moving, frequently shifts course or meanders periodically then site selection is more complex (Fig. 1.13). Assuming that the bridge will not span the entire floodplain width, some form of river training works may be needed to stabilise the channel and ensure that the flow approaches the bridge as planned, or does not miss the waterway opening altogether during flood. Training works are usually expensive to

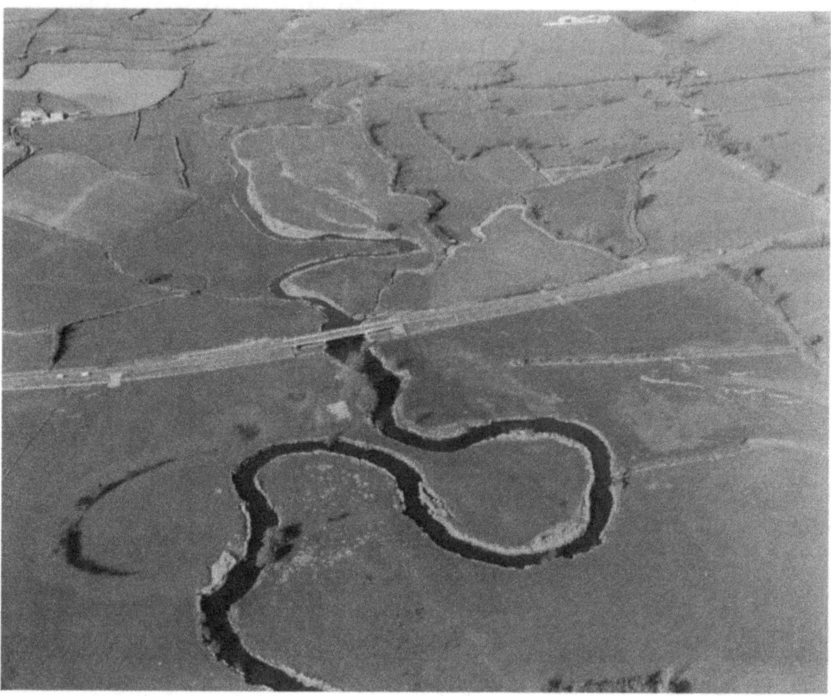

Fig. 1.13 Axe Bridge, Colyford, Devon, illustrating the problem of lateral shifting of the channel and route selection. The meander (bottom centre) looks as though it will be cut off at some time, possibly within the life of the bridge. When this happens the reduced channel length could increase velocities and cause degradation (see Fig. 8.8). An old ox-bow can be seen on the left. (Reproduced by permission of Devon Library Service)

construct and maintain, so the aim should be to minimise such works by using any natural rock outcrops or other permanent controls, locating the crossing at a node that forms the crossover point in a sinusoidal meander, or possibly constructing the crossing at a bend if this is the only relatively stable feature. At a bend the channel may be typically about 1.5–2.0 times the average depth and less likely to shift (Neill, 1973; Farraday and Charlton, 1983). Sharp bends can be indicative of the presence of relatively resistant local bed material, which is why the bend is there.

Farraday and Charlton (1983) suggested that with wide floodplains it is generally cheaper to confine meandering rivers with training works and construct road embankments approaching the bridge than to bridge the entire inundated width. They quoted examples of the latter option being 1.12–2.94 times more expensive for floodplains between 1463 m and 3658 m wide respectively. As described above, theoretically there is an optimum waterway opening width that will minimise the cost of both the bridge and the river training works. It was also suggested that it may be better to consider going around very unstable alluvial fans, but if they cannot be avoided then one should at least try to select an alignment near the inlet.

1.4.2 *Examination of existing data*

Existing data that are already available or easy to obtain should be examined before embarking on an expensive field investigation. Some typical data and sources were given by Richardson *et al.* (1990) and the Highways Agency (1994) for the USA and UK respectively. The sort of data required are as follows.

TOPOGRAPHIC MAPS

To determine channel and floodplain width, to help identify possible crossing sites, to obtain the channel gradient, and to indicate floodplain use. Forested areas upstream may suggest debris problems. Old editions of the maps may assist in identifying changes in channel alignment and land use, and hence problems with channel shifting and long-term degradation or aggradation (see Section 8.2.2). The land use may help decide the maximum permissible backwater.

AIRPHOTOS AND SATELLITE PHOTOS

Largely as for topographic maps.

CHARTS AND TIDE TABLES

For navigable waters charts may be available showing widths and depths, while tide tables will help determine the range of depths in estuaries. Any information regarding wave heights should also be obtained.

GEOLOGICAL MAPS, SOIL MAPS AND GEOLOGICAL MEMOIRS

May provide some details about the local geology and the likelihood of the channel migrating or of scour being a problem.

CONSTRUCTION DETAILS OF EXISTING BRIDGES

May yield some data regarding the geology, depth of foundations required, the design flood and the waterway dimensions. Have these bridges been successful? Are all of their spans used? Are the bridges bypassed during flood? Are their decks high enough? Has serious scour occurred? Have river training works been added subsequently? Is there another bridge or structure (see Section 4.3.4) that will interfere hydraulically with the proposed crossing or vice versa?

HYDROLOGICAL DATA

Such as gauging station records, annual maxima, stage–discharge relationship, and flow–duration curve. These are needed to help identify the design flood and the maximum likely water level, and to plan construction work.

METEOROLOGICAL DATA

Rainfall depths and intensities, snowfall and snow-melt, temperature range and wind speed. These data may help in assessing the possibility of flash flood, and of ice formation and ice loading on the superstructure. If appropriate, the wind speed can be used to assess potential wave heights. Both ice and waves may have to be allowed for when determining the height of the deck.

RIVER CHANNEL DATA

Things such as the roughness of the channel and floodplains (say for use in the Manning equation) are never easy to determine at the best of times and may have to be inferred initially from photographs or from preliminary visits to the site.

In hydraulic calculations it is really the slope of the water surface or energy line over a distance of around 10–20 channel widths that is required, not the physical slope of the bed. However, this can often be obtained only from observations of floods, not from maps. Similarly it may be difficult to determine the width and depth of the river channel from maps. There are equations and diagrams that can be used to obtain typical dimensions (see Section 8.6) but it is much better to base even preliminary calculations on real observations.

SPECIAL CONSIDERATIONS

Such as river regulation, ice flows, logging, or navigation that will require additional clearance between the flood level and the bridge superstructure. Environmental considerations are rightly becoming increasingly important and may determine whether or not river training works, spur dykes (see Section 7.4) or bypass channels can be constructed, and may also influence the appearance and design of the bridge and its highway embankments. Any environmental restraints should be fully explored at the outset, particularly in sensitive areas.

1.4.3 *Field investigations*

These are typically used to confirm the accuracy of the data above and to provide missing information: for example, the location of rock outcrops and hydraulic control points within the river, such as rapids. Samples of bed and bank material can be obtained for analysis to determine composition, grading and D_{50} value, and to assess scour and erosion potential (Table 1.3). In this respect it is worth looking for evidence of scour, especially at bends where the maximum flow velocity may be around 1.5 times the mean velocity of the river (in a straight channel the maximum is around 1.25 times the mean). If appropriate and possible, measure the actual velocity at the bends and relate this to the depth of the bed; this may give a guide as to what will

Table 1.3 (a) Approximate erosion threshold velocities, V_E. If water velocities exceed V_E then erosion is predicted

Bank material	*Approximate diameter, D (mm)*	V_E *(m/s)*
Clay	< 0.002	> 1.2
Silt	0.002–0.06	0.2
Fine sand	0.06–0.2	0.3
Medium sand	0.2–0.6	0.5
Coarse sand	0.6–2	0.6
Sandy loam		0.6
Loam		0.9
Gravel	2–60	1.0
Pebbles/stones	>60	3.0

After Netlon Ltd

(b) Tentative guide to competent mean velocities for the erosion of cohesive materials. The competent mean velocity (V_{SC}) is the velocity at which the flow is just competent to move the exposed bed material

Depth of flow (m)	*Low values: easily erodible material (m/s)*	*Average values (m/s)*	*High values: resistant material (m/s)*
1.5	0.6	1.0	1.8
3	0.65	1.2	2.0
6	0.7	1.3	2.3
15	0.8	1.5	2.6

After Neill, 1973; reproduced with permission, University of Toronto Press

Notes:

1. This table is to be regarded as a rough guide only, in the absence of data based on local experience. Account must be taken of the expected condition of the material after exposure to weathering and saturation.
2. It is not considered advisable to relate the suggested low, average and high values to soil shear strength or other conventional indices because of the predominating effects of weathering and saturation on the erodibility of many cohesive soils.

happen in the opening. The same argument applies to the noses of groynes and spur dykes, where the maximum velocity may be around 2.0 times the average.

Evidence of flood levels may be apparent from trash deposits, ice scars or stains on existing structures. If the available hydrological data are limited, interviews with local residents can often provide rudimentary information regarding the frequency and height of floods. The opportunity should also be used to visualise flood flows both before and after the construction of the proposed crossing (Figs 8.8 and 8.16). How will the flow pattern be affected? Is bypass flow around the bridge possible? Should relief openings be provided on the floodplain(s), or would they be an unnecessary expense? Are flood flows spread evenly across the full width of the floodplain? Or are parts of the floodplain used for flood storage with zero conveyance? This will affect the calculation of the bridge opening ratio and the estimation of the backwater. In the bridge approach, is the direction of flow on the floodplain the same as in the main channel, or different? What is the optimum orientation for the piers and abutments?

1.4.4 General hydraulic design philosophy

In addition to deciding the relative length of the bridge opening and any highway embankments, the designer will often have some idea as to whether the bridge will be a single or multispan structure from the channel and floodplain width and a knowledge of the economic importance of the crossing. What may be less obvious is whether or not a 'conventional' design with the approach embankments above the highest expected flood level (so traffic is not disrupted) is better than the alternative of keeping the approaches low so that anything larger than the design flood spills over them with relatively little damage (Chapter 6). It is highly unlikely that one of the minimum-energy waterways described in Section 7.5 would be adopted, but at problem sites with certain characteristics this may be an option.

1.4.5 Preliminary design of the bridge waterway

From the above it should be possible to identify one preferred location for the waterway. The design procedure can be shortened if it is already known what the pier arrangement will be, otherwise an educated guess is needed. Pier spacing depends upon economics, ground conditions, navigation requirements, backwater restrictions, potential scour, and aesthetics. Of course, from the hydraulic perspective a single-span opening is preferable, but if piers are unavoidable then use as few as possible. Spillthrough abutments are better than vertical-wall types and will help to reduce afflux and scour (see Chapters 4 and 8).

As a starting point it can be assumed that the width of the opening (b) is either equal to the width between the river banks (B) or is between the values obtained from the two equations below, according to the importance of the crossing.

$$\text{Lacey's surface width of an alluvial channel} = 4.75Q^{1/2} \quad (1.5)$$

$$\text{minimum suggested trial opening width} = 3.20Q^{1/2} \quad (1.6)$$

where Q is the design flood discharge (m^3/s). Equation 1.5 gives the regime channel width (m) measured along the water surface at 90° to the banks (see Section 8.6). Both equations may overestimate the waterway width when the discharge is small or when a deep, non-alluvial channel is involved.

Multiplying the width (b) by the depth of flow during the design flood gives the approximate net cross-sectional area of the opening (a). From b, the distance between the abutments can be obtained by adding the width of the piers. At this point some allowance will have to be made for the contraction of the flow as it passes through the openings. This depends upon the size and shape of the piers and abutments and their orientation to the flow. Obviously, these guesses have to be refined during later rounds of calculation. Skew will have already been allowed for, provided that the waterway dimensions are measured at right angles to the principal direction of flow.

Dividing the design discharge (Q) by a gives the mean opening velocity, V. The head loss and afflux are proportional to the velocity head, so a large opening velocity may indicate potential problems with flooding upstream. It may also suggest that scour will occur at the bridge. A very general indication of the possibility of scour can be obtained by comparing V with the velocities shown in Table 1.3. It is possible that the channel bed in the bridge opening will continue to scour until V in the enlarged opening is just less than the indicated value.

With cohesionless material a preliminary estimate of the potential scour depth (d_{SC}) caused by the bridge contraction can be obtained from the mean competent velocity (V_{SC}) in Fig. 1.14. This is the velocity at which the flow is just competent to move the exposed bed material of median diameter (D_{50}) by weight at the scoured depth (d_{SC}), as discussed in more detail in Section 8.4.1. If piers are located in the river channel then the approximate pier scour depth (d_{SP}) shown in Fig. 1.15 should be added to d_{SC}. Note part (b) of the diagram, which is needed when the approach flow is at an angle to the piers. The use of these diagrams is illustrated in Example 1.1, where the preliminary design of a bridge waterway is undertaken and the approximate depth of d_{SC} and d_{SP} is calculated.

Once a firm estimate of the waterway dimensions has been obtained, a more detailed calculation of the depth of scour will have to be undertaken (Chapter 8), the afflux and backwater calculated (Chapters 2–6), and any necessary river training, flow improvement or protection works designed (Chapters 7 and 8). This is an iterative process involving several interrelated variables. For example, if scour occurs in the waterway then the increase in the cross-sectional area of flow will reduce the velocity, so

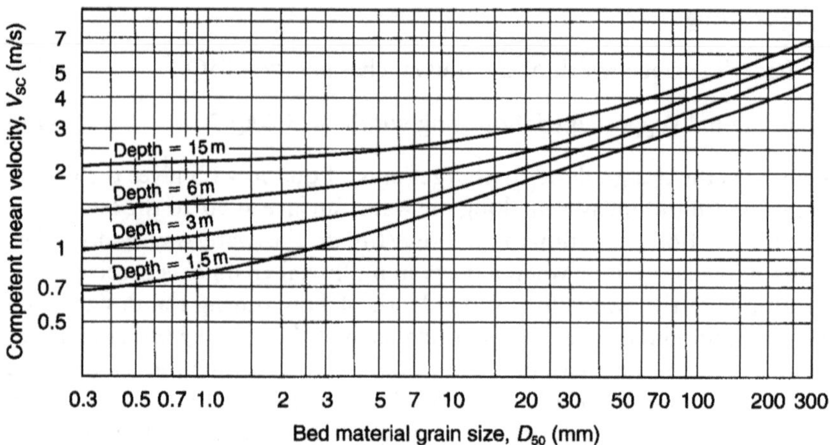

Fig. 1.14 Variation of competent mean velocity (V_{SC}) with the median diameter by weight (D_{50}) and depth of flow. The competent mean velocity is that at which the flow is just competent to move the exposed bed material at the scoured depth. (After Neill, 1973. Reproduced by permission of University of Toronto Press)

the original dimensions may require modification, which affects the afflux. However, at some point an acceptable waterway should be obtained. The process is quicker if any bridge piers have been included, otherwise they may have to be added during the next stage and some of the calculations repeated.

Once the size of the bridge openings is known, the design should be reviewed. Have all of the factors listed earlier been considered? Has the correct balance been obtained between the length of the approach embankments and the span of the bridge? Has the correct balance been struck between minimising the waterway cost and the use of river training works to offset any undesirable consequences? Consult the checklist in Section 8.5.2. If the design passes this review then the next stage, the preliminary design of the structure itself, can be started. If not, some redesign is necessary.

Subsequently additional data and calculation will be needed to complete the hydraulic design: for example, some techniques for calculating backwater have their own specific requirements, which may necessitate additional

Fig. 1.15 (a) Approximate pier scour depth, d_{SP}, for various pier shapes aligned to the approach flow, where b_P is the pier width perpendicular to the flow. If the depth of flow exceeds $5b_P$ then d_{SP} should be increased by 50%. (b) If the approach flow is at an angle to the pier then multiply d_{SP} from part (a) by the factor shown in the table. (After Neill, 1973. Reproduced by permission of University of Toronto Press)

(a)

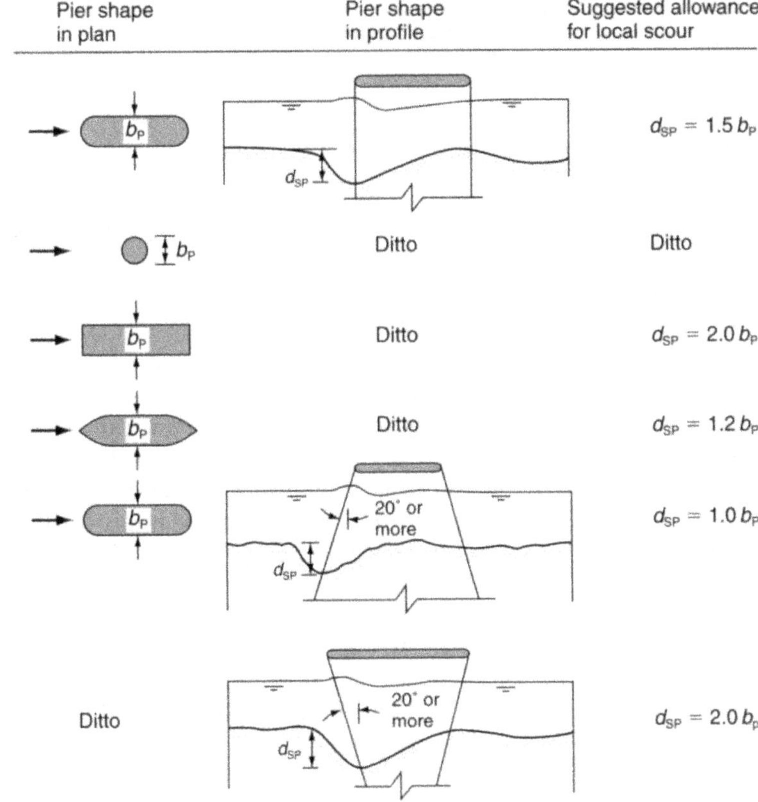

(b)

Multiplying factors for local scour at skewed piers*
(to be applied to local scour allowances of part a).

Angle of attack	Length-to-width ratio of pier in plan 4	8	12
0°	1.0	1.0	1.0
15°	1.5	2.0	2.5
30°	2.0	2.5	3.5
45°	2.5	3.5	4.5

*The table is intended to indicate the approximate range only. Design depths for severely skewed piers, where the use of these is unavoidable, should preferably be determined by means of special model tests. The values quoted are based approximately on graphs by Laursen (1962).

survey work and computation. Similarly, more detailed information may be needed to assess contraction and local scour (i.e. at piers and abutments), plus any long-term changes in bed elevation, as described in Sections 3.11 and 8.2.2.

1.4.6 *Preliminary structural design of the bridge*

This includes deciding the level of the deck and the foundation details, which may be related to the maximum scour depth (see Section 8.5.3 and Table 8.9). If the bridge deck will be submerged during flood, buoyancy and hydrodynamic forces must be allowed for when designing the superstructure (Appendix A), and the afflux calculated accordingly. The possibility of additional loading due to debris and ice jams should be assessed. In some circumstances ship collision should be included.

Submergence of the opening may cause a very large vertical contraction from the soffit of the deck so that the cross-sectional area of flow is considerably reduced and waterway velocities significantly increased (Sections 2.4 and 7.2). The feasibility of using entrance rounding to improve flow efficiency and reduce scour should be considered (Fig. 1.1).

Having established the basic structural form of the bridge, ensure that any river training works considered necessary during the hydraulic design are still appropriate and effective. If the structural requirements are such that the integrity of the hydraulic design has been compromised, then it is necessary to loop back to an earlier stage (Fig. 1.12). If everything is satisfactory, a costing of the bridge, approaches and training works can be obtained.

1.4.7 *Consider alternatives*

Having found one acceptable design, the opportunity should be taken to revisit the original data and to evaluate whether or not an alternative location or different design approach would yield a cheaper and/or more effective alternative. If the existing design proves to be superior, the design can be finalised. Alternatively, with complex sites or prestigious projects it may be decided to undertake physical or computer modelling to confirm the detailed and long-term flow behaviour before committing to the final design stage.

Example 1.1

A preliminary hydraulic design is required for a bridge to cross a 60 m wide main channel with floodplains 150 m wide on each side when the discharge is 490 m^3/s. The corresponding flow depth is 2.6 m in the main channel and, since the river has banks 1.0 m high, 1.6 m on the immediately adjoining floodplain. The diameter of the sand bed material is D_{50} = 1.0 mm.

Lacey (maximum waterway) width = $4.75Q^{1/2} = 4.75 \times 490^{1/2}$
$= 105\,\text{m}$ (1.5)

approximate minimum waterway width = $3.20Q^{1/2} = 3.20 \times 490^{1/2}$
$= 71\,\text{m}$ (1.6)

Try an opening 80 m wide between abutments with three piers each 1.2 m wide at 20 m centres. This gives a net opening width $b = 76.4$ m, which allows for some contraction of the flow. If the depth (Y) is approximately 2.6 m and $b = 76.4$ m then the net cross-sectional area of flow in the opening, $a = 2.6 \times 76.4 = 199\,\text{m}^2$. Thus the opening velocity $V = 490/199 = 2.5$ m/s. Now consider the potential contraction scour depth, d_{SC}.

As a first iteration, with $D_{50} = 1$ mm and $Y = 2.6$ m from Fig. 1.14 the competent mean velocity (V_{SC}) is about 1.0 m/s, suggesting that the opening area will increase until $V = V_{SC}$ when $a = Q/V_{SC} = 490/1.0 = 490\,\text{m}^2$ and the average depth of flow $Y = a/b = 490/76.4 = 6.4$ m. With $Y = 6.4$ m the competent mean velocity increases to about 1.6 m/s.

Second iteration: say $V = 1.35$ m/s when $a = 490/1.35 = 363\,\text{m}^2$ giving $Y = 363/76.4 = 4.8$ m. With $D_{50} = 1$ mm both $V_{SC} = 1.35$ m/s and $Y = 4.8$ m are broadly consistent with Fig. 1.14 so assume the average contraction scour depth in the opening is about $d_{SC} = (4.8 - 2.6) = 2.2$ m. At the piers the scour depth will be larger as a result of the locally increased velocity and vortices. Say the piers have a width $b_P = 1.2$ m, have round noses, and are aligned to the approach flow (therefore no correction for the angle of attack is needed). From Fig. 1.15 the local pier scour depth $d_{SP} = 1.5\,b_P = 1.5 \times 1.2 = 1.8$ m. Adding this to the contraction scour depth gives a combined depth $d_S = 4.0$ m.

These very approximate scour depths can be compared with the results of the more detailed calculations shown in Example 8.5 and Fig. 8.24.

References

Ackers, P. (1992) Gerald Lacey Memorial Lecture, Canal and river regime in theory and practice: 1929–92. *Proceedings of the Institution of Civil Engineers, Water Maritime and Energy*, **96**, September, 167–178.

d'Aubuisson de Voisins, J.F. (1840) *Traité de hydraulique* (Treatise on Hydraulics), 2nd edn, Piois, Levraut et Cie, Paris.

Blench, T. (1939) A new theory of turbulent flow of liquids of small viscosity. *Journal of the Institution of Civil Engineers*, Paper 5185.

Bradley, J.N. (1978) *Hydraulics of Bridge Waterways,* 2nd edn, US Department of Transportation/Federal Highways Administration, Washington DC.

Brandon, T.W. (ed.) (1987) *Water Practice Manuals, 8, River Engineering – Part I, Design Principles*, The Institution of Water and Environmental Management, London.

Brandon, T.W. (ed.) (1989). *Water Practice Manuals, 8, River Engineering – Part II, Structures and Coastal Defence Works*, The Institution of Water and Environmental Management, London.

Brice, J.C. (1984) Assessment of channel stability at bridge sites, in *Transportation Research Record 950,* Second Bridge Engineering Conference, Vol. 2, Transportation Research Board/National Research Council, Washington DC, pp. 163–171.

Bryan, B.A., Simon, A., Outlaw, G.S. and Thomas, R. (1995) Methods for assessing the channel conditions related to scour-critical conditions at bridges in Tennessee. Final report project no. TN-RES1012, United States Geological Survey, Nashville.

Criswell, H. (1968) East Devon Floods 10th–11th July, 1968. Report to Roads Committee, Devon County Council, Exeter.

Farraday, R.V. and Charlton, F.G. (1983) *Hydraulic Factors in Bridge Design*, Hydraulics Research Station, Wallingford, England.

Fowler, D. (1990) Glanrhyd report says BR had no warning. *New Civil Engineer*, 3 May, 5.

Fowler, D. and Russell, L. (1989) Seven die as scour undermines US bridge. *New Civil Engineer*, 6 April, 7–8.

George, A.B. (1982) Devon floods and the waterways of bridges. *Proceedings of the Institution of Civil Engineers, Part 2*, **73**, 125–134; discussion, **73**, 687–692.

Hamill, L. and O'Leary, A.M. (1985) The hydraulic performance and failure of bridges in Devon and Cornwall. *Municipal Engineer*, **2**, August, 213–221.

Highways Agency (1994) *Design Manual for Roads Bridges*, Vol. 1, Section 3, Part 6, BA59/94, *The Design of Highway Bridges for Hydraulic Action*, HMSO, London.

HMSO (1975) *The Reservoirs Act*, Her Majesty's Stationery Office, London.

Holford, I. (1977) *The Guinness Book of Weather Facts and Feats*, Guinness Superlatives, Enfield, Middlesex.

Inglis, C. (1949) The behaviour and control of rivers and canals. Research Publication

No. 13, Central Water Power Irrigation and Navigation Report, Poona Research Station, India.

Kennedy, R.G. (1895) The prevention of silting in irrigation canals. *Proceedings of the Institution of Civil Engineers*, **119**, 281–290.

Lacey, G. (1929–30) Stable channels in alluvium. *Proceedings of the Institution of Civil Engineers*, **229**, 258–384.

Lacey, G. (1933–34) Uniform flow in alluvial rivers and canals. *Proceedings of the Institution of Civil Engineers*, **237**, 421–453.

Lacey, G. (1939) Regime flow in incoherent alluvium. Publication 20, Central Board of Irrigation and Power, India.

Lane, E.W. (1921) Experiments on the flow of water through contractions in an open channel. *Transactions of the American Society of Civil Engineers*, **83**, 1149–1219.

Laursen, E.M. (1962) Scour at bridge crossings. *Transactions of the American Society of Civil Engineers*, **127**, part 1, 166–180.

Montague, S. and Fowler, D. (1989) Rains wash away Inverness rail bridge. *New Civil Engineer*, 9 February, 5.

Nagler, F.A. (1918) Obstruction of bridge piers to the flow of water. *Transactions of the American Society of Civil Engineers*, **82**, 334–395.

Natural Environment Research Council (NERC) (1975) *Flood Studies Report*. Vol I: *Hydrological Studies*. Vol II: *Meteorological Studies*. Vol III: *Flood Routing Studies*. Vol IV: *Hydrological Data*. Vol V: *Maps*. NERC, London.

Neill, C.R. (ed.) (1973) *Guide to Bridge Hydraulics*, Roads and Transportation Association of Canada/University of Toronto Press, Toronto.

Netlon Ltd. *Maritime and Waterway Engineering with Tensar and Netlon Geogrids*, Blackburn, England.

Novak, P., Moffat, A.I.B., Nalluri, C. and Narayanan, R. (1996) *Hydraulic Structures*, 2nd edn, Unwin Hyman, London.

Overman, M. (1975) *Man, The Bridge Builder*, Priory Press, Hove, England.

Pease, J. (1989) BR fails to notice rail bridge scour. *New Civil Engineer*, 21 September, 7.

Railtrack (1995) Scour and flooding – managing the risk. Railway group standard, GC/RT 5143, Railtrack, London.

Richardson, E.V., Simons, D.B. and Julien, P.Y. (1990) Highways in the river environment. Report no. FHWA-HI-90–016, National Highways Institute/Federal Highways Administration, McLean, VA.

Richardson, E.V., Harrison, L.J., Richardson J.R. and Davis S.R. (1993) *Evaluating Scour at Bridges*, 2nd edn. Publication no. FHWA-IP-90–017, Hydraulic Engineering Circular No. 18, National Highways Institute/Federal Highways Administration, McLean, VA.

Ruddock, E. (1979) *Arch Bridges and Their Builders 1735–1835*, Cambridge University Press, Cambridge.

Smith, D.W. (1976) Bridge failures. *Proceedings of the Institution of Civil Engineers,* Part 1, 60, August, 367–382.

Smith, D.W. (1977) Why do bridges fail? *Civil Engineering*, American Society of Civil Engineers, November, 58–62.

Watson, R. (1990) Hundreds of bridges to undergo scour tests. *New Civil Engineer*, 9 August, 5.

White, K.R., Minor, J. and Derucher, K.N. (1992) *Bridge Maintenance, Inspection and Evaluation*, 2nd edn, Marcel Dekker Inc., New York.

Yarnell, D.L. (1934) Bridge piers as channel obstructions. Technical Bulletin no. 442, November, US Dept. of Agriculture, Washington DC.

2 How a bridge affects river flow

2.1 Introduction

When a bridge is placed in a river it forms a narrowing of the natural channel and an obstacle to the flow. This results in a loss of energy as the flow contracts, passes through the bridge and then, most significantly, re-expands back to the full channel width. To provide the additional head necessary to overcome the energy loss the upstream water level increases above that which would be usually experienced without the bridge. This additional head is called the afflux, and its variation with distance upstream is called the backwater profile. The smaller the opening, the greater the afflux and backwater.

One situation where a knowledge of bridge hydraulics is essential is when a new bridge is to be built that obstructs the main river channel and/or encroaches onto the floodplain. The construction of a hydraulically inefficient bridge could cause flooding upstream, or exacerbate that which already occurs. This would be extremely damaging and expensive if a large number of properties or factories were flooded, so it is important that a hydraulically efficient structure is designed and the backwater calculated accurately.

Sometimes the analysis may involve an existing bridge where flooding already occurs and it is necessary to determine what proportion of this is attributable to the afflux from the structure. First appearances can be deceptive. In many situations flooding, in the form of overbank flow, would occur even if the bridge was not there. For instance, the small arch bridge just visible in the centre of Fig. 2.1 looks as though it should be the cause of the flooding but, although the afflux increases the depth and width of water on the floodplain, it is actually the inadequacy of the river channel that is largely to blame.

When investigating whether or not a bridge is (or will be) the primary cause of flooding the hydraulic capacity of the main river channel without the bridge (Q_R) should be compared with the capacity of the bridge waterway (Q_W) and the design flood (Q_{DF}). Then as a rough guide:

Fig. 2.1 Upstream of Canns Mill Bridge during a 1 in 5 year flood (see also Fig. 2.15). The bridge is in the centre of the photo. The width of the floodwater is about 55 m while the bridge span is 4.3 m. The bridge appears to be responsible for the flooding, but only exacerbates it. The primary cause is the low conveyance of the downstream channel. Much of the water on the upstream floodplain is static overbank storage, a fact that must be recognised in any hydraulic analysis. (Photo courtesy of G.A. McInally)

- if $Q_R < Q_W$ the bridge is relatively blameless;
- if $Q_R < Q_{DF}$ inundation of the floodplains would occur without the bridge;
- if $Q_W < Q_R$ the bridge forms an obstacle to flow and may cause or exacerbate flooding;
- if $Q_W < Q_{DF}$ the waterway is underdesigned;
- if $Q_W > Q_{DF}$ the waterway is overdesigned or has a margin of safety.

Although the above relationships are fairly logical, the extent of any flooding actually induced by the bridge will depend upon such factors as the height of the banks and their freeboard, whether normal or abnormal stages exist, the Froude number, and the severity of the contraction.

The problem of analysing flow through a bridge often appears deceptively easy, perhaps because it has been oversimplified by assuming a horizontal channel and flow at the normal depth parallel to the bed. In reality the bed level will vary considerably so that the depth becomes almost meaningless, while the flow may not be at the normal depth. Additionally, some complex hydraulic phenomena are involved and there are many different types of flow that can occur at any particular bridge site, while identical bridges may perform differently in different locations.

This chapter describes how the flow in a river is affected as it passes through a bridge, and defines some basic variables such as afflux, head loss and energy loss. This is important: for example, engineers not familiar with bridge hydraulics have been known to define the afflux incorrectly. Chapter 3 describes the principal factors that significantly affect the hydraulic performance of a bridge, while Chapters 4 and 5 outline various methods that can be employed to obtain the afflux.

2.2 What happens when water flows through a bridge

If water flows through a constriction – that is, a section narrower than the natural river channel – the water level is increased upstream of the constriction compared with that which would otherwise exist (Kindsvater and Carter, 1955; Tracy and Carter, 1955). The backwater extends upstream to section 0, at which point the constriction has no effect on the water level, and uniform flow exists (Fig. 2.2a). The distance between the constriction and section 0 depends upon such factors as the geometry, roughness and slope of the channel, and can be calculated using a backwater analysis (e.g. French, 1986; Hydrologic Engineering Center, 1990; Chadwick and Morfett, 1993). In the reach affected by the backwater the depth will be greater than normal, so the velocity and energy loss are less than would otherwise occur. Unless the constriction is very severe the flow is usually subcritical, with gradually varied flow upstream and downstream of the structure and rapidly varying flow at the bridge.

When the opening is running free (i.e. not submerged) the water surface is drawn down as it approaches the opening (of width or span b). This zone of drawdown approximates a semicircle of radius b radiating from the centreline of the opening at the upstream face (Fig. 2.2b). Thus the maximum afflux is generally assumed to occur on the centreline of the channel at one opening width (span) upstream of the upstream face of the constriction (Kindsvater *et al.* 1953; Bradley, 1978). This is a simplification: the maximum afflux tends to be nearer the upstream face when the contraction is slight and further upstream when it is severe (Fig. 4.33). Additionally, with wide flooded valleys (of water surface width B) section 1 may be better located around $0.5(B - b)$ upstream of the constriction. Nevertheless, for simplicity and consistency the maximum afflux will be assumed to occur one span upstream from the bridge face at section 1.

After passing through section 1 the water surface is drawn down as it accelerates through the opening, passing through normal depth at section 2 at (or near) the upstream face of the bridge. The body of water in the centre of the channel experiences the greatest acceleration, while deceleration occurs along the outer boundaries. This is significant for two reasons: first, it results in the region of drawdown being roughly semicircular; second, sepa-

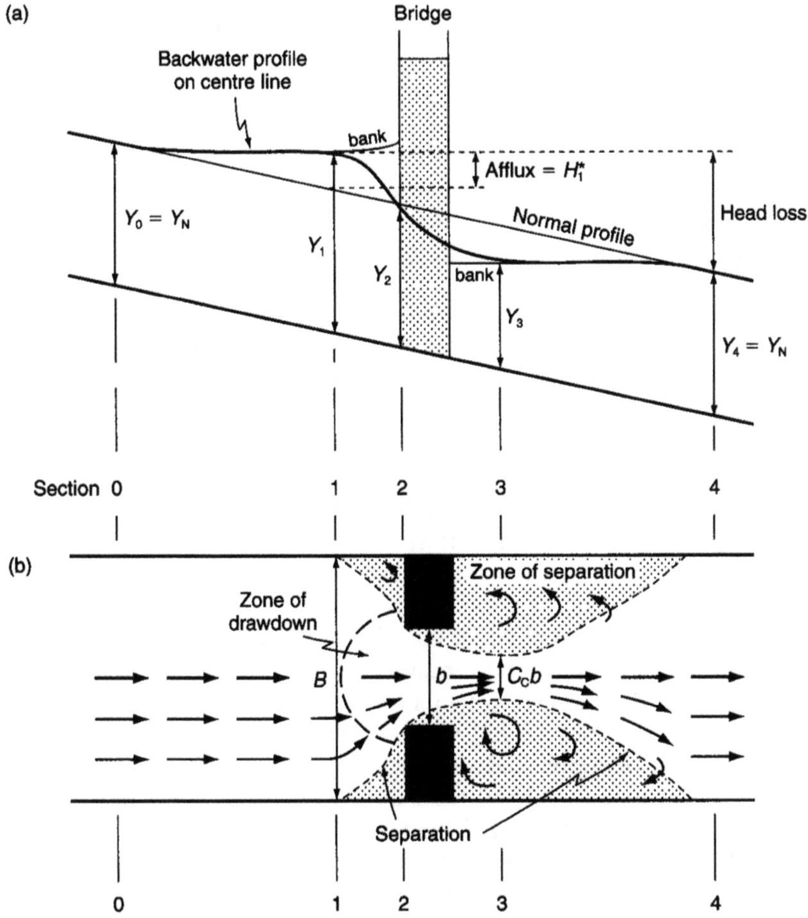

Fig. 2.2 (a) Diagrammatic longitudinal section of uniform flow at normal depth (Y_N) in a river channel with (superimposed) the surface profile arising from the introduction of a bridge. (b) Plan view showing how the flow separates and forms a vena contracta of width C_Cb. For clarity the diagrams are not to scale.

ration occurs in the corners between the edge of the channel and the upstream face of the constriction. The separation zone is delineated by a dashed line in Fig. 2.2b. Eddies may form in the separation zone, especially when the opening has a small span relative to the width of the river. The size of the zone depends upon the upstream channel characteristics and the geometry of the constriction. An important point is that the water level in the separation zone is higher than that in the region of drawdown, so the water level at the edge of the floodplain may be significantly higher than at the centre of the channel. This transverse hydraulic gradient is needed to drive water off the floodplain and through the opening, especially where there are wide, rough floodplains. Skewed openings also cause transverse gradients, the water being trapped in one corner against the bridge face (see Fig. 3.8 and Section 3.7). This is most pronounced when the channel slope is steep.

Although water approaching the bridge along the centre of the channel can pass through the opening unimpeded, as the distance from the centreline increases the streamlines have to curve ever more sharply inwards. Water flowing along the upstream abutments or approach embankments has to curve particularly sharply, and this can interfere with the main central jet passing through the opening, a process that is very troublesome in extreme cases (Section 7.4 and Fig. 7.11). Another result of the inward curvature is that the 'live' stream continues to contract as it passes through the opening so that it reaches a minimum width and depth at section 3 (effectively a vena contracta, as for an orifice). Section 3 may be located either in the opening or some distance downstream of it, the actual position depending upon the length of the waterway and the downstream conditions, but for convenience it is usually assumed to be at the downstream face of the constriction (Fig. 4.34). The width of the live stream can be estimated as $C_C b$, where C_C is the dimensionless coefficient of contraction and b is the width of the opening (m). Both the location and width of the vena contracta significantly influence the energy loss and hydraulic performance (see Section 3.4).

The expansion of the live stream starts at the vena contracta and continues until the flow fills the full width of the channel and normal conditions have been re-established at section 4. There is no absolute rule regarding the location of section 4: the distance depends upon the geometry of the channel, its roughness and the flow characteristics. HEC-2 suggests an expansion of the jet at the rate of 1:4 (width:length) so the section would be located at least $2(B - b)$ from the downstream face (Hydrologic Engineering Center, 1990), but this is questionable (Kaatz and James, 1997). Note that some computer models (e.g. WSPRO) assume that section 4 is one span (b) downstream, so there are many variations. However, for consistency it is always assumed in this book that section 4 is far enough downstream for normal depth conditions to have been re-established and for the flow to be unaffected by the bridge.

Downstream of the structure there is a large zone of separation (Fig. 2.2b). In all of the zones of separation eddying occurs between the live stream and the side of the channel, but this can be very pronounced downstream of the opening. Here, under appropriate conditions, the flow can be very turbulent with large eddies, and it is possible for the flow direction to be backwards towards the constriction: that is, from right to left in Fig. 2.2. The line along which separation occurs represents a shear boundary. As the jet expands, the shear causes a deceleration of the jet, its width increasing in the direction of flow to maintain continuity of discharge. The creation of eddies along the boundary results in lateral mixing of the two bodies of water until the diffused jet occupies the whole channel. This process is aided by the existence of a deep, slower-moving body of water downstream. The transfer of momentum from the jet to the separation zone also results in the water in this zone being accelerated, entrained and carried downstream by the mixing process.

The region of expanding flow is important since the greatest energy loss usually occurs between the vena contracta and the section where the expansion ends. A rule of thumb is that the energy loss due to the expansion is twice that of the contraction. In the expansion the high rate of production of turbulence in the shear zones results in a large loss of energy that, when added to the boundary shear loss, may lead to a total energy loss greater than the initial kinetic energy of the jet (Laursen, 1970). To overcome this energy loss the water level increases upstream of the structure, which is why afflux occurs.

The presence of wide, densely vegetated floodplains can significantly increase the backwater arising from a constriction, perhaps even doubling it (Laursen, 1970; Bradley, 1978; Kaatz and James, 1997). In addition to the process described above, there is now a significant flow from the floodplain into the main channel in order to pass through the opening, followed by the expansion of the flow back onto the floodplain downstream. Laursen called this accretion and abstraction respectively. In both cases an additional head of water is needed to drive the tranverse flow. This is an important factor to remember when selecting a method to calculate the afflux and when assessing the accuracy of the results (see also Section 4.6).

Figure 2.2a shows the water depth relative to a uniformly sloping bed. In real situations the bed is often irregular so it is better to work in terms of the elevation above a datum (such as ordnance datum) as in Fig. 2.3. This shows diagrammatically the longitudinal profile at a bridge site resulting from uniform flow: that is, when the flow in the channel without the bridge is constant at the normal depth (Y_N) as calculated from the Manning equation, for example. For uniform flow to occur the discharge must be constant, and the channel, within a sufficiently long reach, must have a uniform cross-section, a uniform surface roughness and a uniform gradient. Skogerboe *et al.* (1973) pointed out that these stringent conditions suggest that uniform flow at a bridge site may be the exception rather than the rule. Non-uniform flow occurs if there is downstream control of water levels due to a severe bend, flood conditions at a confluence between two streams, vegetation, tides or any sort of obstruction, including another structure. In this case the usual water surface profile is a shallow curve, termed the abnormal profile, lying above the normal depth line (Fig. 2.4). The backwater profile is obtained by adding the bridge afflux to the abnormal profile. Obviously this is more complicated since the abnormal profile is not parallel to the bed. Again, working to a datum below bed level makes things easier.

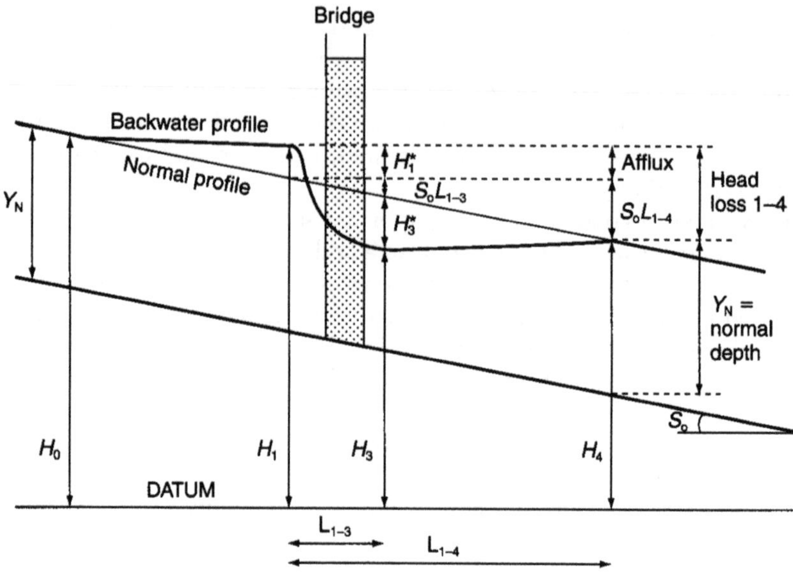

Fig. 2.3 Definition of afflux ($H_1^* = Y_1 - Y_N$) and piezometric head loss for uniform flow at normal depth with the elevation of the water surface measured above a datum. The corresponding water depths at the sections are Y_1, Y_2, Y_3, etc. Note that the difference in water level between sections 1 and 3, $\Delta h = H_1^* + S_O L_{1-3} + H_3^*$.

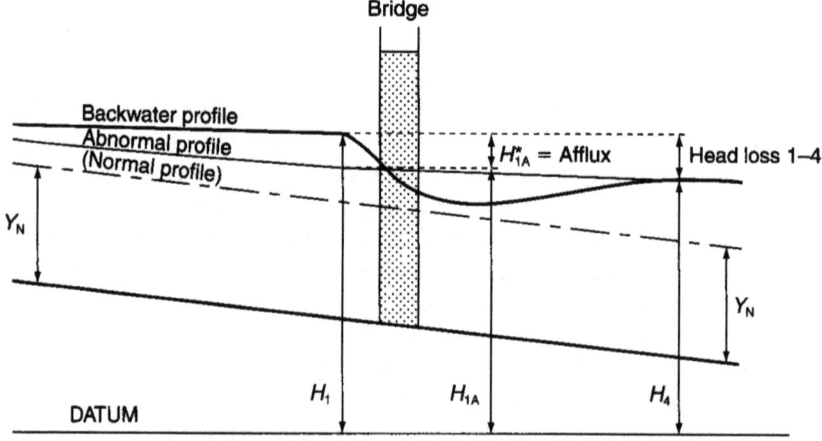

Fig. 2.4 Definition of afflux and piezometric head loss for non-uniform flow at abnormal depth. The afflux (H_{1A}^*) must be obtained from the difference in the elevation of the water surface at section 1 with and without the bridge, i.e. H_1 and H_{1A} respectively. The elevation of the water surface is measured from a datum, the corresponding water depths at the sections being Y_1, Y_{1A}, Y_4, etc.

It should be appreciated that the bridge does not alter the elevation of the water surface or the energy line at section 0, nor at section 4. By definition (in this book) these two sections are located outside the reach affected by the structure. It is only between these two sections that the values are changed as a result of the obstruction to flow.

2.3 Afflux, piezometric head loss and energy loss

Regardless of whether it is a proposed or existing bridge that is to be analysed, most hydraulic investigations require an estimate of the change in water level caused by the structure. This means that the afflux and the piezometric head loss are required. The energy loss includes a consideration of the change in both the velocity head and the piezometric head. These variables are defined below, while methods that can be used to evaluate them are described in Chapters 3–5.

2.3.1 The uniform flow condition

Because it is the simplest, the uniform flow condition will be used initially to define and illustrate the afflux and head loss across a bridge. For the situation shown in Fig. 2.2a, at any particular discharge:

$$\text{maximum afflux, } H_1^* = Y_1 - Y_N \qquad (2.1)$$

where Y_1 is the water depth (m) at section 1 and Y_N is the normal depth (m). In reality the bed is likely to have a variable slope and the depth of flow will not be constant, so a more widely applicable definition is shown in Fig. 2.3. Here the elevation of the water surface above a datum is used. Again the maximum afflux (H_1^*) is the difference in water level with and without the bridge, but it is now apparent that

$$H_1^* = H_1 - (H_4 + S_O L_{1\text{-}4}) \qquad (2.2)$$

where H_1 and H_4 are the elevation of the water surface (m) at sections 1 and 4 respectively, S_O is the dimensionless gradient of the channel and $L_{1\text{-}4}$ is the plan distance (m) between sections 1 and 4. With existing bridges the difference in bed level ($S_O L_{1\text{-}4}$) can be measured. Note that ($H_1 - H_3$) is greater than the value obtained from equation 2.2 but is not the true afflux, simply the difference in water level across the constriction. In channel flow with a transverse slope towards the centreline, the maximum difference may be between the bank water levels at sections 2 and 3 (Fig. 2.2a). If the water level rises above the top of the waterway opening, then the maximum head difference may occur between the centre of the opening at section 2 (where recovery of the velocity head occurs) and section 3. The latter may be of interest with respect to hydrostatic loading of the structure.

The piezometric head loss is the difference in the elevation of the water surface between two points, which must be specified. With a constant bed slope as in the diagrams, the fall in bed level depends upon the distance between the two points. The true head loss across the constriction is that measured between section 1 and section 4, where the normal depth has been recovered (Fig. 2.3). Thus

$$\begin{aligned} \text{head loss 1–4} &= H_1 - H_4 \\ &= H_1^* + S_O L_{1\text{–}4} \end{aligned} \tag{2.3}$$

Equation 2.3 illustrates that the afflux and head loss are related, and that the head loss increases as $L_{1\text{–}4}$ increases (but the afflux does not). If normal depth is not recovered at section 4, say as a result of a large energy loss, then the head loss ($H_1 - H_4$) increases.

Figure 2.5 shows how the gradient of the energy line varies in the vicinity of a bridge. From section 0 to 1 a backwater occurs so the flow is deeper and slower than normal, which reduces the energy losses and decreases the energy gradient. The reverse happens between sections 1 and 3, while between sections 3 and 4 the flow is expanding against an adverse hydraulic gradient (i.e. an increasing depth of water) so the energy losses are correspondingly high giving a steep energy gradient. For simplicity the energy gradient between the sections has been drawn as a straight line.

The energy loss, or total head loss, is the difference in the elevation of

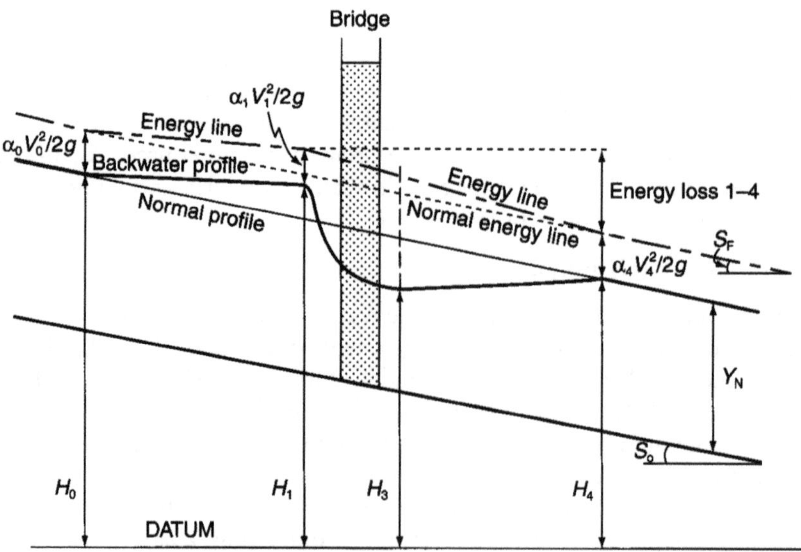

Fig. 2.5 Diagrammatic illustration of the variation of the slope of the energy line near a bridge, and the definition of the energy loss for uniform flow at normal depth. The increased depth upstream of the bridge results in lower velocities and a shallower energy or friction gradient (S_F) than downstream, where the reverse happens. The energy loss for non-uniform flow is defined in a similar manner.

the energy line between two points, which again must be specified. The true energy loss across a bridge is measured between sections 1 and 4:

$$\text{energy loss 1–4} = (H_1 + \alpha_1 V_1^2/2g) - (H_4 + \alpha_4 V_4^2/2g) \quad (2.4)$$

where α is the dimensionless velocity head coefficient to allow for the non-uniform velocity over the area of flow, V is the mean velocity of flow (m/s), and g is the acceleration due to gravity (m/s^2).

The calculation of either the head loss or the energy loss is far from easy: different (but correct) answers can be obtained according to whether it is calculated along the centreline of the channel or elsewhere. Additionally, H, α and V all vary both longitudinally and transversely in the river channel, particularly in compound channels during flood, so obtaining accurate values is difficult. Estimating the value of α is not a simple matter, and can be done with relative accuracy only if the variation of velocity over a particular cross-section is known. Unfortunately, this information is usually not available so the values of α and V have to be guessed or estimated crudely (see Section 3.10 and Fig. 4.20). Remember that the energy line must always fall in the direction of flow, so if calculations result in anything else this is a good indication that the values assigned to the variables are incorrect or that a mistake has been made.

The energy loss caused by a bridge can be assumed to arise from three main things:

- contraction of the flow caused by the abutments, noses of the piers and, when the opening is submerged, the soffit or deck of the bridge (15%);
- friction between the water and the surfaces of the piers, abutments and, when the opening is submerged, the soffit of the bridge (20%);
- expansion of the live stream downstream of the bridge (65%).

The figures in brackets give a very approximate indication of the relative size of the three energy losses, which can vary significantly. These losses are governed by the shape, length and alignment of the abutments and piers (and the deck when submerged), the severity of the constriction, and by the velocity of flow. Generally the energy loss, and hence the afflux, is proportional to the opening velocity head. If the abutments are not in the river, the friction loss due to the central piers can be negligible compared with that arising from the contraction and expansion of the flow, and so can be ignored in many cases. These factors are considered in more detail later. Also of importance is the type of flow through the waterway opening, as described in Section 2.4.

2.3.2 *The non-uniform flow condition*

With non-uniform flow and abnormal stages (suffix A below) that are greater than the normal depth, the calculation of the afflux is not quite so simple. Without the bridge the slope of the bed and the gradient of the water surface are not equal, so the depth of water at the site varies with distance upstream of section 4 (Fig. 2.4). Consequently if an existing bridge is being investigated this abnormal profile will have to be computed from a backwater analysis; if the bridge has not yet been constructed it can be either measured in the field or calculated. The geometry of this longitudinal profile will change with the discharge, but at any particular discharge

$$\text{maximum afflux, } H^*_{1A} = H_1 - H_{1A} \tag{2.5}$$

where H_1 is the elevation of the water surface at section 1 with the bridge, and H_{1A} is the elevation at section 1 without the bridge, both measured above an arbitrary datum level. The head loss between sections 1 and 4 is given by

$$\text{head loss 1–4} = H_1 - H_4 \tag{2.6}$$

where H_4 is the elevation above datum of the water surface at section 4, which is the same with or without the bridge. There is now no simple relationship between afflux and head loss.

The energy loss again can be calculated from equation 2.4, and is caused by the same factors as described earlier. However, with an abnormal stage the flow is deeper and slower than normal so the energy loss and afflux may be reduced, but deck submergence may be more of a problem. Under these conditions a large waterway and/or a significant head difference across the structure may be needed to pass the design discharge.

When analysing any bridge, regardless of whether normal or abnormal stages exist, it is essential that the types of flow occurring at the site are correctly identified.

2.4 Classification of flow types at a bridge

There are many types of flow that can occur at a bridge site depending upon the upstream and downstream stage, the discharge, the severity of the constriction, and its geometry. Similarly, identical bridges at different sites may experience different types of flow. The flow types have varying characteristics so different equations have to be used to calculate the discharge, afflux and energy loss. Thus it is important to be able to identify the flow types that will occur. These are described below and illustrated in Fig. 2.6.

Type 1

The opening is submerged or drowned at both the upstream and downstream face, with the waterway flowing completely full. This can be referred to as the drowned orifice condition and evaluated using an equation of this type (equations 2.9 and 2.10). This condition occurs quite frequently when the design discharge is exceeded at sites with abnormal stages, shallow

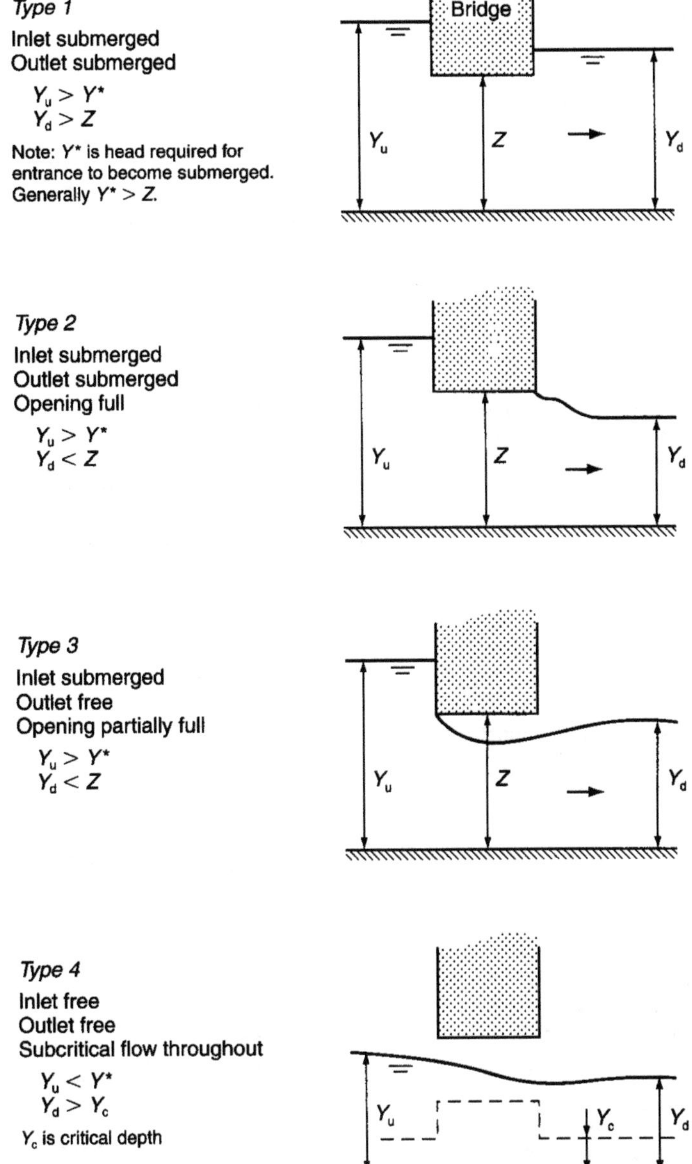

Fig. 2.6 Classification of some common types of flow through bridges. Y^* is the stage needed to submerge the waterway opening; usually $Y^* > 1.1Z$. The critical depth is Y_C, which is higher in the narrow opening.

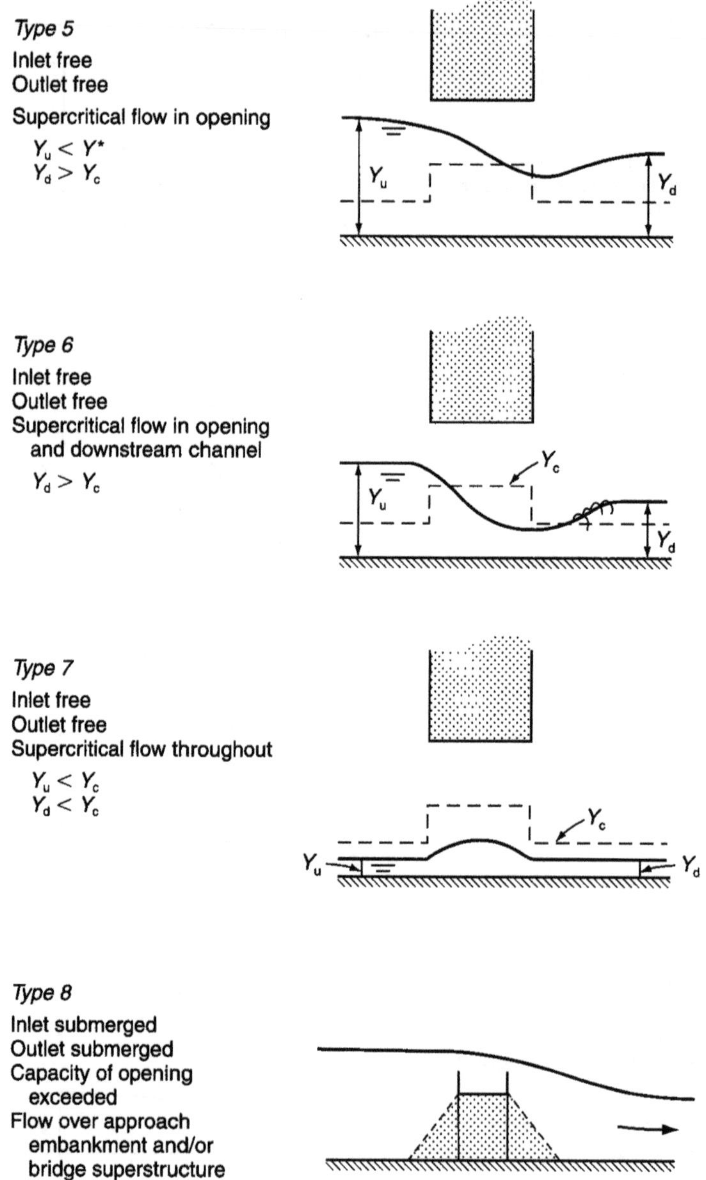

Fig. 2.6 Classification of some common types of flow through bridges (*cont.*) Y^* is the stage needed to submerge the waterway opening; usually $Y^* > 1.1Z$. The critical depth is Y_C, which is higher in the narrow opening.

channel or friction slopes, and low normal depth Froude numbers (eg $F_N < 0.25$). When conducting a numerical analysis of drowned orifice flow it is essential that the opening is truly submerged at both faces. This may only occur when both water levels exceed $1.1Z$, where Z is the height of the opening.

This type of flow is significant because it can result in a large afflux and head loss, and so cause extensive flooding upstream, as illustrated in Section 2.6. It is inefficient, partly because the waterway is flowing full resulting in a large boundary friction loss, and partly because (unlike type 3 sluice gate flow) the water in the downstream channel is deep enough to prevent the waterway from discharging freely. Equation 2.9 shows that with drowned orifice flow the geometry of the opening (C_d, a_W), the difference in head across the structure (ΔH) and the depth of water in the downstream channel (Y_d) all affect the discharge (Q), so there is a combination of structure and channel control.

Type 2

This is the transitional case between types 1 and 3. The flow is similar to type 1 except that the outlet is not submerged, although the waterway is still running full. This means that the friction loss is still relatively high.

Type 3

The upstream face of the opening is submerged but the water level is below soffit level at the downstream face. This is frequently referred to as sluice gate flow, and can be evaluated using an equation of this type (equation 2.8). The discharge through the opening depends upon the upstream water level (Y_u) and the geometry of the opening, so structure control is dominant. The downstream water level is irrelevant. The waterway is running only partially full so the friction loss is reduced compared with type 2 flow. This condition frequently occurs when the capacity of the opening is exceeded in channels with relatively steep bed or energy gradients and $F_N >$ 0.25, or with flatter channels in the transition to flow types 2 and 1 (Hamill, 1997).

The head required for the opening to become permanently submerged is denoted by Y^* in Fig. 2.6 with $Y^* > 1.1Z$. Around this stage the upstream water level is usually extremely turbulent and fluctuates considerably, resulting in the opening running alternately free and submerged (Fig. 2.7). In this photograph, despite the horizontal soffit, the opening is submerged at the sides but not at the centre, where the velocity head is largest. After becoming permanently submerged there can be a rapid increase in upstream water level (Fig. 2.14). A helpful rule of thumb when trying to decide whether or not sluice gate flow will occur at a site is that laboratory tests indicated that in channel flow (type 4) normally about one-half of the total fall of the

Fig. 2.7 There was some concern regarding the safety of this bridge at Camelford in June 1993. Note the extreme turbulence as the opening starts to drown. This illustrates the difficulty of measuring upstream water levels accurately and in determining the height required for an opening to ensure a minimum freeboard of 0.6 m. (Reproduced by permission of Newquay Press Service)

water surface from section 1 to 3 occurs between section 1 and the upstream face of the bridge (Kindsvater and Carter, 1955). Often the maximum flood level at sections 1 and 3 can be determined from trash marks.

Type 4

This is the subcritical channel flow condition that is most commonly encountered. The water level is below the top of the opening at both the upstream and downstream face of the bridge. The flow can be either at the normal depth or at an abnormal stage, but the depth is always greater than the critical value (Y_C) that represents the start of supercritical flow.

Type 5

In this channel flow condition the water surface passes through the critical depth in the opening indicating supercritical flow. Note that the critical depth is larger in the opening than in the river channel because the opening is narrower. Thus supercritical flow often occurs first in the opening. This type of flow can be very difficult to detect from a visual inspection of the site, because if the flow is only weakly supercritical there will be no obvious hydraulic jump. In some instances it is possible to calculate the critical depth (Section 3.3.1) corresponding to a particular discharge and compare this with observations made in the field, if there are any. Alternatively, it is possible to calculate the width of constriction needed for flow to occur at the critical depth (Section 3.3).

Type 5 flow is more likely to occur (than type 4) when F_N or the channel slope is relatively large, the constriction is relatively severe, or perhaps when the opening is located eccentrically in the channel. Flow types 5, 6 and 7 are significant because supercritical flow in the waterway means that the upstream water level is now independent of the conditions downstream.

Type 6

This is similar to type 5 except that the flow is supercritical in both the opening and the channel immediately downstream of the bridge (Fig. 2.8). In the more extreme cases ($F_N > 2$) it may be possible to see clearly that the flow is supercritical because there may be a well-developed hydraulic jump. The conditions associated with this type of flow are similar to those described for type 5, only more severe. The problem of scour may now be quite acute.

Fig. 2.8 Type 6 supercritical flow in the opening and downstream channel. This bridge is eccentric to the channel so there is a considerable flow along the upstream face of the right hand abutment (on the left in the photo). This pushes the jet towards the left abutment, narrows the vena contracta, and results in high velocities (see Fig. 7.11).

Type 7

Supercritical flow now exists throughout the channel and opening, and a loss of energy results in an increase in water level. This type of flow is rarely encountered but could occur in mountainous regions or in smooth, steep, artificial channels.

Type 8

This type of flow occurs only when the capacity of the opening is grossly exceeded and water spills over the bridge deck and/or the approach embankments. Because all of the flow does not now pass through the opening, this condition is not consistent with those described above (which is why it was placed last). A different method of analysis is required, as described in Chapter 6.

The eight types of flow above give rise to the important generic classification of subcritical and supercritical flow. This is important not only because the hydraulic behaviours of these two types of flow are different, but also because many model studies of bridge hydraulics have not included supercritical flow at all. Crucially, it is usually more difficult to analyse the supercritical condition and produce appropriate designs, while the phenomenon known as 'choking' can result in a much larger afflux than expected (see Section 3.3.3).

Another important generic classification is that of drowned orifice, sluice gate and open channel flow with equations 2.7–2.10 (or similar) being used to analyse the stage–discharge relationship. These equations are different in form, so it is essential to know what general type of flow is being dealt with. For example, Fig. 2.9 illustrates how the velocity profile in a laboratory channel (diagram a) is modified by the flow contraction that occurs at the entrance to a bridge opening, in this case an arch. Even in channel flow the contraction has a radial component that deflects the flow downwards so that the maximum relative velocity occurs nearer to the bed (b and c). Once the stage is above soffit level, the contraction is most pronounced in sluice gate flow (d). Many bridges have a scour pit just downstream because the jet has a large velocity and a downward component as it emerges from the opening.

A knowledge of the flow type is important when designing works to improve the hydraulic performance of a bridge and reduce flooding (see Chapter 7). For type 1 drowned orifice flow with downstream control and an abnormal stage, some form of channel improvement below the bridge may be most effective. This can result in an increased conveyance and reduced stages at the bridge site. On the other hand, if a rounded entrance to the bridge waterway is to be employed, this may be ineffective in channel flow, fairly effective with type 1 drowned orifice flow, but very effective with type 3 sluice gate flow. This is because with type 3 flow there is a large contraction from the soffit and a very uneven distribution of velocity within the

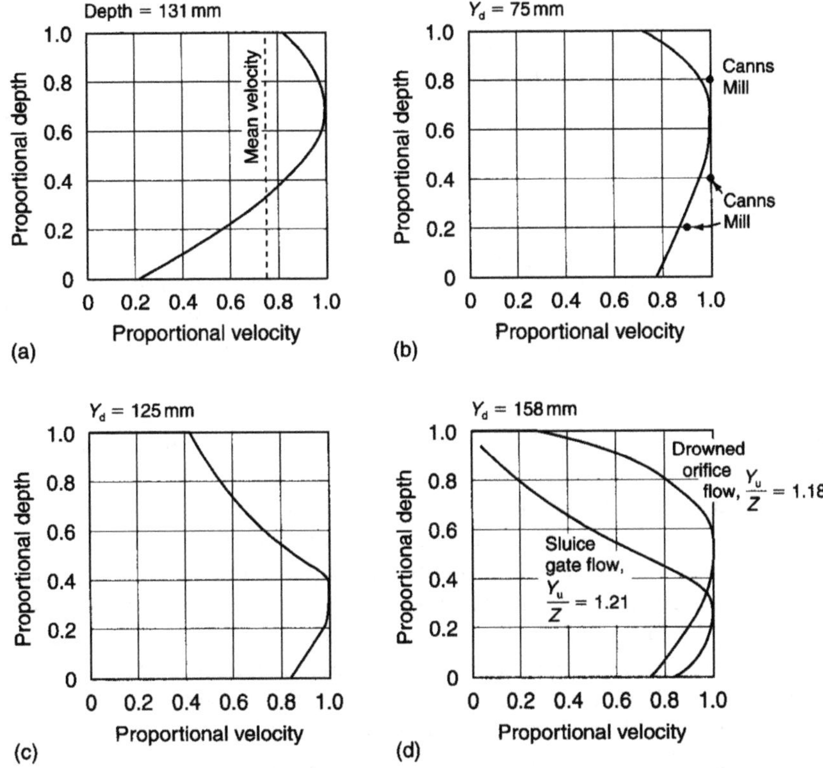

Fig. 2.9 Variation of proportional velocity with proportional depth: (a) in an unconstricted 450 mm wide laboratory channel; (b) in the jet emerging from a 300 mm span model arch bridge with $Y_u/Z = 0.54$ (three field measurements from Canns Mill Bridge are also shown); (c) in the jet emerging from the 300 mm arch with $Y_u/Z = 0.98$; (d) in the jet emerging from the 300 mm arch with the opening submerged and experiencing drowned orifice flow and sluice gate flow. Note how the radial contraction from the arch results in the maximum velocity moving progressively closer to the bed.

opening, whereas with type 1 the whole opening is already being used and there is a more uniform velocity distribution. This is apparent in Fig. 2.9d, which shows that in sluice gate flow only about 50% of the jet emerging from the opening has a centreline proportional velocity of 0.7 or more, whereas in drowned orifice flow about 85% of the jet has a proportional velocity over 0.7. Thus it pays to know what type of flow is being dealt with.

2.5 Channel control and structure control

This is another generic classification that incorporates some of the factors described above. The classification depends upon whether the stage and discharge at the bridge site are controlled by the channel or by the structure. An example of channel control (i.e. channel characteristics dominating) is a low stage at a bridge opening that is almost as wide as the river so that the flow is practically unaffected. The stage–discharge relationship for a channel experiencing uniform flow can be predicted using the Manning equation:

$$Q = \frac{A}{n} R^{2/3} S_F^{1/2} \qquad (2.7)$$

where Q is the discharge (m^3/s), A (m^2) is the cross-sectional area of flow in the channel, R (m) is its hydraulic radius (= cross-sectional area/wetted perimeter), n ($s/m^{1/3}$) is the Manning roughness coefficient, and S_F is the (dimensionless) slope of the energy line (friction gradient). In uniform flow S_F equals the bed slope, S_O, but in non-uniform flow it does not. In American texts using English units the $(1/n)$ of equations 2.7 and 3.6 becomes $(1.49/n)$.

Two features of channel control are that the stage–discharge curve is concave downwards (Fig. 2.10), and that the bridge opening ratio (M) is an important factor in determining the hydraulic performance of a bridge, as described in Section 3.2. Channel control can occur with normal or abnormal stages, the latter being where a backwater from further downstream controls water levels at the bridge site. A feature of abnormal stage channel control is that the water level downstream of the bridge often increases more rapidly than that upstream. Under these conditions the channel characteristics significantly influence the afflux and head loss.

Structure control occurs if an opening is very narrow and/or low so that the constriction itself controls the flow and determines the upstream water level. In sluice gate flow where the water level rises above the top of the opening, submerging or drowning the waterway, the opening ratio (M) becomes unimportant so the discharge (Q) can be calculated from equation 2.8:

$$Q = C_d a_W \left[2g \left(Y_u - \frac{Z}{2} + \frac{\alpha_u V_u^2}{2g} \right) \right]^{1/2} \qquad (2.8)$$

where C_d is a dimensionless coefficient of discharge, a_W is the total cross–sectional area (m^2) of the opening flowing full, g is the acceleration due to gravity (9.81 m/s^2), Y_u is the water depth (m) on the centreline at an upstream cross-section (either the bridge face or one span upstream depending upon which has the largest stage), Z is the vertical height (m) of the opening from mean bed level (that is, the height to the crown of an arch), α_u is the dimensionless velocity distribution coefficient, and V_u is the mean upstream approach velocity (m/s). The discharge through the bridge is determined mainly by Y_u and the geometry of the waterway (i.e. C_d, a_W, Z), so the bridge controls the flow. In structure control the discharge and

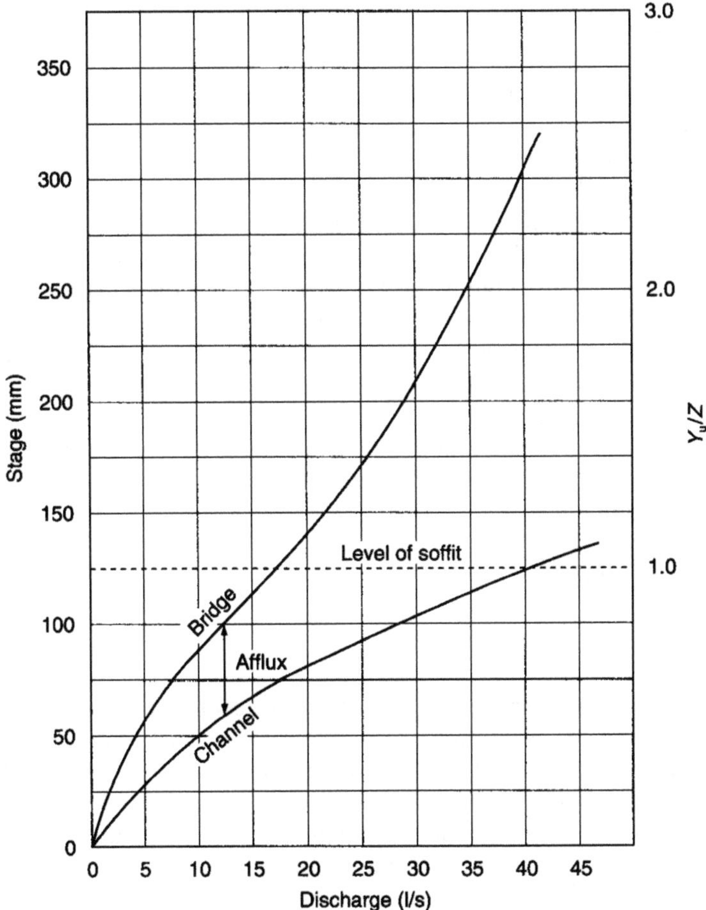

Fig. 2.10 Stage–discharge curves for an unconstricted 450 mm wide laboratory channel and a 250 mm span model bridge with a rectangular opening. The channel curve is concave downwards while, after submergence, the bridge curve is concave upwards. These curvatures are characteristic of channel and structure control. The vertical difference between the curves is the afflux.

upstream water level are independent of the conditions in the downstream channel. However, when sluice gate flow is just becoming established, the conditions in the channel may still have an influence on the upstream stage, but this influence diminishes as Y_u increases.

A feature of structure control is that the upstream stage (Y_u) increases more quickly than that downstream (the two are not related). Another is that the stage–discharge curve is usually concave upwards after the upstream water level has risen above the soffit of the bridge; channel flow is concave downwards (Fig. 2.10). The curvature can be a useful way of identifying the flow: at some sites it may not always be clear whether the flow is controlled by the channel or by the structure, and there may be an indistinct transition zone. This often occurs on a rising or falling stage when the flow alternates between channel flow and sluice gate flow, and neither is really established. The vertical difference between the two stage–discharge lines is the afflux caused by the bridge.

The shape of the lines in Fig. 2.10 is easy to explain. With respect to equation 2.7, $A = BY$ where B is the width of the channel (m) and Y is the depth of flow (m); if a wide rectangular channel is assumed so that R can be approximated by Y, then ignoring the other variables the equation reduces to $Q \propto Y^{5/3}$. On the other hand, with equation 2.8 it can be assumed that the area of the opening, a_w, is constant so basically $Q \propto Y^{1/2}$. Spend a moment plotting the graphs of these two relationships and they will have the curvature described above.

When both the upstream and downstream water levels are above the top of the opening the flow is of the drowned orifice type and can be described by an equation of the form

$$Q = C_d a_W \, (2g\Delta H)^{1/2} \tag{2.9}$$

$$\text{where } \Delta H = \left(Y_u + \frac{\alpha_u V_u^2}{2g} - Y_d\right) \tag{2.10}$$

and Y_d is the downstream water depth (m) at the centre of the opening. The conditions associated with drowned orifice flow are such that the velocity head is often negligibly small so the differential head can be taken as $\Delta H = (Y_u - Y_d)$. There is always some degree of both structure and channel control since both the geometry of the opening (i.e. C_d, a_W) and ΔH (which includes Y_d) affect the discharge through the bridge.

Typical values of the discharge coefficients (C_d) for use with equations 2.8 and 2.9 are shown in Figs. 2.11 and 2.12. These have been derived from several sources (Bradley, 1978; Hamill and McInally, 1990; Hamill, 1997). The diagrams show the range of values applicable to various types of bridge; it would be wrong to assume that at any stage there is a single value that can be adopted. Remember that the transition from channel flow to submerged flow is always unpredictable until $Y_u \gg 1.1Z$ or $Z/Y_u < 0.9$.

As a further example of a few of the things discussed in this chapter, imagine a laboratory channel that is supplied with water by a pump. The slope of the channel can be altered, while the depth of flow can be adjusted by raising or lowering a tailgate at the end of the channel. Suppose that initially the tailgate and slope are set so that uniform channel flow at a constant normal depth is obtained as in Figs 2.2a and 2.3. If everything remains the same except that the tailgate is raised to create a backwater, there will now be an abnormal stage as in Fig. 2.4. It is the same bridge operating at

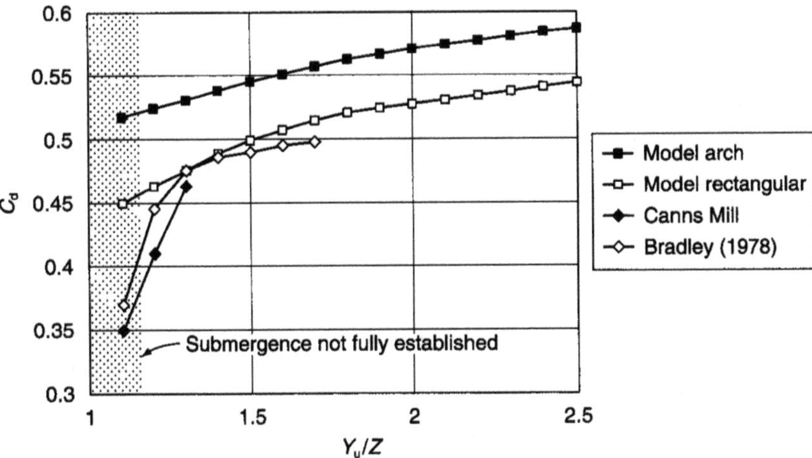

Fig. 2.11 Coefficients of discharge (C_d) for sluice gate flow. Below $Y_u/Z = 1.1$ submerged flow is not established, and the results are unreliable. The results are from Bradley (1978), Hamill and McInally (1990), and Hamill (1997).

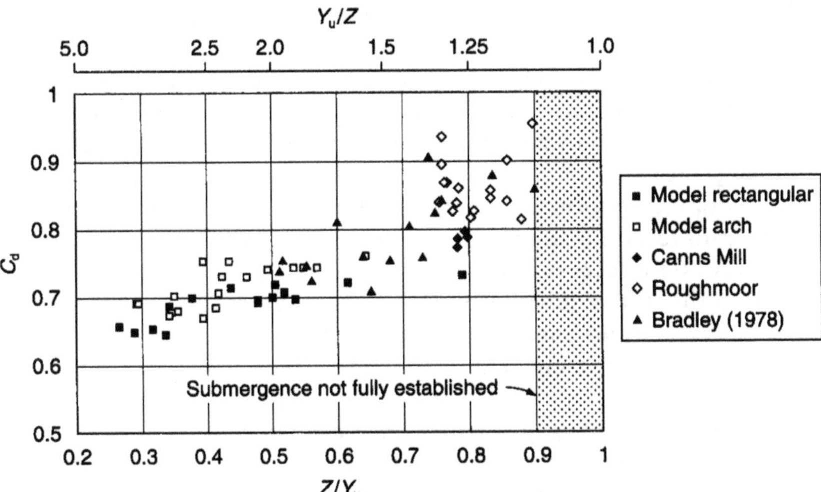

Fig. 2.12 Coefficients of discharge (C_d) for drowned orifice flow. Below $Y_u/Z = 1.1$ or $Z/Y_u = 0.9$ submerged flow is not established and the results are unreliable. The results are from Bradley (1978), Hamill and McInally (1990) and Hamill (1997).

the same discharge, but the afflux is different and the method of analysis would have to be different. Note that under these conditions, if the depth of flow and bed slope are used to calculate the discharge from the Manning equation, the answer would be too large. With non-uniform flow, the slope of the energy line (the friction gradient) must be used in the Manning equation, not the bed slope.

Continuing the above example, if the tailgate is raised progressively, eventually the bridge opening will become submerged by the backwater so that type 1 drowned orifice flow is established. This would require yet another method of analysis. As observed earlier, identical bridges experiencing the same discharge may exhibit very different types of flow and hydraulic behaviour.

2.6 Case study: Canns Mill Bridge

Canns Mill Bridge is located on the River Dalch in Devon (Fig. 2.1). It is a single-span segmental arch structure with a waterway opening approximately 4.28 m wide, 1.8 m high and 3.35 m long. The segmental arch springs from vertical abutments approximately 0.6 m high. During the course of the investigation a number of floods caused slight scouring of the bed so that the height of the arch above mean bed level increased to 1.9 m (which explains why a different but equally correct figure may be used later).

The channel approaching the bridge is quite straight and steep, having a gradient of about 1:60. However, the downstream channel has an adverse gradient as a result of a scour pit: that is, it slopes towards the bridge for the first 12 m or so before beginning to fall in the direction of flow (Fig. 2.13). After this the river channel is tortuous and overgrown so that it provides a control on water levels at the bridge during flood, resulting in abnormal stages.

The hydraulic performance of the bridge was investigated by locating pressure transducers in stilling tubes at four locations, as shown in Figs 2.13 and 2.15. All transducer readings were checked against staff gauges located adjacent to the stilling tubes. The transducer readings were recorded at 15 minute intervals by an intelligent outstation (McInally and Hamill, 1987). Additional manual measurements were made at the faces of the bridge. The river discharge was measured by velocity meter gauging from the bridge, and the results checked against a mathematical model of the site.

The largest event recorded was about 15 m^3/s, which caused the water level to rise 0.76 m above the upstream soffit of the bridge, equivalent to $Y_u/Z = 1.4$. This level is indicated by the white mark on the bridge face in Fig. 2.15. Typically a shallow standing wave formed against the face, moving further upstream as the degree of submergence increased. At the downstream face the water level was probably around 0.2 m above the downstream soffit of the bridge (which is higher than the upstream soffit). There was significant overbank flow both upstream and downstream of the bridge, but no bypass flow around it. This study provided some reliable and accurate water level and discharge data (Hamill and McInally, 1990).

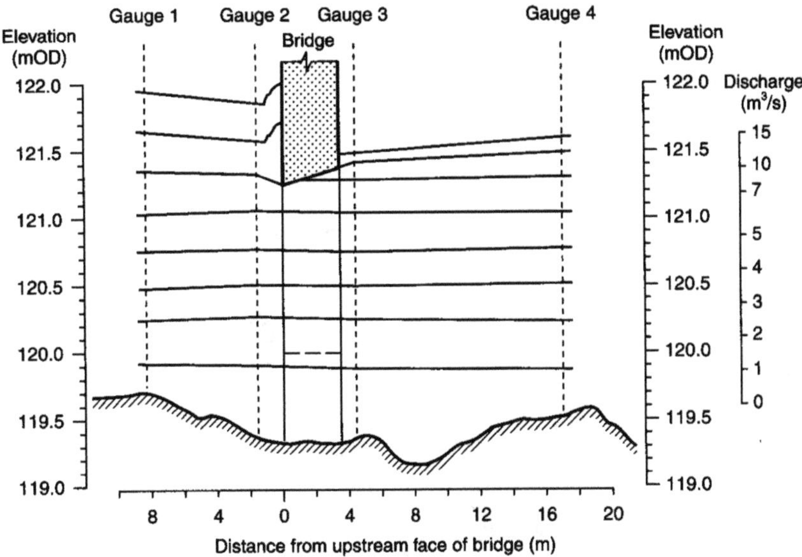

Fig. 2.13 Longitudinal section through Canns Mill bridge showing the profile of the water surface at various discharges. Until submergence the site is controlled by an abnormal stage arising downstream, so the depth initially increases more rapidly downstream than upstream.

The longitudinal profile of the water surface observed during a series of floods in November 1986 is shown in Fig. 2.13. At low flows the water surface has a relatively uniform gradient throughout the site. At medium stages (type 4 flow) a backwater from further downstream reduces the slope of the water surface and the depth is greater downstream than upstream, indicating typical non-uniform flow behaviour associated with abnormal stages and channel control. The normal depth Froude number at Canns Mill is difficult to estimate accurately because of the compound channel (see Section 3.3) but is around 0.3, more or less on the boundary between sluice gate and drowned orifice flow. Thus at higher stages with the opening permanently submerged flow types 3, 2, and 1 occur. With structure control established, there is a larger increase in depth upstream than downstream.

The hydraulic performance of the bridge is illustrated nicely by the relationship between the head loss and the discharge (Fig. 2.14 and Table 2.1). The head loss is small until just before the opening becomes permanently submerged, after which it increases rapidy. The line gets steeper as type 3 flow changes to type 2 and then type 1. Type 1 and 2 flow usually result in a relatively large afflux and head loss.

Prior to submergence of the waterway, the river channel rather than the bridge was the primary cause of flooding, after which the afflux added up

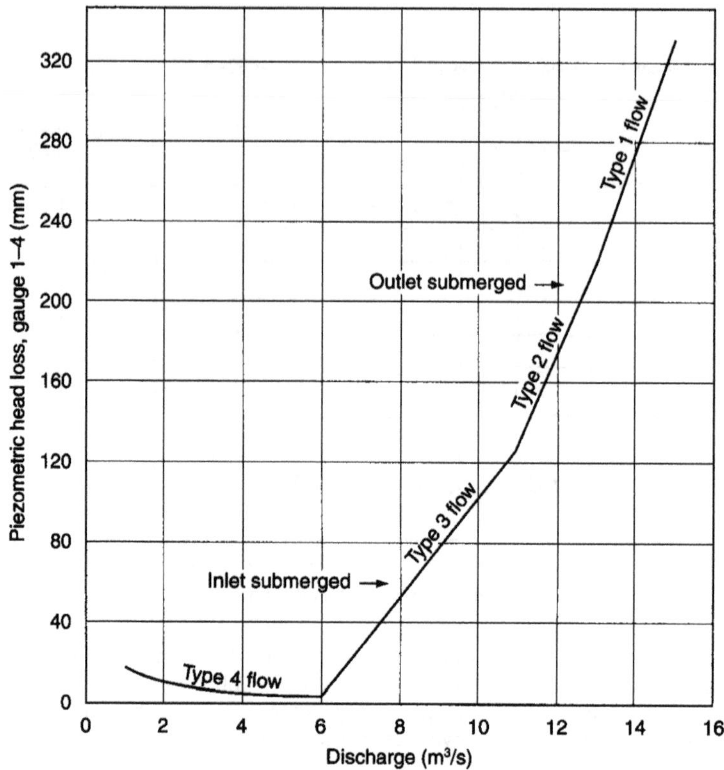

Fig. 2.14 The relationship between the piezometric head loss from gauge 1 to 4 and the discharge at Canns Mill. The head loss increases rapidy after the opening submerges and flow types 3, 2 and 1 occur.

Table 2.1 Observed head loss and afflux at Canns Mill according to flow type

Discharge (m^3/s)	*Head loss 1–4* (*mm*)	*Afflux* (*mm*)	*Flow type*
2.4	8	3	Open channel – channel control
4.5	3	5	
5.8	3	11	
7.0	27	17	
8.3	60	45	Sluice gate – structure control
9.5	90	72	
10.8	123	115	
12.5	200	175	Drowned orifice
13.8	264	220	
15.0	330	270	

Fig. 2.15 Canns Mill after the flood, showing the water level gauges and flood height (the white mark on the bridge parapet).

to 270 mm to the upstream stage. Thus severe flooding would occur without the bridge. This again demonstrates the need to fully understand the nature of a site and the prevailing flow conditions before considering either improvement works or a replacement bridge.

References

Bradley, J.N. (1978) *Hydraulics of Bridge Waterways*, 2nd edn, US Department of Transportation/Federal Highways Administration, Washington DC.

Chadwick, A. and Morfett, J. (1993) *Hydraulics in Civil and Environmental Engineering*, 2nd edn, E & FN Spon, London.

French, R.H. (1986) *Open-Channel Hydraulics*, International Student Edition, McGraw-Hill, Singapore.

Hamill, L. (1997) Improved flow through bridge waterways by entrance rounding. *Proceedings of the Institution of Civil Engineers, Municipal Engineer*, **121**, March, 7–21.

Hamill, L. and McInally, G.A. (1990) The hydraulic performance of two arch bridges during flood. *Municipal Engineer*, 7, October, 241–256.

Hydrologic Engineering Center (HEC) (1990) *HEC-2 Water Surface Profile User's Manual*, US Army Corps of Engineers, Davis, CA.

Kaatz, K.J. and James, W.P. (1997) Analysis of alternatives for computing backwater at bridges. *American Society of Civil Engineers, Journal of Hydraulic Engineering*, 123(9), 784–792.

Kindsvater, C.E. and Carter, R.W. (1955) Tranquil flow through open-channel constrictions. *Transactions of the American Society of Civil Engineers*, 120, 955–992.

Kindsvater, C.E., Carter, R.W. and Tracy, H.J. (1953) *Computation of peak discharge at contractions. Circular 284*, United States Geological Survey, Washington DC.

Laursen, E.M. (1970) Bridge backwater in wide valleys. *Proceedings of the American Society of Civil Engineers, Journal of the Hydraulics Division*, 96(HY4), 1019–1038.

McInally, G.A. and Hamill, L. (1987) Telemetry systems for research and practice. *Proceedings of the British Hydrology Society*, National Hydrology Symposium 1987, pp. 27.1–27.6.

Skogerboe, G.V., Barrett, J.W.H., Walker, W.R. and Austin, L.H. (1973) Comparison of bridge backwater relations. *Proceedings of the American Society of Civil Engineers, Journal of the Hydraulics Division*, 99(HY6), 921–938.

Tracy, H.J. and Carter, R.W. (1955) Backwater effects of open channel constrictions. *Transactions of the American Society of Civil Engineers*, 120, 993–1018.

3 Factors that affect the hydraulic performance of a bridge

3.1 Introduction

This chapter reviews the main factors that affect the hydraulic performance of a bridge, explains why they are important, and illustrates their effect. Many of these variables are related to the geometry of the structure and are summarised in Fig. 3.1. It is these factors in combination with the types of flow encountered (as described in Chapter 2) that give each bridge its own unique hydraulic characteristics.

Some of the factors described below are quite difficult to estimate numerically with any degree of accuracy. However, this fact should not necessarily deter anyone from starting an analysis. Sometimes it is better to have an inaccurate estimate of (say) the afflux that occurs at a particular discharge than no estimate at all. The important thing is to realise that the estimate is inaccurate so that the appropriate allowances can be made. Possibly the estimate can be refined at a later date when better field data become available. If such data already exist, then provided care is taken to prove and verify the mathematical models adopted in the analysis, the solutions obtained can be surprisingly robust.

3.2 The bridge opening ratio, M

The bridge opening ratio (M) is basically a measure of the severity of the constriction (Fig. 3.1a). In channel flow this is perhaps the most important of all the variables: the smaller the opening, the greater the obstacle to flow, the greater the afflux, and the smaller the discharge through the opening for a given stage. However, once the waterway opening is submerged M ceases to be important since the dimensions of the opening and the degree of submergence of the waterway determine the stage–discharge relationship (Section 2.5).

With channel flow, the flow in the middle of the river can go straight through a centrally located bridge opening whereas the flow in the sides of the channel or on the floodplains will have to bend inwards and contract to

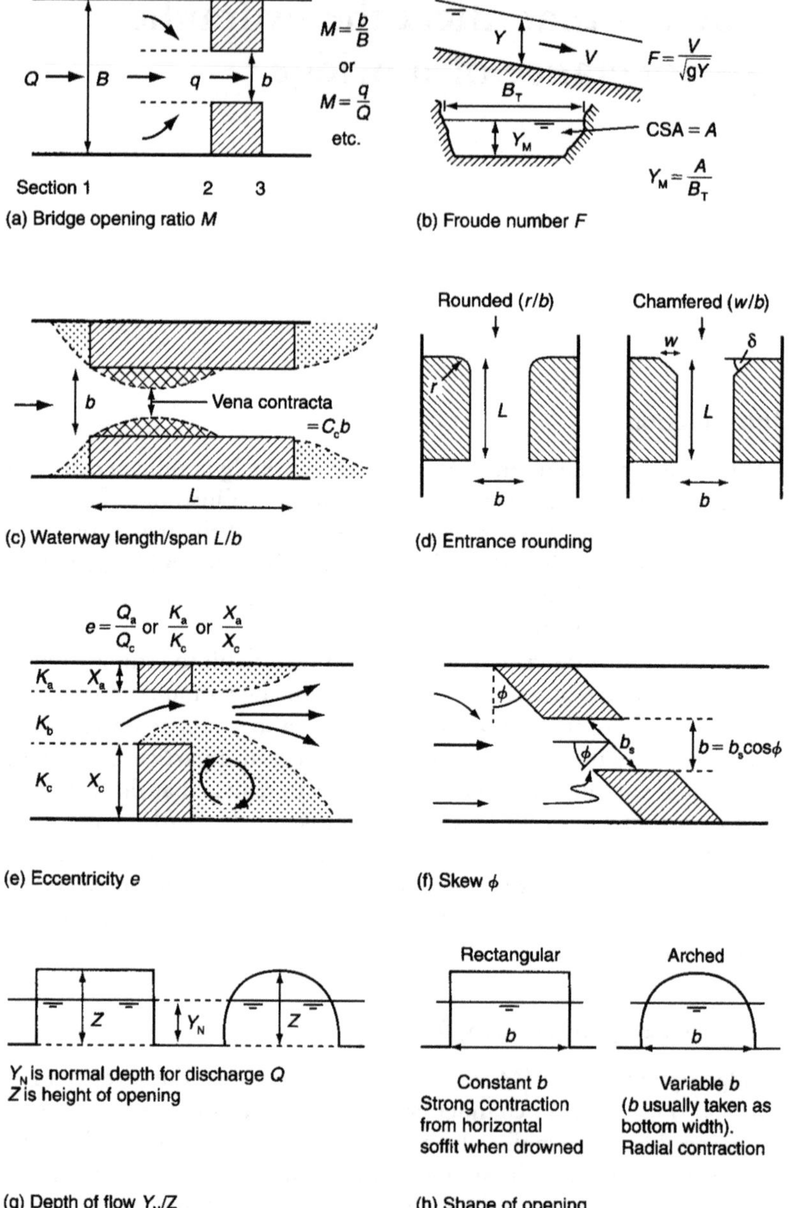

Fig. 3.1 Summary of the principal hydraulic variables affecting the hydraulic performance of a bridge: (a) bridge opening ratio, M; (b) Froude number, F; (c) waterway length/span, L/b; (d) entrance rounding; (e) eccentricity, e; (f) skew, ϕ; (g) depth of flow, Y_N/Z; (h) shape of opening.

pass through. The opening ratio represents the ratio of flow that can pass straight through the bridge opening without having to contract, q, to the total flow in the river, Q (Kindsvater *et al.*, 1953; Kindsvater and Carter, 1955, Tracy and Carter, 1955; Bradley 1978), so:

$$M = \frac{q}{Q} \tag{3.1}$$

Thus M is basically the ratio of the discharge (q) through a width (b) of section 1 compared with the total discharge (Q) through the full channel width (B) of section 1. A value of 1.0 means that the bridge does not represent a constriction and that all of the flow can pass through the opening unimpeded without having to contract.

Usually M is calculated by assuming normal depth (Y_N) at section 1, where Q is the total discharge through the section, and q is the discharge through an area of channel at section 1 equivalent to the opening area at normal depth (A_{N2}) projected upstream to section 1. Another way of putting this is that q is calculated assuming that normal depth occurs across the width (b) of the waterway opening giving a cross-sectional area of flow of A_{N2} ($= Y_N\, b$ for a rectangular opening). Note that it is not A_2, the actual drawn-down area of flow in the waterway, that is used. Similarly, when calculating q (it is almost impossible to measure directly) assume that the area A_{N2} has the bed roughness of the appropriate part of section 1, since the waterway opening has been projected upstream to section 1.

If at section 1 it is assumed that both the normal depth (Y_N) and the associated mean velocity (V_N) are constant across the full width of the channel, equation 3.1 can be rewritten as

$$M = \frac{q}{Q} = \frac{A_{N2} \times V_N}{A_{N1} \times V_N} \tag{3.2}$$

If $A_{N2} = a$ and $A_{N1} = A$ then

$$M = \frac{a}{A} \tag{3.3}$$

If both the opening and the channel are rectangular, so $a = b \times Y_N$, and $A = B \times Y_N$ then

$$M = \frac{b}{B} \tag{3.4}$$

At high flood stages in real, compound river channels the depth of flow and velocity vary greatly across the width of the channel so neither equation 3.3 nor 3.4 adequately describes M. Under these conditions a popular way to calculate M is to split the channel into subsections wherever there is a large change in hydraulic radius (changes in roughness are not important unless the flow is shallow) and then calculate the hydraulic conveyances. Provided all parts of the channel (i.e. the main channel and floodplains) have the same longitudinal energy (friction) gradient, S_F, the opening ratio becomes

$$M = \frac{K_b}{K} \tag{3.5}$$

where K_b is the conveyance of that part of the approach channel equivalent to the opening width (b) projected upstream to section 1 and K is the conveyance of the whole channel at section 1. Conveyance is defined in terms of the Manning equation (equation 2.7) so, in metric units, for any channel or channel subsection (denoted by subscript i below) its conveyance is

$$K_i = \frac{Q_i}{(S_{Fi})^{1/2}} = \frac{A_i R_i^{2/3}}{n_i} \tag{3.6}$$

The conveyance is usually calculated from the expression on the right of equation 3.6 where A_i (m^2) and R_i (m) are respectively the cross-sectional area and hydraulic radius, and n_i is the Manning roughness coefficient ($s/m^{1/3}$). Example 3.2 at the end of the chapter provides an illustration.

The apparent simplicity of the opening ratio (e.g. q/Q or b/B) has little basis in reality, and it is often necessary to be pragmatic. For example, if the main channel meanders upstream of the bridge, then when projecting the waterway upstream to section 1 the opening should be superimposed on the main part of the channel (Matthai, 1967). Common sense should be applied rather than simplistic rules. Similarly, instead of using the normal depth, M can be evaluated from observed water levels upstream of the bridge. These levels would include the afflux, but if this is uniform across the width of the channel the value of M obtained from equation 3.3 would be the same. The same logic applies where abnormal stages occur, the flow is not at the normal depth, and the water surface is not parallel to the bed (Fig. 2.4). In this situation the calculation of the abnormal profile without the bridge may necessitate making many subjective decisions, so using the observed conditions at section 1 with the bridge in place may be just as good.

For a multispan bridge with piers (unless they are unusually wide) most methods of analysis in Chapter 4 use the gross opening area, which includes the area occupied by the piers, when calculating q (or K_b) and hence M. The effect of the piers is allowed for later by introducing a coefficient. This is more practical than trying to decide which part of the flow can pass through the many waterways unimpeded and what part contracts around relatively slender piers. The exception to this is a multiple-opening contraction that has 'interior embankments' within the river channel (Fig. 4.16). Here, each opening is considered individually by separating the approach channel into subchannels, one to each opening.

The calculation of the opening ratio becomes more complex with arched waterways whose width (b) changes with stage. However, M can still be evaluated directly using equation 3.1 or 3.5 or, alternatively, by assuming a rectangular upstream channel with flow at the normal depth an expression can be developed from equation 3.3 (Biery and Delleur, 1962). For a semi-circular opening of radius r with the arches springing from bed level, with a depth in the waterway of Y_S above the springs of the arch (Fig. 3.2a), working in radians :

$$M = \frac{Y_S(r^2 - Y_S^2)^{1/2} + r^2 \sin^{-1}(Y_{S/r})}{BY_N} \tag{3.7}$$

For a semicircular arch with the springline above the bed, the area of the flow beneath the springs should be added to the numerator (Fig. 3.2b). With segmental arches the centre of curvature of the arch, O, is at some depth, d, below spring level (Fig. 3.2c). If the height of the water surface above the centre of curvature is h, then for a segmental arch with the springline at bed level, working in radians:

$$M = \frac{r^2[\sin^{-1}(h/r) - \sin^{-1}(d/r)] + h(r^2 - h^2)^{1/2} - d(r^2 - d^2)^{1/2}}{BY_N} \tag{3.8}$$

As before, if the springline is above the bed (Fig. 3.2d) then the area of the flow beneath the springs should be added to the numerator. Example 3.1 illustrates the use of the equation.

Instead of calculating the area of flow, a practical alternative is to produce a scale drawing of the opening and measure the area using a planimeter or a digitiser. This is often the easiest way with elliptical openings or arches that have a unique shape.

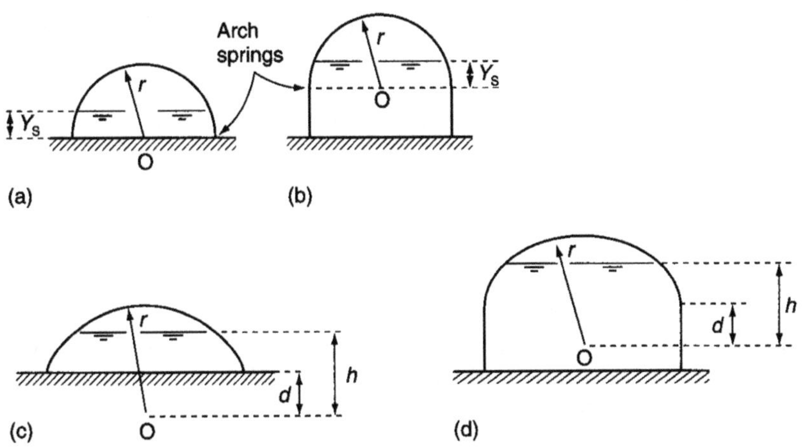

Fig. 3.2 Definition of the variables needed to calculate the area of flow through some common types of arched waterway: (a) semicircular arch with springs at bed level; (b) semicircular arch with springs above bed level; (c) segmental arch with springs at bed level; (d) segmental arch with springs above bed level.

In natural river channels the value of M is not constant and changes with stage (Fig. 3.3). As the stage rises, the capacity of the main channel will be exceeded and flow will commence on the floodplains. When this happens the values of q, K_b and a may remain relatively constant while Q, K and A continue to increase, resulting in a change in gradient of the stage–M curve. With an arch bridge there is an added complexity because the width of the opening also varies with stage so b is not constant. Figures 3.3 and 3.11 can be compared to see the difference in M that results from applying equations 3.1 and 3.8 respectively. The latter uses a simpler definition of M, and assumes that the channel has a constant width of 5.5 m and that there is no overbank flow prior to the submergence of the opening.

From the comments above it will be apparent that the opening ratio is quite difficult to evaluate accurately. Often the field data required are unavailable. For an accurate estimate it is important that the person doing the calculations observes a flood at the site in question. Unfortunately this rarely happens. For example, at Canns Mill the channel is compound in nature (Fig. 3.3a) with a hedgebank and then pasture on the left, and a hedge or barbed wire fence bordering open pasture on the right. There is flow over both floodplains during flood, with ultimately all of the flow passing through the bridge opening (Fig. 2.1). However, during a 15 m^3/s flood, at about 10 m and 15 m from the left and right banks the water on the floodplains is stagnant and does not contribute to flow (this may also happen where there are levees, as in Fig. 8.16). If this is ignored when using equations 3.3 and 3.4 the value of M obtained is far too small: $b/B = 0.08$ and $a/A = 0.14$. With zero flow on parts of the floodplains S_F cannot be the same as in the main channel, so equations 3.5 and 3.6 are not appropriate. Thus equation 3.1 is the only reliable method of obtaining M and indicates that the true value is around 0.55 (Fig. 3.3b and Example 3.3).

Forty-seven undergraduate students at the University of Plymouth were given the task of calculating the opening ratio at Canns Mill. All conducted a technically correct analysis, but there was a very large range of answers and almost all produced values that were too small. This was mainly a consequence of never having seen the site or having had the opportunity to study the flow. It serves as a reminder that in many cases only a rough estimate of M can be obtained, which has repercussions for the accuracy of any subsequent calculations.

In open channel flow the opening ratio is very important. With a wide opening (large M) the flow in the river is barely affected. If the opening is narrow compared with the river (small M), the flow entering the waterway will have a significant transverse velocity resulting in a relatively narrow live stream at the vena contracta (Fig. 2.2b) and a small coefficient of contraction, C_C (and a small coefficient of discharge, C, since friction losses are small and C is basically a coefficient of contraction). Thus a fundamental relationship exists between M and C_C or C in open channel flow, as illustrated by the base curves of Fig. 4.3a. It is apparent from the diagram that

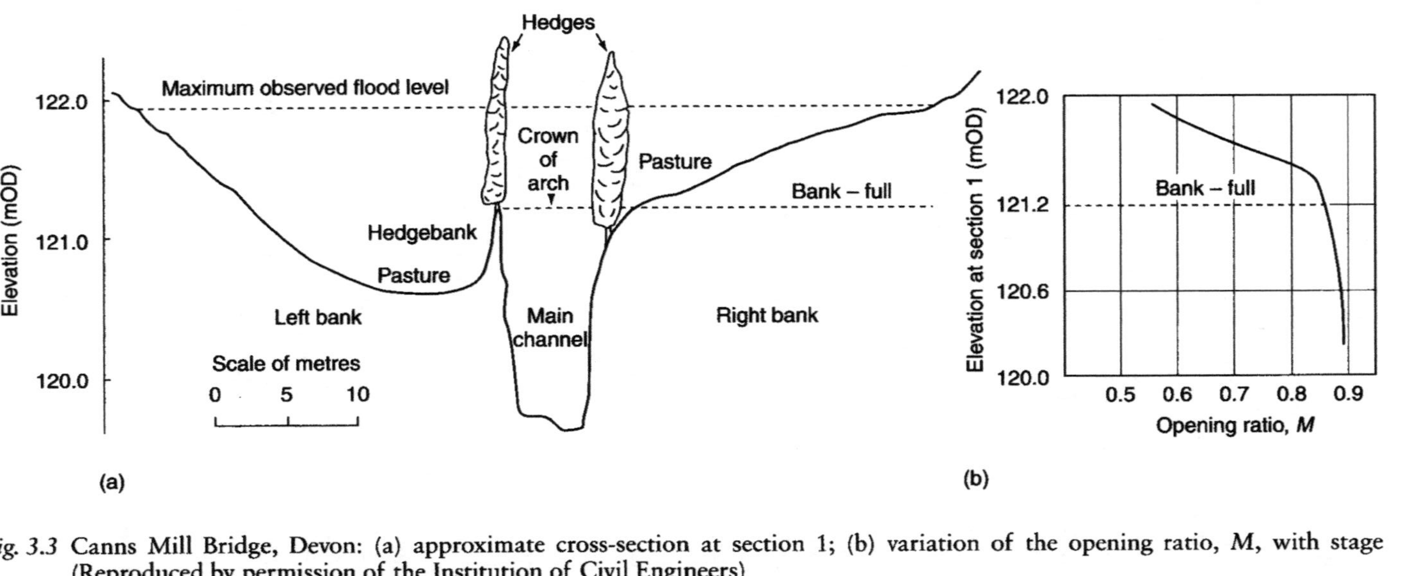

Fig. 3.3 Canns Mill Bridge, Devon: (a) approximate cross-section at section 1; (b) variation of the opening ratio, M, with stage (Reproduced by permission of the Institution of Civil Engineers)

the value of M can change C' and C and hence the discharge through the waterway by up to 30%.

The opening ratio is also instrumental in determining the type of flow at a particular site. Waterways that are narrow and have a small opening ratio are more likely to experience supercritical flow in either the opening or the downstream channel (flow types 5 and 6 in Fig. 2.6). A narrow opening also makes it more likely that the capacity of the opening will be less than the floods that occur, so increasing the possibility of flooding upstream and/or that the bridge will operate with the waterway submerged (flow types 1 to 3).

The opening ratio ($M = q/Q$) is also referred to as the bridge opening ratio or the channel opening ratio. In this book only equations 3.1–3.8 are used to define and evaluate the opening ratio, so a value of 1.0 always means that the flow is unobstructed. Note that the contraction ratio, the channel contraction ratio or the blockage ratio (often symbolised by m) is usually defined as $(1 - q/Q)$ or $(1 - b/B)$. With this definition a value of zero means that the bridge has no effect on the flow, while a value above zero gives an indication of the quantity of flow that cannot pass unimpeded through the opening: that is, the amount that flows in from the sides of the channel. Both of these alternative definitions are in common usage, so whenever a value is quoted in other literature it is essential to check which has been adopted.

3.3 Froude number (F), subcritical and supercritical flow

Whether or not the flow is subcritical or supercritical is instrumental in determining the hydraulic performance of a bridge and the type of flow that will be encountered (Section 2.4). In open channel flow F also affects the discharge through the bridge opening: for example, compared with the standard value of $F = 0.5$ in Fig. 4.3b, when $F = 0.2$ there is roughly a 6% decrease in Q and when $F = 0.8$ there is about a 10% increase. This relationship is independent of the opening ratio, M.

Flow at the critical depth ($F = 1.0$) can be used to optimise the performance of a waterway, as described in Section 7.5, so a knowledge of the critical (or limiting) contraction that will cause this condition is important. A waterway narrower than the critical contraction may result in an unexpectedly large afflux due to the phenomenon known as choking. This is explained below.

3.3.1 *Froude number*

The Froude number of the flow in a particular channel depends upon the velocity and depth, thus:

$$F = \frac{V}{(gY)^{1/2}} \tag{3.9}$$

where V is a characteristic velocity of flow (m/s) and Y is a characteristic depth (m). F itself is dimensionless. If $F = 1$, the flow is in the critial state. If $F < 1$ the flow is subcritical, and if $F > 1$ it is supercritical. The denominator of the Froude number is the celerity of an elementary gravity wave in shallow water. The significance of this is that in subcritical flow a gravity wave (or disturbance to the flow) can propagate upstream ($(gY)^{1/2} > V$) so that the upstream reach is in hydraulic communication with the downstream reach. Therefore with subcritical flow through a bridge the control is downstream of the constriction and this would be the starting point for a backwater analysis, and the calculations would progress in an upstream direction. In supercritical flow a gravity wave cannot propagate upstream because $V > (gY)^{1/2}$ so the effect of any disturbance is swept downstream. Consequently the upstream reach is not in hydraulic communication with that downstream. With supercritical flow the control is upstream so the calculations for the backwater analysis proceed in a downstream direction.

The Froude number is more important than just being an indicator of where a backwater analysis should begin. The Froude number of the downstream control section (section 4 in Fig. 2.2) gives an indication of the ease with which water can flow away from the constriction. For example, a low Froude number of 0.1 might indicate a deep, slow flow with the possibility of an abnormal stage at the bridge site. The relatively low velocities would probably result in a correspondingly small afflux. If the opening submerges, then drowned orifice flow might be expected. On the other hand, a high Froude number of 0.8 would suggest a very steep channel with high flow velocities, so the afflux might be relatively large. There would also be the likelihood of supercritical flow in the opening or downstream channel, with it being less likely that the opening would submerge. If it did, sluice gate flow would be more likely than drowned orifice flow. In model tests Hamill (1997) found that sluice gate flow usually occurred when $F_N > 0.25$.

The importance of the Froude number in determining the type of flow encountered at a bridge site was investigated using a hydraulic model of Canns Mill Bridge (see Section 2.6). The study was conducted in a tilting laboratory channel. The model bridge was tested under a wide range of conditions, not just those actually encountered at the prototype site, in order to determine how an identical bridge would perform hydraulically at a different location. Some of the results are shown in Fig. 3.4, which is simplified and for illustration only: it should not be used for design purposes.

The vertical scale of Fig. 3.4 is the dimensionless ratio of the normal depth at section 4 to the arch height, Y_N/Z. The horizontal scale is the dimensionless normal depth Froude number, F_N, at section 4. The contours on the diagram represent the dimensionless ratio of upstream depth at section 1 to the normal depth: that is, Y_u/Y_N. This is effectively a measure of the afflux, a contour value of 1.0 meaning that there is no afflux; a contour value of 2.0 means that the afflux is equal to the normal depth. From the diagram it is easy to see that for a given stage (Y_N/Z) the contour values

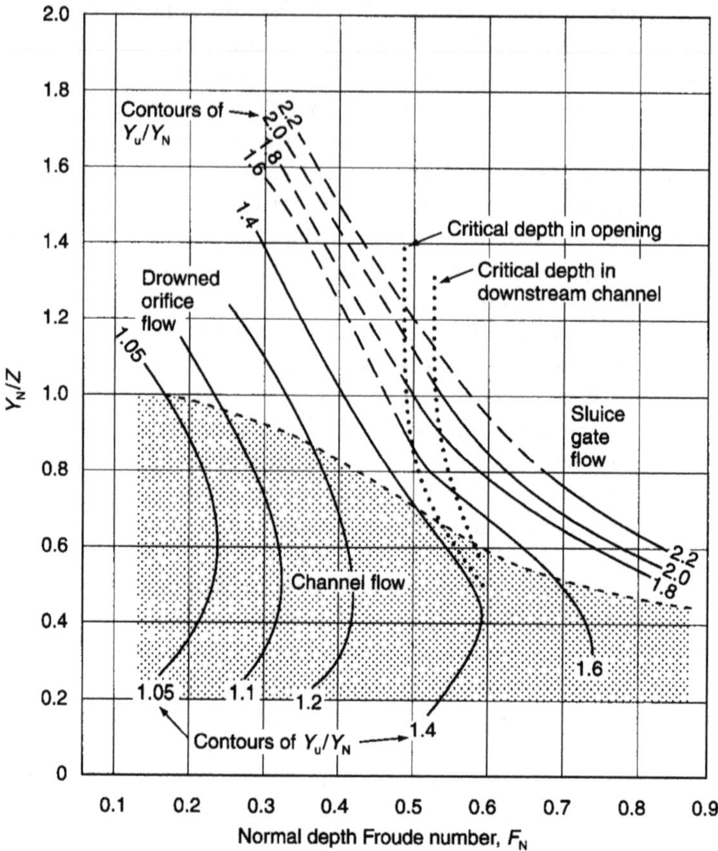

Fig. 3.4 Illustration of the variation in performance of a bridge with depth of flow and Froude number based on a 1:15 scale hydraulic model of Canns Mill. The vertical scale (Y_N/Z) represents normal depth as a proportion of the arch height. The horizontal scale is the normal depth Froude number (F_N). The contours of Y_u/Y_N show the upstream depth as a proportion of normal depth. For a given stage (Y_N/Z) it is apparent that the contour values, which are equivalent to the afflux, increase with increasing F_N. The type of flow experienced is also indicated.

increase from left to right, showing that afflux increases with increasing Froude number. The diagram is also marked to show the types of flow experienced under different conditions. With low Froude numbers, channel flow and drowned orifice flow predominate (flow types 4 and 1 in Fig. 2.6). There is no possibility of critical depth flow occurring in the opening or downstream channel (flow types 5 and 6) until the value of F_N approaches 0.5. At higher Froude numbers channel flow (types 4 to 6) and sluice gate flow (type 3) are possible.

Figure 3.4 has been included not just to illustrate the importance of the Froude number in determining the type of flow that occurs at a particular site, but also to emphasise once again that identical bridges can have a very different hydraulic performance at different sites. Consequently the types of flow that will occur at any given site must be identified before trying to assess the afflux, head loss and energy loss. It should also be appreciated that Fig. 3.4 represents only one opening ratio, and that changing this would also affect the type of flow experienced.

The Froude number is another parameter that at first appears easy to calculate, but which in practice involves some considerable margin for error. Indeed, for anything other than a rectangular channel the value of F may be unreliable. For instance, determining accurately the value of the characteristic hydraulic depth (Y) can be difficult in natural channels where the depth varies considerably. Normally Y is taken as the mean depth (Y_M), which is the flow area (A) divided by the width of the top (free) water surface (B_T) so $Y_M = A/B_T$ (Fig 3.1b). Thus

$$F = \frac{V}{(gA/B_T)^{1/2}} \tag{3.10}$$

Of course, $V = Q/A$ so another way of writing this is

$$F = \left(\frac{\alpha Q^2 B_T}{gA^3}\right)^{1/2} \tag{3.11}$$

This time α, the dimensionless velocity distribution coefficient, has been included to allow for the non-uniform velocity in the channel (see below). Since the value of this coefficient is often unknown or near 1.0, it is frequently assumed to be 1.0 and omitted. However, near bridges and in compound channels it may have a value of around 2.0 (see Section 3.10).

It is apparent from equation 3.11 that the width of the water surface (B_T) is influential in determining the Froude number, so a large change in F will occur immediately overbank flow commences. The variation of F with river stage in natural compound channels of irregular cross-section with overbank flow is complex, and the equations above are inadequate. Under such conditions the compound section Froude number may be more appropriate, details of which were given by French (1986).

Another problem is that some methods of analysing the hydraulic performance of a bridge (most notably the USGS method in Section 4.2) involve calculating the Froude number within the bridge waterway (F_3). This value is not easy to estimate accurately because the exact cross-sectional area of flow (A_3) and depth (Y_3) are often unknown or difficult to calculate or measure. With irregularly shaped or trapezoidal openings having a surface water width of b_T, the mean value of Y_3 may be calculated as $Y_{M3} = A_3/b_T$ (the subscript M is often omitted because it is understood that the mean or most representative value is being used, as in Figs 4.4 and 4.7). However, with arched openings the width of the water surface (b_T) tends to zero as the stage increases. Consequently Y_{M3} and F_3 at first become numerically dubious, then meaningless. The same problem is encountered when trying to calculate the critical depth (Y_C) in an opening. For a rectangular opening:

$$Y_C = \left(\frac{aQ^2}{b^2 g}\right)^{1/3} \tag{3.12}$$

where b is the width of the opening. With an arch the question arises as to what width to use. Hamill (1993) suggested using the bottom width of the arch (i.e. b) when calculating, Y_{M3}, F and Y_C since this eliminates many of the problems. For example, if Q is the discharge through the opening and A_3 is the cross-sectional area of flow at section 3 in the opening, then the mean velocity is $V_3 = Q/A_3$. If the mean depth of flow at the section is calculated as $Y_{M3} = A_3/b$ then $Q = V_3 Y_{M3} b$ is a valid expression. This means that Y_{M3} is not necessarily the same as the actual depth, Y_3, measured on site, but neither is A/b_T. The problem of b tending to zero does not become apparent until the stage reaches about 70% of the arch height, so the use of the bottom width is reasonable (see the notes accompanying Table 3.1). If the stage in the opening rises above 70% of the arch height then the calculation of the Froude number and critical depth in the waterway may be pointless anyway, because the opening must be very near to submerging or submerged.

In Fig. 2.6 the line representing the critical depth (Y_C) is shown dashed. From equation 3.12 it is apparent that Y_C is proportional to the discharge per unit width, so the line is higher in the opening than in the river channel because Q/b is greater than Q/B. If, at any particular discharge, field measurements show that the water level is always above the critical level then subcritical type 4 flow exists. If the water surface cuts the critical depth line then supercritical flow types 5 or 6 exist, depending upon how far downstream the supercritical flow extends. This is fine provided accurate observations of water levels in the opening and downstream channel are available, but for existing bridges this is usually not the case. For new bridges that have yet to be constructed the accurate calculation of the depth in a region of expanding, rapidly varying flow is not straightforward, even using computer software. With important crossings hydraulic model tests may be advisable.

There are some guidelines for determining whether the flow is subcritical or supercritical. Bradley (1978) suggested that the backwater should be calculated assuming the flow to be subcritical throughout. If the result appears unrealistic, the backwater should then be calculated assuming supercritical flow (see Section 4.3.5). Normally one of the results should appear to be 'erratic' so the flow is not of that type. Bradley also stated that if the backwater calculated from supercritical flow has a lower value than that obtained assuming subcritical flow, then the flow is definitely supercritical. Remember, the higher the normal depth Froude number and the more severe the contraction, the more likely it is that supercritical flow will occur.

3.3.2 The limiting or critical contraction

If a wide waterway opening is made progressively narrower, there is a limiting width at which the flow in the opening will no longer be subcritical but critical. This is the limiting or critical contraction. Henderson (1966) considered the limiting contraction required to cause flow at the critical depth between the piers of a bridge, and presented two equations that relate the limiting opening ratio (M_L) to the Froude number (F or F_L) in the channel. These equations are generally used in connection with the flow between bridge piers, but can be cautiously applied to single-span openings. Both assume that the velocity is uniform across section 3 between the piers, and that this is where the most contracted section is located and where critical flow first occurs ($F_3 = 1.0$). The first equation assumes that the specific energy at section 1 equals that at section 3. The limiting bridge opening ratio, $M_L = b/B$, is given by

$$M_L^2 = \frac{27F_1^2}{(2+F_1^2)^3} \tag{3.13}$$

where F_1 is the Froude number at section 1. This equation is attributed to Yarnell (1934), who used it to distinguish between unchoked flow and choked flow (see Sections 3.3.3 and 5.4.1). Chow (1981) presented basically the same equation but equated the energy at section 3 between the piers to that at section 4 while introducing ε to represent the proportion of energy recovered, so:

$$\varepsilon\left(Y_3 + \frac{V_3^2}{2g}\right) = \left(Y_4 + \frac{V_4^2}{2g}\right)$$

With $F_3 = 1.0$, after manipulation this becomes

$$M_L^2 = \frac{27\varepsilon^3 F_4^2}{(2+F_4^2)^3} \tag{3.14}$$

where F_4 is the normal depth Froude number at section 4. Henderson's second equation assumes that the momentum at sections 3 and 4 are equal so:

$$M_L = \frac{(2+1/M_L)^3 F_4^4}{(1+2F_4^2)^3} \tag{3.15}$$

According to Henderson, equation 3.15 is more likely to be accurate because there are no assumptions about energy conservation. This appeared to be confirmed by a limited series of tests conducted in the Civil Engineering Hydraulics Laboratory at the University of Plymouth. Equation 3.13 underestimated the critical value of M_L (e.g. 0.47 when it should be 0.56) with an average error of 16%, whereas equation 3.15 underestimated by 10%. However, it should be emphasised that even in the laboratory it is difficult to determine precisely when critical flow exists, so these results are not very significant.

For convenience, equations 3.14 and 3.15 have been plotted as Fig. 3.5 so it is a simple matter to obtain the limiting opening ratio. Although this is derived in terms of b/B above, because rectangular cross-sections are assumed, generally this is not a very good way of defining M. Usually the area, conveyance or discharge ratio produces a more accurate result.

The significance of critical flow, or supercritical flow, occurring in the waterway is that the methods used to calculate the afflux or backwater in

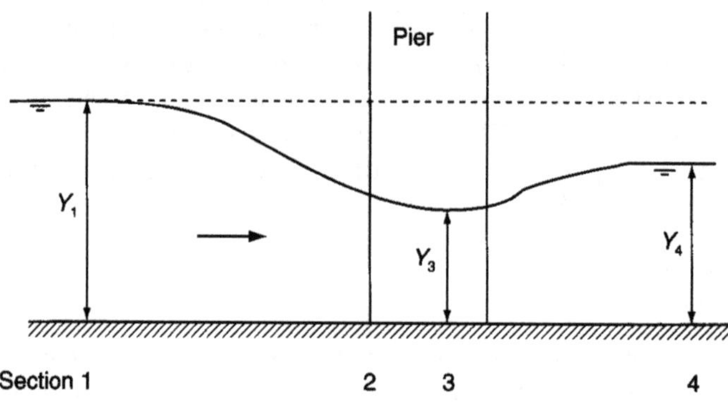

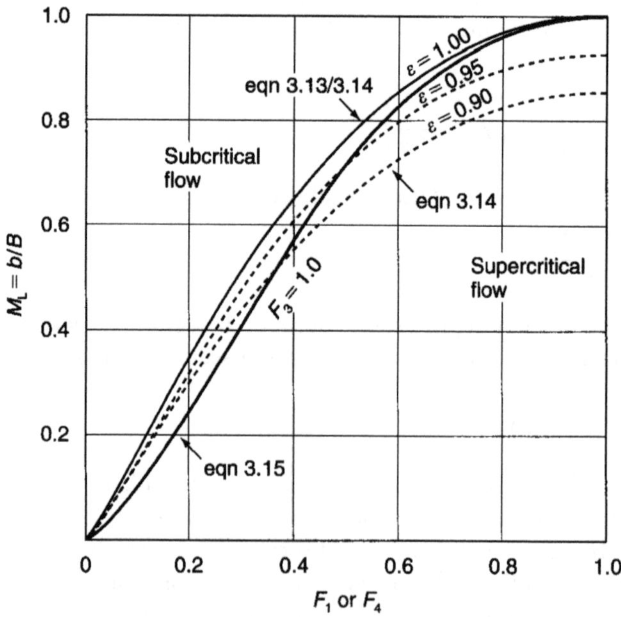

Fig. 3.5 Relationship between the limiting bridge opening ratio (M_L) and the Froude number at either section 1 or 4 when the flow at the contracted section is at the critical depth (Y_C) with $F_3 = 1.0$. The proportion of energy recovered from section 3 to 4 is represented by ε.

subcritical flow are no longer appropriate, so a different technique must be adopted (e.g. Section 4.3.5). In addition, there may be a substantial increase in the afflux because a higher local value of specific energy is needed to enable the transition from subcritical to supercritical flow to occur. This is referred to as choking.

3.3.3 *Choking of a bridge waterway*

If the flow through a bridge waterway chokes this will result in an increase in the upstream water level that is additional to the afflux caused by the energy loss. The concept of a choke has to be explained in terms of specific energy, E_S, which is the energy calculated above bed level, so

$$E_S = Y + \left(\frac{V^2}{2g}\right) \tag{3.16}$$

If a particular discharge (Q) is assumed in a channel of constant width (B), then for any depth (Y) the value of E_S (m) can be calculated from equation 3.16 and plotted in the form of a diagram of Y versus E_S, as in Fig. 3.6. This shows the curve ABC obtained when $Q = 5\,\text{m}^3/\text{s}$ and $B = 10\,\text{m}$. The upper part of the curve (AB) represents subcritical flow, the lower part (BC) supercritical flow, with critical depth (the boundary between the two) coinciding with the minimum value of E_S at B. With the exception of flow at the critical depth, at any value of specific energy two alternative depths of flow are possible, one subcritical on AB and one supercritical on BC. An important point is that if Q and B are constant then the flow in the channel must follow the line ABC of the Y–E_S diagram: the flow cannot move vertically from one limb of the curve to the other, except in the special case of a hydraulic jump. This means that any change in the depth of flow must be accompanied by a change in specific energy, or vice versa.

The Y–E_S curve ABC in Fig. 3.6 is plotted for the condition where the discharge per metre width of channel, $q = Q/B = 5/10 = 0.5\,\text{m}^3/\text{s}$ per m. If the channel width (B) is decreased by a bridge to $b = 2.884\,\text{m}$ then $q = 5/2.884 = 1.734\,\text{m}^3/\text{s}$ per m and another Y–E_S curve, DEF, is necessary to represent the new flow condition and the increased value of q. The curves for $q = 1.111\,\text{m}^3/\text{s}$ per m and $2.500\,\text{m}^3/\text{s}$ per m are also shown. It can be seen that as q increases new Y–E_S curves are obtained to the right of the previous one, while the value of the critical depth (Y_C) increases.

The effect of a bridge opening that causes a sudden decrease in channel width is to initiate a corresponding increase in q so that the flow switches from one Y–E_S curve to another. If the contraction is slight, the flow will accelerate smoothly through the opening and will remain subcritical. For example, the change from 0.500 to $1.111\,\text{m}^3/\text{s}$ per m is represented by the movement G to H on the Y–E_S diagram. This requires no change in specific energy, so it is possible both in theory and reality. However, if the value of b

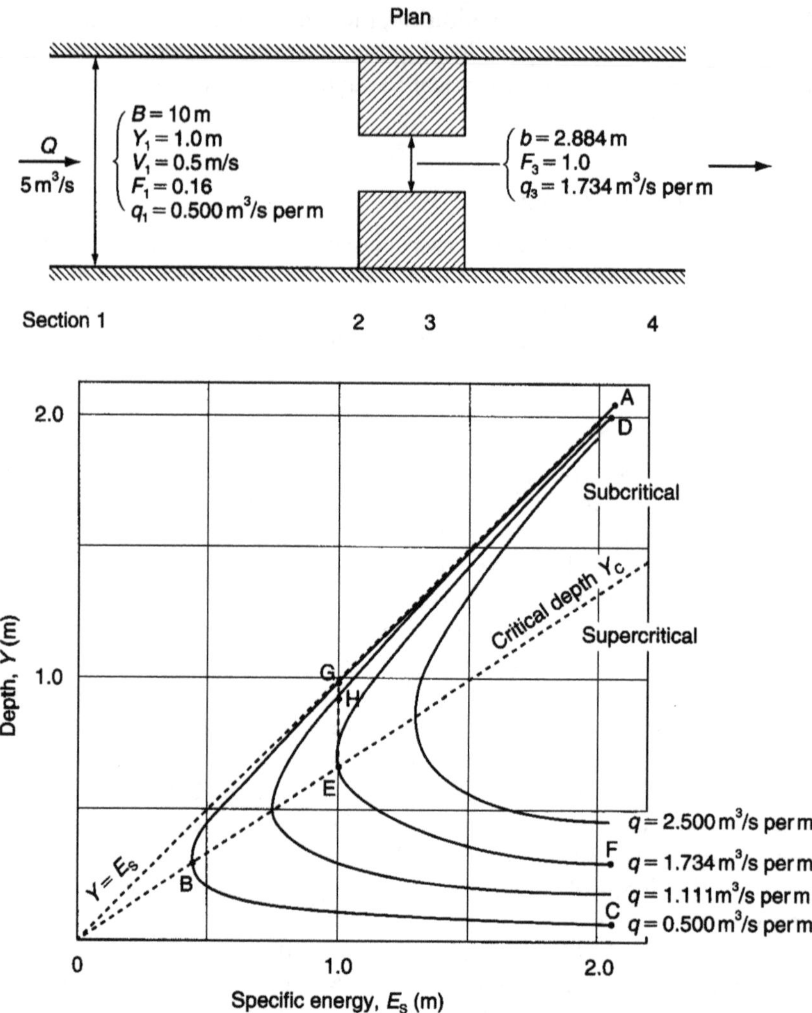

Fig. 3.6 Illustration of choking caused by the need for a local increase in specific energy, E_S. The narrowing of a channel by a constriction leads to an increase in q (from 0.500 m^2/s to 1.734 m^2/s in this example). If the constriction is narrower than this the appropriate E_S curve cannot be reached by dropping vertically from G, necessating an increase in E_S and depth, Y. For example, starting from G the $q = 2.500$ m^2/s curve is inaccessible without an increase in E_S.

is progressively reduced then a point will be reached where the flow in the opening is at the critical depth, which is represented by the movement G to E on the Y–E_S diagram. This is not a problem because the flow can accelerate smoothly from subcritical to critical while the specific energy remains constant. This represents the limiting contraction defined by equation 3.13:

If $V_1 = 0.5\,\mathrm{m/s}$ and $Y_1 = 1.0\,\mathrm{m}$ then

$$F_1 = \frac{V_1}{\sqrt{gY_1}} = \frac{0.5}{\sqrt{9.81 \times 1.0}} = 0.160$$

$$M_L^2 = \frac{27F_1^2}{(2+F_1^2)^3} = \frac{27 \times 0.160^2}{(2+0.160^2)^3} = 0.0832$$

$M_L = 0.2884 = b/B$ where $B = 10\,\mathrm{m}$

so $b = 0.2884 \times 10 = 2.884$ m

The same result can be obtained from Fig. 3.5 with $F_4 = F_1 = 0.160$.

Thus $b = 2.884\,\mathrm{m}$ is the limiting contraction, which is why E is vertically below G. However, consider what happens if b is further reduced, say to $2.000\,\mathrm{m}$ so $q = 2.500\,\mathrm{m^3/s}$ per m. A line drawn vertically down from G now fails to intercept this Y–E_S curve indicating that this flow condition is not possible with $E_S = 1.0\,\mathrm{m}$. In order to intercept the $q = 2.500\,\mathrm{m^3/s}$ per m line a shift to the right is necessary, which means that an increase in specific energy must occur. This can be accomplished only by an increase in the upstream depth, which is called choking. A phenomenon such as choking is logical because if a channel is progressively narrowed there must be some limit at which the water level upstream is affected, irrespective of the fact that critical or supercritical flow occurs.

Choking is significant because it can result in the afflux being larger than expected. Most methods of calculating afflux are valid only when the flow is subcritical, and certainly do not allow for choking. In many ways afflux and choking are similar: the afflux provides the additional head necessary to overcome energy losses in subcritical flow through a constriction, while choking represents the additional head required to increase the specific energy and accelerate the flow through a constriction in the critical or supercritical condition. Choking is also significant when a waterway is deliberately designed to operate at the critical depth (e.g. see Section 7.5).

The occurrence of choking is not always as easy to predict as the example above may suggest. No allowance was made for a non-uniform distribution of velocity or energy losses; Fig 3.5 illustrates how this alone can throw doubt upon the value of the critical contraction. Then there is always the problem of debris caught on the piers or abutments altering the flow condition.

3.4 Ratio of waterway length to span, *L/b*

In general, long waterways are more efficient than short ones, so the length of the opening is another factor that affects the hydraulic performance of a bridge. This is assessed in terms of the length ratio (L/b), which is the ratio of the waterway length between the upstream and downstream faces of the constriction, L, to the width or span of the opening, b. With an arch bridge the bottom width is usually adopted so that L/b remains a constant for a particular structure (Fig. 3.1c). To be classed as a bridge this ratio should be less than 1.0, waterways with larger values being classed as culverts. However, the distinction is rather arbitrary.

Kindsvater and Carter (1955) showed that reducing the waterway length from $L/b = 1$ to $L/b = 0.15$ can decrease the discharge through a waterway by as much as 15%. Thus the ratio L/b appears in Fig. 4.3a (for example) in recognition of the fact that the waterway length is one of the more important variables controlling the hydraulic performance of a bridge.

The reason why long waterways are more efficient is that the contraction and subsequent expansion of the flow is more controlled in a long waterway, so the energy loss is reduced. This can be explained as follows. As the flow enters the opening it contracts and forms a vena contracta. With a short waterway the contraction will continue unsuppressed into the downstream channel (as in Fig. 2.2b). The expansion between the vena contracta and section 4 is then relatively large, and takes place in a region with an adverse (positive) pressure gradient, so there is a large energy loss. With a long waterway the vena contracta will be inside the opening, and the flow will have expanded to the full width of the waterway before emerging into the downstream channel (Fig. 3.1c). This effectively means that the water between the separation boundary of the live stream and the abutment wall (shown cross-hatched) is trapped. It cannot be replaced from downstream, but it will be entrained by the jet and removed. Thus the depth of water in this zone is less than that in the jet, and the consequent negative pressure gradient aids the expansion of the jet and reduces the energy loss. When the jet leaves the opening it is wider than it would otherwise have been, so the expansion and energy loss between sections 3 and 4 are reduced. The double expansion still results in an energy loss, but it is smaller than for a short waterway so there is an overall improvement in hydraulic performance.

The length ratio is one of the few variables that is easy to calculate and which has a constant value once a bridge has been built, unlike the opening ratio and Froude number, which vary with river stage.

3.5 Entrance rounding

One method of improving the hydraulic efficiency of a waterway opening is to round the upstream edges so that the geometry of its entrance more closely resembles the shape of the boundary streamlines (Fig. 3.1d). Entrance rounding reduces the contraction of the live stream and increases the width of the vena contracta, and hence increases the coefficient of discharge, such as C_d in equations 2.8 and 2.9. Since the discharge for a given stage is directly proportional to C_d, this means that the bridge can pass larger floods for the same water level.

Kindsvater *et al.* (1953) and Kindsvater and Carter (1955) showed that as the relative radius of curvature, r/b, increases so the improvement in the hydraulic performance of a constriction operating with open channel flow increases, but at a diminishing rate until a maximum value is reached. The maximum value corresponds to a value of the coefficient of contraction (C_C) approaching unity. Under these conditions the difference between C_C and C (i.e. the coefficient of discharge in open channel flow) represents the boundary friction loss. Predictably, the greatest benefit arises with narrow openings (small M). For example, with $r/b = 0.14$ and $M = 0.2$ the improvement in C, and hence Q, is about 20% (Fig. 4.3c). This probably represents about the maximum improvement that can be gained from entrance rounding in channel flow. It should be appreciated that improvements on this scale are not automatic: the type of flow, Froude number and bridge geometry all influence the value, and in most cases it will be much less than 20%. Hamill (1997) found that in channel flow the increase in discharge or reduction in afflux that could be obtained through entrance rounding was usually less than 5%, although much larger improvements that increased with the degree of submergence were obtained once the opening had drowned (Section 7.2.2).

An alternative to using a rounded entrance is to use a chamfer (bevel), which has the advantage that it is easier to construct. In channel flow this virtually amounts to angled wingwalls. The degree of chamfer is indicated by the angle (δ) and its length measured as w/b (Fig. 3.1d). When the radius of a rounded entrance is small a chamfer may be as effective, but rounding becomes superior as the radius increases. Kindsvater *et al.* (1953) found that with $w/b = 0.12$ and $M = 0.2$ a 30° chamfer used with channel flow resulted in a maximum 9% increase in discharge, a 45° chamfer 15%, and a 60° chamfer 29% (Fig. 4.3 d to f). The improvement is smaller with larger M values, but the 60° chamfer/wingwall configuration is always best since this minimises the transverse contraction.

Kindsvater *et al.* (1953) found that rounding the downstream corners had a negligible effect in laboratory tests. Hamill (1997) studied the contraction from the soffit of rectangular and arched openings and concluded that this was where rounding or a chamfer was most effective, since there is usually a large vertical contraction from the deck of a bridge, particularly in sluice gate flow. This type of deck rounding is apparent in Fig. 1.1. Entrance rounding is considered in more detail in Chapter 7, along with other ways to improve hydraulic performance.

3.6 Eccentricity, *e*

If a bridge opening is eccentrically located in the river channel, as shown in Fig. 3.1e, this can affect the flow through the constriction. To take an extreme example, if $X_a = 0$ there is no contraction on that side of the channel so the flow on the left bank can pass through the constriction unimpeded. This would increase the width of the live stream at the vena contracta (Fig. 2.2b) and increase the coefficient of discharge. The reverse is also true. If one abutment is very long (e.g. X_c in Fig. 3.1e), so that the water has to flow along it prior to passing through the opening, the resulting contraction may be very large, reducing the width of the live stream and decreasing the coefficient of discharge. One consequence of this may be an increased likelihood of supercritical flow in the waterway (Fig. 2.8). Another may be that a very large eddy will form in the downstream separation zone.

The amount of eccentricity can be quantified as $e = X_a/X_c$ (Fig. 3.1e), but it is best defined in terms of the discharge or conveyances of the approach section, so $e = Q_a/Q_c$ or $e = K_a/K_c$ where the numerator is always the smaller of the two values so $e < 1.0$ (Kindsvater and Carter, 1955; Matthai, 1967; Bradley, 1978). Generally if e is larger than 0.12–0.20 the effect of the eccentricity on the coefficient of discharge can be ignored. Even with smaller values the resulting reduction in discharge will probably be less than 5% (Table 4.1).

3.7 Skew, ϕ

A simple definition of skew is shown in Fig. 3.1f. In this example the longitudinal centreline of the bridge and its approach embankments is at an angle ϕ to the banks of the channel and the direction of flow, although the waterway opening itself is parallel to the flow. For a normal or perpendicular crossing $\phi = 0°$. However, Fig. 3.7 illustrates that there are three possible types of skew in addition to a normal crossing:

- normal crossing – embankments perpendicular to the flow, waterway parallel to the flow;
- skew 1 – embankments skewed to the flow, waterway parallel to the flow;
- skew 2 – embankments skewed to the flow, waterway skewed to the flow;
- skew 3 – embankments perpendicular to the flow, waterway skewed to the flow.

Of these skew 3 is unusual, although not unknown: it may occur unintentionally if the river shifts its course after the bridge has been constructed, but will not be considered further. Both skew 1 and 2 are common and will be considered in some detail. Note that the USGS method of analysis in Section 4.2 treats skew in a slightly different way from that described here.

Skew has many effects, some of which are not immediately obvious. The hydraulic consequences of skew 1 and 2 compared with a normal crossing are summarised below and then explained fully.

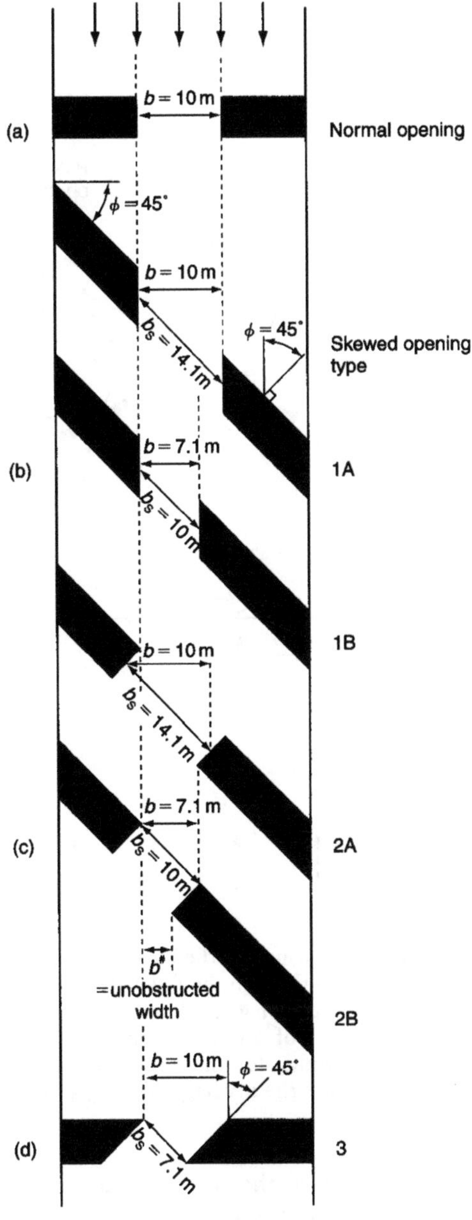

Fig. 3.7 Comparison of normal and skewed crossings and the definition of skew types and the skew angle ϕ: (a) normal crossing; (b) type 1 skew with abutments parallel to the flow; (c) type 2 skew with abutments at 90° to the approach embankments; (d) type 3 skew with the abutments at an angle ϕ to both the flow and approach embankments. Note that with skew type 2 the unobstructed opening width normal to the flow ($b^{\#}$) is reduced compared with type 1.

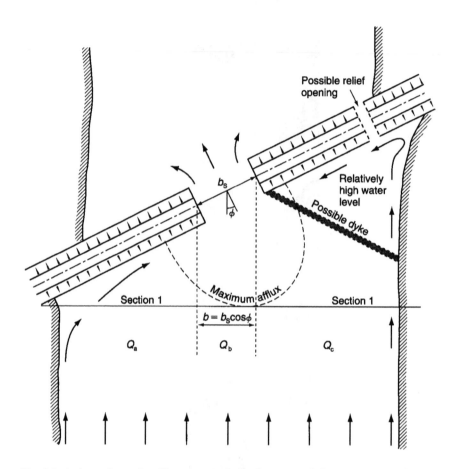

Fig. 3.8 A skewed crossing illustrating: (a) the location of the maximum afflux; (b) the higher water level on the right bank where the water is 'trapped' in the corner; (c) the possible use of a relief opening to reduce the raised water level; (d) the possible use of a dyke to reduce the effect of skew and aid flow through the opening, (e) the projection of the skewed opening upstream to section 1 and the discharge Q_a, Q_b and Q_c in the channel sub-sections.

- The water levels recorded at the two upstream corners of the channel will be different, being higher on the side where the water is 'trapped', such as in the right corner of Fig. 3.8.
- In channel flow there is a reduction in the effective width of the waterway with an associated change in the opening ratio (M) and in hydraulic performance.
- Once the waterway is submerged there is a change in hydraulic performance, and the flow emerging from the opening is angled towards one bank.

3.7.1 Transverse variation of water level

In Fig. 3.8 the water flowing along the right-hand boundary floodplain is trapped in the corner of the channel against the upstream face of the bridge or embankment, and may have to flow backwards to escape. This means that the water in this region is effectively static and, owing to the recovery of the velocity head, the depth is larger than in the centre of the channel. On the left bank the flow along the floodplain is deflected towards the centre by the embankment, so it accelerates smoothly into the opening while experiencing a reduction in depth as the velocity head increases. Thus there is a general slope of the water surface from right to left.

Because of the 'superelevation' of the flow, when a skewed opening first submerges it will do so on the side where the water is trapped in the corner between the upstream face and the channel boundary. It is not unusual for one side of the opening to be submerged while the other side is still operating in open channel flow. Thus the transition between the two may take place over a larger range of Y_u/Z than for a normal crossing.

The river is most likely to burst its banks in the corner of the channel where the water surface is highest. However, unless the channel is wide, steep and the velocity head relatively large the cross-fall may not be particularly significant. Nevertheless, the hydraulic performance of a skewed crossing may be improved by constructing an embankment or dyke across the 'troublesome' floodplain to deflect the flow into the opening. This dyke is more or less the mirror image of the bridge embankment on the opposite floodplain. An alternative is to provide a relief opening through the embankment (Fig. 3.8). An opening (e.g. a pipe) may also be needed at this location if a dyke is contructed, so that the area between the dyke and the bridge embankment can drain if flooded.

3.7.2 Reduction in width

In Figs 3.7 and 3.8 the opening width (b) is measured perpendicular to the direction of flow, and it is this width that is projected upstream to section 1, whereas b_S is the skewed width between the abutments or piers measured along the longitudinal centreline of the highway embankments (except for skew 3). To aid the comparison of the waterway size and ease of flow through the opening, Fig. 3.7 is drawn so that the upstream corners of the left-hand abutments align.

It is generally assumed that skew reduces the efficiency of a waterway, but the true effect is often obscured by confusion regarding the effective 'span'. For example, consider skew 1A in Fig. 3.7 where $b = 10.0\,\text{m}$, as for the normal crossing. The skew angle $\phi = 45°$ so $b_S = b/\cos 45° = 14.1\,\text{m}$. The clear distance between the abutments has increased, so there is a corresponding increase in discharge capacity (although not in direct proportion to b_S), as shown in Fig. 3.9a. Conversely, for skew 1B with $\phi = 45°$ and $b_S = 10\,\text{m}$ the

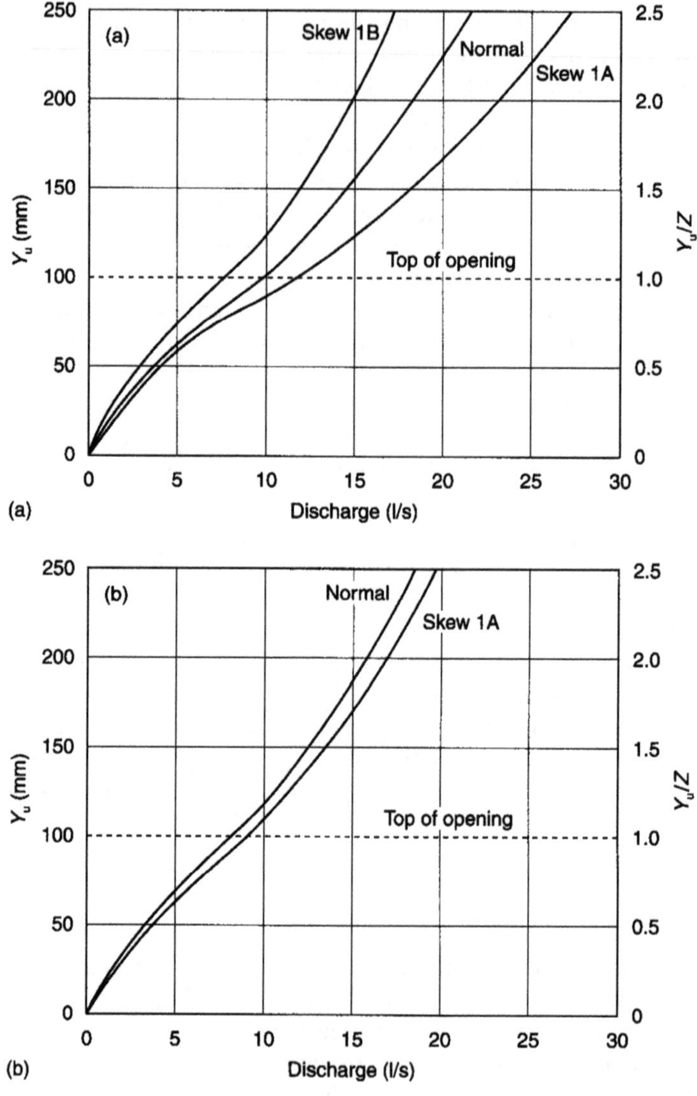

Fig. 3.9 Illustration of the effect of type 1 skew on the stage–discharge curve of model bridges at a slope of 1:50. For the normal bridge $b = b_S = 200$ mm. With $\phi = 45°$, skew 1A has $b = 200$ mm and $b_S = 283$ mm; skew 1B has $b_S = 200$ mm and $b = 141$ mm. (a) Rectangular opening; (b) arched opening.

projected width perpendicular to the flow is $b = b_S \cos 45° = 7.1$ m, which represents a reduction of 29% compared with the 10 m wide normal crossing. As a result, the waterway has a reduced discharge capability (Fig. 3.9a). With skew type 2B there is also a marked reduction in the unobstructed span ($b^{\#}$) between the abutments parallel to the flow. This is one reason why skewed waterways are generally assumed to be inefficient. However, provided the appropriate allowances are made, such as increasing b_S slightly and considering the possibility of scour near the points of the abutments, there is no real need to avoid skewed crossings (particularly with $\phi < 20°$). Thus the question arises: what size of skewed opening actually gives the same performace as a normal crossing?

For skew 1 waterways Bradley (1978) presented a chart (Fig. 4.24) that showed the skewed width of opening (b_S) required to produce the same backwater as a normal crossing of width b. For example, if $b = 10$ m, $\phi = 45°$ and $M = 0.45$ then from the chart $b_S \cos 45°/b = 0.87$, giving $b_S = 12.3$ m. This is between the dimensions indicated in diagrams 1A and 1B of Fig. 3.7. Bradley suggested that with $\phi < 20°$ there were no objectionable results regardless of abutment geometry, but bigger angles may result in large eddies, reduced efficiency and scour.

In open channel flow, Kindsvater *et al.* (1953) and Kindsvater and Carter (1955) suggested that skew (or angularity) of less than 20° has a relatively small effect in most cases, usually causing less than a 5% reduction in discharge. However, the effect of skew becomes larger as as the angle increases, and is most apparent when M is near 1.0. For skew 2, Matthai (1967) showed that when $M = 0.95$ a 45° skew can reduce the discharge through an opening by up to 28%, but this reduces to 0% when $M = 0.36$ (Fig. 4.3g). When M is small other hydraulic factors predominate. Note that in the USGS method of analysis the projected opening width at section 1 (b) is the same as the skewed width between the abutments (b_S) so $b = b_S$ regardless of the skew angle (Fig. 4.14a).

To summarise, a skewed waterway is usually regarded as having a smaller effective width than a normal crossing so the value of Q_b or K_b is reduced and a decrease in M is experienced. These factors generally result in a small loss of hydraulic performance, but this can be as much as 28% of the discharge capacity in extreme cases (but see below).

3.7.3 *Change in submerged performance*

After the opening has submerged, the orientation of the crossing with respect to the direction of the approach flow is much less significant, since the situation is analagous to discharge through an orifice, and M is irrelevant. The flow contracts strongly from the deck and enters the opening at right angles to the face of the bridge, so the emerging jet is basically perpendicular to the plane of the entrance. This is true even with skew 1 openings, where the waterway is parallel to the sides of the channel. In a narrow channel the angled jet can lead to severe scouring of the bank if it has not been protected. With skew 2 openings the direction of the jet is more obvious. Note that the zone of drawdown, as delineated by the dashed line showing the maximum afflux in Fig. 3.8, is also based on the line of the upstream bridge face.

Given that the jet in submerged flow emerges perpendicularly to the bridge face, it is reasonable to expect that the discharge capacity of the waterway is related to the skewed width (b_S) of the opening. For example, assume a rectangular opening of height Z, with the dimensions shown in Fig. 3.7, and then compare the following:

Normal crossing	opening area along the line of the bridge face = 10Z projected opening area at section 1 = 10Z
Skew 1A	opening area along the line of the bridge face = 14.1Z projected opening area at section 1 = 10Z
Skew 1B	opening area along the line of the bridge face = 10Z projected opening area at section 1 = 7.1Z

Skew 1A has the largest all-round dimensions and could be expected to have the largest discharge for a given stage. In model tests involving rectangular openings this was found to be the case (Fig. 3.9a). This does not imply that skewed waterways are beneficial; it merely demonstrates that it is not easy to predict their performance because the effective span of the waterway is not always apparent. Conversely, skew 1B has the same area at the face (10Z) as a normal crossing but a reduced projected area (7.1Z) so it has a worse stage–discharge performance.

The geometry of a skewed, arched opening is more complex than that of a rectangle. If a semicircular waterway is cut at an angle to give a type 1 skew then the entrance to the opening is no longer a semicircle. Additionally, with an arch there is a strong radial contraction towards the centre. This tends to make the effect of skew on the flow through arches difficult to identify and measure, and as yet there appears to have been no satisfactory study of the effect of skew on arch bridges, although there have been attempts (Husain and Rao, 1966). However, Fig. 3.9b compares a normal semicircular arch ($b = b_S = 200\,\text{mm}$) with a 45° skewed 1A arch with $b = 200\,\text{mm}$ and $b_S = 283\,\text{mm}$. The larger value of b_S results in an increased discharge, but the increase is smaller than that obtained with the corresponding rectangular opening in Fig. 3.9a.

3.8 Depth of flow, *Y*

For a given discharge, flow can occur over a wide range of depths depending upon the slope and geometry of the channel, and on whether the flow is uniform or non-uniform. The values of many variables, such as the Froude number (F), conveyance (K) and opening ratio (M), are functions of the depth. Additionally, the depth of flow relative to the height of the bridge opening can influence both the type of flow that occurs at a bridge site and the hydraulic performance of the structure: it may determine whether or not a waterway becomes drowned during flood. For example, if the value of

Table 3.1 Geometric properties of semicircular arches

Proportional depth, Y_u/Z	*Surface width,* $B_T(m)$	*Cross-sectional area* $A(m^2)$
0	$1.000 \times D$	0
0.1	0.995	$0.064 \times \pi D^2/4$
0.2	0.980	0.125
0.3	0.954	0.188
0.4	0.917	0.248
0.5	0.866	0.305
0.6	0.800	0.358
0.7	0.714	0.406
0.8	0.600	0.448
0.9	0.436	0.481
1.0	0	0.500

Notes:
1. Froude number, $F = V/(gY_M)^{1/2}$ where the mean depth, $Y_M = A/B_T$.
2. For example, if $V = 1.5$ m/s, $Y_u = 1.75$ m and the diameter of a semicircular arch, $D = 5.0$ m then $Z = 2.5$ m and $Y_u/Z = 0.7$. From the table $A = 0.406 \times \pi \times 5.0^2/4 = 7.97$ m^2 and $B_T = 0.714 \times 5.0 = 3.57$ m. Thus $Y_M = 7.97/3.57 = 2.23$ m and $F_3 = 0.32$.
3. There is an apparent anomaly with $Y_M = A/B_T = 2.23$ m and $Y_u = 1.75$ m, which is compounded by the fact that B_T decreases to zero when $Y_u/Z = 1.0$. Section 3.3 suggested an alternative approach with $Y_M = A/b = 7.97/5.0 = 1.59$ m. This gives $F_3 = 0.38$. These figures illustrate the problem of obtaining a 'correct' value of F, particularly with arched waterways.

Y_N in an unconstricted channel is similar to the height of a bridge opening (Z) then it is likely that the waterway will become submerged (Fig. 3.1g).

The ratios Y_N/Z and Y_u/Z are used in Figs 3.4 and 3.9 to illustrate hydraulic performance. With arched openings the ratios are valuable in defining changes in geometry (Table 3.1). The listing of the water surface widths and proportional cross-sectional areas is useful when evaluating M and F_3.

3.9 Shape of the waterway opening

The shape of the waterway may affect the hydraulic performance of a bridge (Fig. 3.1h). For instance, a rectangular opening with a width twice its height ($b = 2Z$) has a 27% larger cross-sectional area than the equivalent semicircular arch. This means that at any given stage a rectangular opening will probably have a larger discharge and a smaller afflux than an arch of the same span. These differences are likely to be most significant when the waterway is running nearly full or full.

The distribution of velocity and the pattern of flow through a waterway will also vary according to shape. For example, the upstream edge of an arch can impart a distinct radial contraction (lateral and vertical) to the flow, even before the waterway becomes drowned. This is apparent from

Fig. 2.9 which shows how the point of maximum velocity moves nearer to the bed with rising stage. With arches the gradually closing roof can produce a more gentle transition between channel flow and sluice gate flow than a rectangular opening where (theoretically) the waterway becomes drowned very abruptly when the water level rises above the horizontal upstream soffit. In practice, the water surface may be lower in the centre of the opening (because of the higher velocity head) than at the sides, so this transition is not always as sudden as expected (Fig. 2.7). Nevertheless, a very large vertical contraction can occur from the horizontal soffit after the opening has submerged and $Y_u > 1.1Z$. Section 7.2 shows that rounding the soffit of the opening can significantly reduce the contraction and improve hydraulic performance.

Differences in the shape of the waterway might also be expected to affect the values of the coefficients of contraction and discharge. Matthai (1967) presented four different graphs of the coefficient C' corresponding to the four different types of waterway listed in Section 4.2.1. The coefficients of discharge of arched and rectangular waterways in Fig. 2.11 are also different.

3.10 Channel roughness and shape

These parameters do have an influence on the hydraulic performance of a bridge, but it is usually small. Generally they are not considered directly, although they are taken into account indirectly in the calculation of the opening ratio (in terms of the discharge or conveyance, equations 3.1 and 3.5) and the Froude number.

Channel roughness affects the differential head across the opening (Δh in Fig. 2.3), which increases with increasing roughness. Heavy vegetation on the downstream floodplain interferes with the natural expansion of the jet and causes a large afflux. The existence of rough, scrub- or tree-covered floodplains also causes practical difficulties in estimating roughness and friction loss, in addition to any limitation arising from the way in which these parameters are incorporated into the hydraulic analysis of bridge performance. Conversely, with smooth channels and relatively severe contractions the friction loss arising from boundary roughness is insignificant compared with the differential head across the constriction. In many cases the friction loss in the opening is negligible or assumed to be so (see Section 4.2.1).

The relative roughness and shape of a river channel also affect the distribution of velocity and discharge within it. For convenience, it is often assumed that there is a uniform velocity over the entire width of the channel (as in equation 3.2). Unfortunately this is often unrealistic, particularly when the flow is overbank in a compound channel where the velocity can vary from almost nothing on the floodplains or at the banks to a large value in the deeper or central parts of the main channel. To allow for this the velocity distribution coefficient, or kinetic energy correction factor, α, is used to obtain the weighted average velocity head ($aV^2/2g$) at a cross-section (Chow, 1981; French, 1986).

In a man-made channel with a uniform cross-sectional area a seldom exceeds a value of 1.15. Chow pointed out that a tends to be lower (1.10 minimum) with large, prismatic, straight channels and when the flow is of considerable depth. The value is higher (1.20 maximum) for small channels and when the conditions are the reverse of those just described. Natural streams typically have a value of around 1.30, ranging from as low as 1.15 to a maximum of 1.50 depending upon the conditions. When the flow is over the floodplains, values range from 1.5 to 2.0 with 1.75 being a typical figure. Compound sections and natural rivers that comprise several distinct subchannels and/or floodplains, obstructions, bridges, weirs, and pronounced irregularities in alignment can all result in values greater than 2.0. High values are also possible in closed conduits, which may have relevance under some circumstances to bridge waterways running full. However, large values near 2.0 should be used sparingly, since they cannot be justified very often. It is very difficult to give accurate guidelines regarding a values so those just quoted must be regarded as extremely approximate; if possible the true value should be established by calculation, as described below.

During a flood the mean velocity and hence the velocity head may be quite large (if $V = 3\,\text{m/s}$ then $V^2/2g = 0.46\,\text{m}$), so neglecting a or using an inappropriate value can significantly affect the analysis. This problem often manifests itself in an energy line that looks wrong, possibly because its elevation increases in the downstream direction. If the water surface has a shallow slope, then any errors in the calculated velocity heads at the various sections often become apparent in this way.

The value of the dimensionless coefficient α for a particular cross-section can be obtained by splitting the channel into N subsections and then using the equation

$$a = \frac{\sum_{i=1}^{N} (K_i^3/A_i^2)}{(K^3/A^2)} \tag{3.17}$$

where K_i is the conveyance of the ith subsection (m^3/s), A_i is the area of the ith subsection (m^2), K is the conveyance of the total cross-section, and A is the area of the total section. Alternatively, if the velocity has been measured with a current meter, the following forms of the equation may be used:

$$a = \frac{\sum_{i=1}^{N} (Q_i V_i^2)}{(QV^2)} \tag{3.18}$$

or

$$a = \frac{\sum_{i=1}^{N} (A_i V_i^3)}{(AV^3)} \tag{3.19}$$

where Q_i and V_i are the discharge (m^3/s) and velocity (m/s) in the *i*th subsection, Q is the total discharge of the river, and V is the average velocity of the whole section – that is, Q/A. Example 3.2 illustrates the use of the equations.

3.11 Scour

Scour is not really a hydraulic variable in the same way as some of the others described above, but it can affect the afflux or backwater at a bridge site, so it has been included here as a reminder. In some ways a scoured waterway may be considered as the natural equivalent of the minimum-energy waterway described in Section 7.5. Scour itself is considered in much more detail in Chapter 8.

If a bridge causes significant scour over a length of the river channel then the elevation of both the bed and the water surface will be reduced if the depth of flow remains the same. Consequently the head loss between the scoured sections and section 4 (where normal depth occurs and scour is less likely) will decrease, as will the afflux. One method of allowing for this is to estimate the scour depths and length of reach affected at different discharges, and then adjust the measured bed levels accordingly. Another technique is to calculate the increase in cross-sectional area caused by scour, ΔA_S, and express this as a proportion of the cross-sectional area of flow at section 2 at normal depth, A_{N2} (Bradley, 1978). The scour correction factor, S_C^*, can then be obtained from Fig 3.10 and used to adjust either the normal afflux, H_1^* (Fig. 2.3), or the drawdown of the water surface at section 3, H_3^*. Using this technique the afflux with scour (H_{1S}^*) is

$$H_{1S}^* = S_C^* H_1^* \tag{3.20}$$

For example, if scour increases the cross-sectional area from $10\,m^2$ to $15\,m^2$, $\Delta A_S/A_{N2} = 5/10 = 0.5$, which gives $S_C^* = 0.52$ and $H_{1S}^* = 0.52H_1^*$. If the opening area increases from $10\,m^2$ to $20\,m^2$ as a result of scour then $\Delta A_S/A_{N2} = 10/10 = 1.0$ and $S_C^* = 0.31$ so $H_{1S}^* = 0.31\ H_1^*$. In both cases the afflux is substantially reduced. If the scour depth is large enough then it is possible that zero afflux, or even a negative afflux, may be experienced (Laursen, 1984).

If a bridge has a sound foundation and can withstand high water velocities, then scour may usefully enlarge the waterway during a flood and enable the structure to pass a bigger discharge with a smaller backwater than would otherwise be the case. However, attempts to improve the capacity of a waterway by excavating the bed to enlarge it will be successful only if the increased area can be maintained and the stability of the river channel is undisturbed. Frequently the excavation silts up again, so other alternatives should also be considered (see Chapter 7). When scour is considered in Chapter 8 it is generally assumed to be a local phonomenon so that the water surface elevation remains the same while the depth of flow increases. This is unlike the assumptions described above.

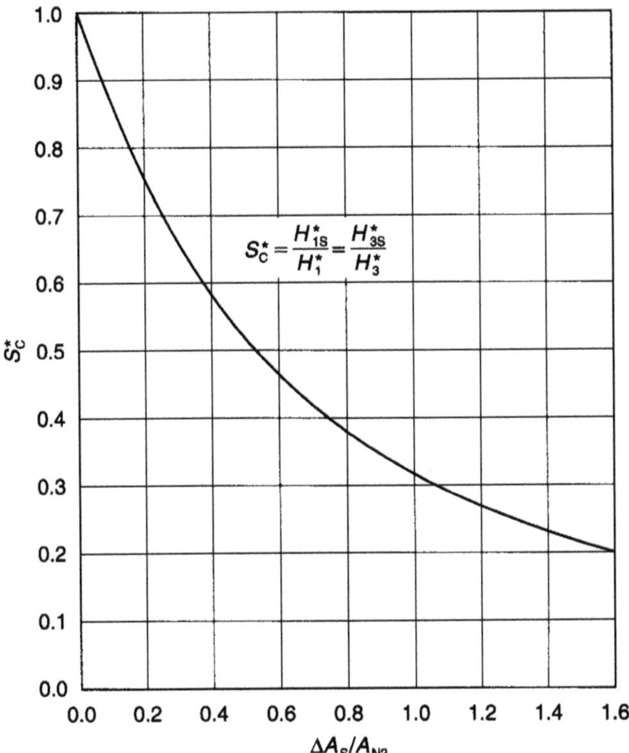

Fig. 3.10 Variation of the scour correction factor, S_C^*, with the proportional increase in scoured area, $\Delta A_S/A_{N2}$. By calculating the depth and area of scour the value of $\Delta A_S/A_{N2}$ and thus S_C^* can be obtained, hence the maximum afflux corrected for scour is $H_{1S}^* = S_C^* H_1^*$. (After Bradley, 1978)

3.12 Examples

Example 3.1

The variation of the bridge opening ratio (M) with stage is required for Canns Mill Bridge (see Section 2.6). The bridge is a segmental arch structure with the springings above bed level (Fig. 3.11a). The relevant dimensions are:

width of bridge opening, $b = 4.28$ m
width of upstream channel, $B = 5.5$ m
height of springs above bed = 0.60 m
height of crown above bed = 1.80 m
radius of arch, $r = 2.70$ m

It would be possible to calculate M from $b/B = 4.28/5.50 = 0.78$, but this is not really appropriate for an arch bridge since the waterway area and width

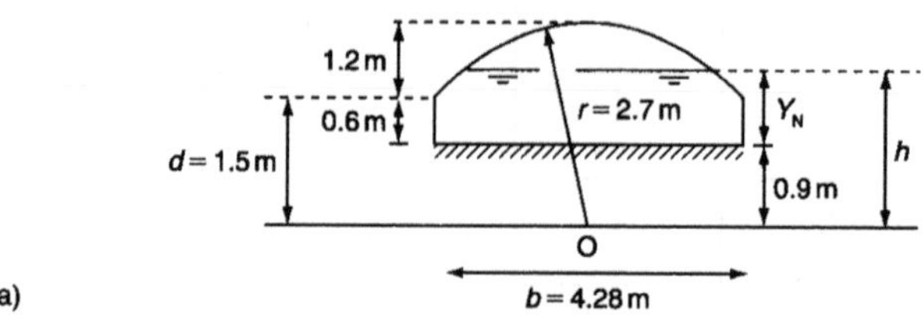

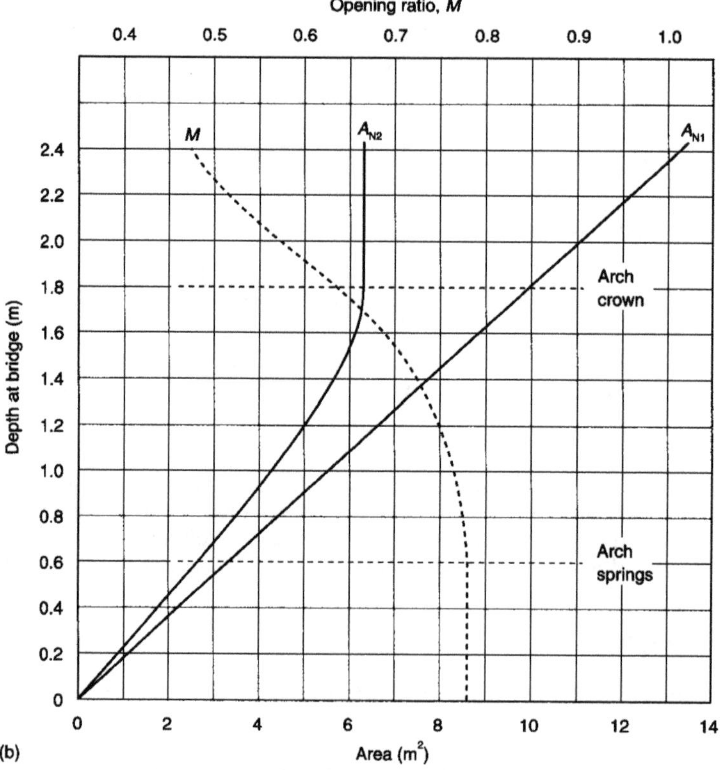

Fig. 3.11 (a) Approximate dimensions of Canns Mill Bridge and (b) the corresponding variation of opening ratio $M = A_{N2}/A_{N1}$ with stage obtained from equation 3.8 assuming a constant channel width of 5.5 m.

change with stage. Equation 3.3 is more appropriate, where $M = a/A = A_{N2}/A_{N1}$ and where A_{N1} and A_{N2} are the cross-sectional areas of sections 1 (upstream) and 2 (opening area) at normal depth (Y_N) so that $A_{N1} = BY_N$. When the water level is below the springline of the arch $A_{N2} = bY_N$, and this applies up to $Y_N = 0.6$ m when $A_{N2} = 4.28 \times 0.60 = 2.57\,m^2$. After this 2.57 m^2 must be added to the area calculated from the numerator of Equation 3.8. Referring to Figs 3.2d and 3.11a:

distance between bed and centre of curvature O = 2.70 − 1.80 = 0.90 m
distance between O and arch springs, $d = 0.90 + 0.60 = 1.50$ m
height of water surface above O, $h = Y_N + 0.90$ m

The area of flow between the bed and the water surface for any depth of flow, Y_N, in the segmental arch can be calculated from the numerator of equation 3.8, written here with $\pi/180$ included to convert to radians. For instance, if $Y_N = 1.65$ m, then $h = 1.65 + 0.90 = 2.55$ m and

$$A_{N2} = \left\{\frac{2.70^2 \times \pi}{180}\left[\sin^{-1}\left(\frac{2.55}{2.70}\right) - \sin^{-1}\left(\frac{1.50}{2.70}\right)\right]\right.$$

$$\left. + 2.55(2.70^2 - 2.55^2)^{1/2} - 1.50(2.70^2 - 1.50^2)^{1/2}\right\} + 2.57$$

$$= \{0.127[70.812 - 33.749] + 2.263 - 3.368\} + 2.570$$

$$= 6.17 \text{ m}^2$$

Now $A_{N1} = 5.50 \times 1.65 = 9.08$ m^2
Thus $M = A_{N2}/A_{N1} = 6.17/9.08 = 0.68$

Notes

1. M is not constant but varies with stage, so this calculation has to be repeated with other values of Y_N. The resulting variation of A_{N1}, A_{N2} and M is shown in Fig. 3.11b. It is apparent that A_{N1} increases linearly, while A_{N2} curves upward to reach a constant value of 6.36 m^2, which corresponds to the full waterway opening. As the stage rises and the crown begins to close, M reduces to around 0.64. After this M decreases sharply as A_{N2} is constant whereas A_{N1} continues to increase.
2. If the opening had been rectangular, the line representing A_{N2} would have been straight so that A_{N2}/A_{N1} would have a constant value equal to b/B.
3. The calculation of the cross-sectional areas (channel and opening) is often best accomplished with a digitiser since in reality they are usually irregular. A simple graph of area against stage can be produced, eliminating the need for the calculations above.
4. This example over-simplifies: see Fig. 3.3 and Example 3.3.

Example 3.2

A bridge has a single rectangular opening 10 m wide, which is the same width as the main river channel at low stages (Fig. 3.12). However, during flood the bridge obstructs the flow over the floodplains. The dimensions and Manning roughness coefficients are shown in the diagram. Assuming uniform flow, that the longitudinal slope of the channel and floodplains is 1 in 1000, and a depth of 4 m in the main channel, estimate the following: (a) the

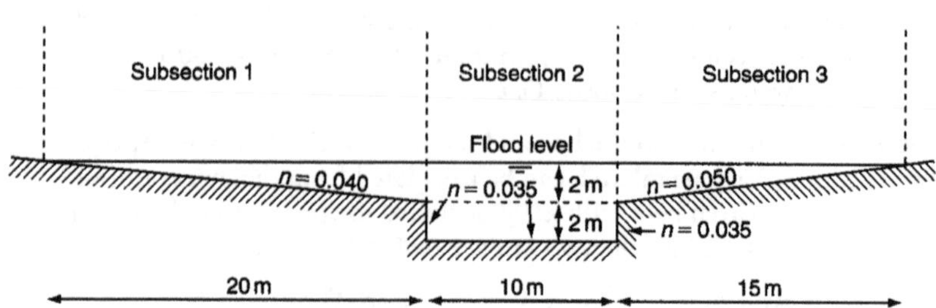

Fig. 3.12 A compound channel split into three subsections by dotted vertical lines. The roughness of the various parts of the channel is shown by the values of Manning's *n*. The conveyance of the subsections is calculated in Example 3.2.

conveyance of the upstream cross-section, *K*; (b) the velocity distribution coefficient, *α*; (c) the bridge opening ratio, *M*.

(a) The total conveyance of the upstream section 1 was denoted by K in equation 3.5. The value of K can be obtained by evaluating the right-hand part of equation 3.6 with respect to each of the three subsections shown in Fig. 3.12 and then adding the values, thus:

$$K = K_1 + K_2 + K_3 \tag{3.21}$$

$$\text{where } K_i = Q_i / S_{Fi}^{1/2} = (1/n_i)\, A_i\, R_i^{2/3} \tag{3.6}$$

As before, the subscript i refers to the ith subsection, of which there are a total of three in this example ($N = 3$). The method is best illustrated by means of a table, but with familiarity most of the arithmetic can be done on a calculator or computer speadsheet without writing out the intermediate steps. When estimating the length of the wetted perimeter, P_i, the imaginary lines that represent the water interface between the subsections are ignored.

Subsection	n_i $(s/m^{1/3})$	A_i (m^2)	P_i (m)	R_i (m)	$K_i = (1/n_i)A_iR_i^{2/3}$ (m^3/s)
1	0.040	20.00	20.10	1.00	500
2	0.035	40.00	14.00	2.86	2302
3	0.050	15.00	15.13	0.99	298
Total		75.00			3100

Therefore the total conveyance of the section is 3100 m^3/s. Note that the conveyance is not equal to the discharge, despite the units being the same. The discharge is obtained from the other part of equation 3.6, namely $Q = KS_F^{1/2}$, so that $Q = 3100 \times 0.001^{1/2} = 98.03\, m^3/s$. The discharge in the individual subsections is shown in the table below.

(b) The velocity distribution coefficient can be estimated by calculating the discharge (Q_i) and velocity (V_i) in each of the subsections and then using equation 3.18.

Subsection	$Q_i = K_i S_F^{1/2}$ *(m^3/s)*	$V_i = Q_i A_i$ *(m/s)*	$Q_i V_i^2$
1	15.81	0.79	9.87
2	72.80	1.82	241.14
3	9.42	0.63	3.74
Total	98.03		254.75

The total discharge through the section is 98.03 m^3/s and the total cross-sectional area of flow is 75.00 m^2, giving an average velocity across the whole section, $V = 98.03/75.00 = 1.31$ m/s. The velocity distribution coefficient, $\alpha = \Sigma(Q_i V_i^2)/QV^2 = 254.75/(98.03 \times 1.31^2) = 1.51$.

(c) The bridge opening ratio can be obtained from equation 3.5 where K_b in this example corresponds to K_2:

$$M = K_2/K = 2302/3100 = 0.74$$

Notes

1. The opening ratio above relates to one depth of flow only, and M would be expected to vary with stage. Thus the calculations have to be repeated for different depths.
2. Since the slope of the main channel is the same as that of the floodplains, in this case there is no difference between the ratio of the sectional conveyances and the ratio of the sectional discharges. Thus $q/Q = 72.80/98.03 = 0.74$. If the gradients of the various parts of the channel are different, then the discharge ratio should be used, as in Example 3.3.
3. In this example the simple ratio $b/B = 10/45 = 0.22$, which is significantly different from 0.74. A value of 0.22 represents a severe constriction. This ratio is particularly inaccurate with compound channels and overbank flow, as Example 3.3 illustrates.
4. Channels with a complex geometry require the use of more subsections, a dividing vertical being inserted wherever appropriate. Matthai (1967) suggested a change of subsection where a significant change in R occurs; a change of subsection where there is a significant change in n is needed only if the flow is shallow. If the discharge has been measured on site, then it is unlikely that the gauged value will agree initially with the calculated value. Some judicious modification of the boundary roughnesses or effective longitudinal gradients is usually required to adjust the relative conveyance or subsectional discharges. Matthai suggested that (working across a section from the left to right bank) a curve of cumulative cross-sectional area against distance be drawn. A graph of cumulative conveyance against distance should then be plotted

to the same distance scale. The shape of the cumulative conveyance curve between the plotted points can then be 'shaped' using the cumulative area curve as a guide. By this means a better estimate of the conveyance in various parts of the channel might be obtained.

5. Another point to remember is that when the value of Manning's n is not uniform over the whole perimeter of a subsection then the effective value of n has to be calculated: that is, n_E. This is not difficult, but if the problem has not been encountered before it is possible to make mistakes. Equation 3.22 was recommended (Hydraulics Research, 1988) for British rivers:

$$n_E = \frac{PR^{5/3}}{\sum_{i=1}^{N}\left(\frac{P_i R_i^{5/3}}{n_i}\right)} \quad (3.22)$$

$$n_E = \left[\frac{\sum_{i=1}^{N}(P_i n_i^{1.5})}{P}\right]^{2/3} \quad (3.23)$$

where the variables have the same meaning as before. Equation 3.23 is simpler and used in HEC-2 (Hydrologic Engineering Center, 1990), while an alternative expression is used in Example 3.3. A good introduction to this aspect of hydraulics is French (1986) and Hydraulics Research (1988). Both gave different equations that may be preferred under different conditions.

6. The method of calculating sectional conveyances and discharges used in this example appears in many of the original publications on bridge hydraulics. It is well documented. However, there are more modern methods of calculating conveyance in two-stage channels. Ackers (1992, 1993) stated that the conventional approach over-estimates the conveyance when in flood because of the interaction between the main channel flow and the slower-moving floodplain flows. A new hydraulic design procedure was presented that allows for this effect, especially when extrapolating observed flows to greater depths. If accuracy or extrapolation is necessary this research should be consulted.

Example 3.3

The variation with stage of the bridge opening ratio for Canns Mill is required. The cross-section upstream of the bridge at the maximum discharge of 15 m^3/s is shown in Fig. 3.3a and diagrammatically in Fig. 3.13. This diagram also shows the Manning n values (determined initially by visual inspection of the channel, and then modified on a trial-and-error basis). The slope of the energy line for the main channel is thought to be different from that on the floodplains, where areas of stationary water have been observed, but these slopes are unknown. The stage–discharge relationship is known. Determine the opening ratio at the maximum discharge.

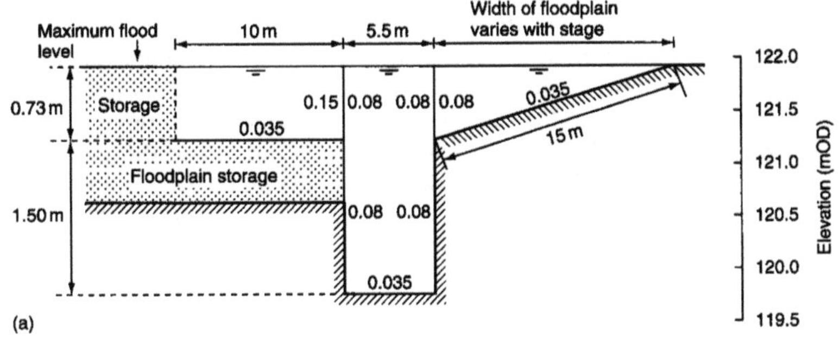

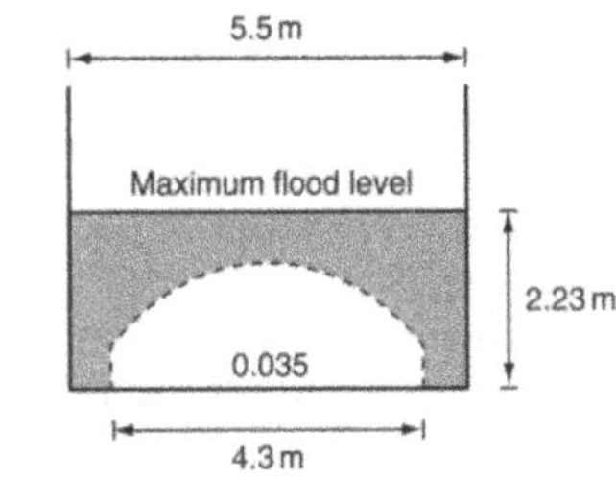

Fig. 3.13 (a) A simplified cross-section of section 1 at Canns Mill (Fig. 3.3a) showing the areas of floodplain storage that do not contribute to flow. The dimensions and Manning's n values are also marked. The discharge in the three channel subsections is calculated in Example 3.3 prior to calculating M. (b) The dimensions of the bridge. The roughness coefficients are as for section 1.

Introduction

One of the most common mistakes made by graduate engineers when analysing hydraulic problems involving bridges is to assume that the flow upstream of the structure is uniform so that the bed slope can be substituted in the Manning equation to obtain the discharge. In reality, there may be an abnormal stage with or without the bridge (Fig. 2.4) so the bed, water surface, and total energy line all have different gradients. Under these conditions the slope of the energy line (friction gradient, S_F) over about 10–20 channel widths should be used. This is often much flatter than the bed slope. Similarly, the friction slope on the floodplain may be less than in the main channel.

Another potential source of error lies in using the maximum flood level (which may be indicated by deposited flood debris) to determine the cross-sectional area of flow. This often introduces significant errors because the flood level shows the maximum flooded cross-section not the maximum cross-sectional area of flow, which may be much less. In other words, there may be no flow through some of the cross-section, the water on parts of the floodplains being stationary (floodplain storage). The only way to ascertain whether or not this is the case is to be on site during a flood to observe what happens. The importance of this is illustrated below.

The approach adopted in this example is to split the compound channel into three subsections: the left floodplain, the main channel, and the right floodplain. It is assumed that the sum of the discharges in the three subsections must equal that in the whole channel, and indeed the quantity passing through the bridge opening since there was no flow around it. At the maximum gauged flow of 15 m^3/s a stage of approximately 2.23 m exists in the main channel at section 1.

Assumptions

LEFT FLOODPLAIN

By observation it is known that at 15 m^3/s there is flow only through the 10 m of the floodplain adjacent to the main river channel; the rest of the inundated area beyond this does not contribute to the flow, and is marked as storage in Fig. 3.13a. It is assumed that the dividing vertical on the left has zero roughness, as is customary.

The top of the hedgebank between the left floodplain and the main channel is about 1.5 m high when measured from the bottom of the main channel (Fig. 3.3a). Because the bank is lower much further upstream, water can enter the floodplain but it cannot return to the main channel and pass through the bridge until the stage exceeds 1.5 m. Thus it is assumed that any water stored below the 1.5 m stage is floodplain storage, over the top of which passes the live part of the stream. Thus at 15 m^3/s a depth of (2.23 − 1.50) = 0.73 m exists over the 10 m wide floodplain. It is not appropriate to assume that this horizontal interface between the stationary water and the live stream has zero roughness (because then practically the entire perimeter of the subsection would be frictionless) so a roughness equivalent to the underlying floodplain, namely 0.035 s/$m^{1/3}$, is adopted (see Hydraulics Research, 1988).

The dividing vertical on the right of the subsection is the bank and hedge through which the water ultimately has to flow, incurring a significant and visible loss of head in the process. This boundary is assumed to be relatively rough, with n = 0.150 s/$m^{1/3}$.

THE MAIN CHANNEL

This is assumed to have n = 0.035 s/$m^{1/3}$ across the 5.5 m width of the channel with both sides having n = 0.08 s/$m^{1/3}$ over the full height, although the bottom part of the channel is perhaps smoother and the top part with the hedge rougher.

To calculate the bridge opening ratio from $M = q/Q$ the waterway has to be projected upstream to section 1, with q representing the flow in this projected subsection of the main channel (i.e. the quantity that can pass through the bridge without contraction), and Q the total discharge, including the floodplains. Thus the main channel itself has to be subdivided

(Fig. 3.13b). It is assumed that the bed has a roughness (0.035 s/m$^{1/3}$) while the imaginary projected boundary of the arch is ignored.

At 15 m^3/s the stage is 2.23 m while the height of the arch is 1.8 m, so the opening is submerged. By calculating the discharge (q) through the projected waterway area (6.30 m^2) the value of M will include the vertical contraction through the bridge. Note that if the width of the opening (4.30 m) was projected upstream over the full 2.23 m depth this would be equivalent to an unsubmerged rectangular opening with an area of 9.59 m^2.

RIGHT FLOODPLAIN

The right floodplain, like the left, is approximately 1.5 m above the bottom of the main channel. It has a transverse slope towards the main channel, so the width of the floodplain varies with stage. At 15 m^3/s the maximum depth at the interface with the main channel is (2.23 − 1.50) = 0.73 m, reducing to zero over a slope distance of 15 m as shown in Fig. 3.13a. The roughnesses of both surfaces are included in the calculations, being 0.080 s/m$^{1/3}$ and 0.035 s/m$^{1/3}$ for the left and bottom boundaries respectively.

Calculations

These are conducted for each of the subsections in turn. Important features are:

(a) the non-uniform roughness around the wetted perimeter necessitating the calculation of the effective Manning roughness (n_E) for the subsection, as follows:

$$n_E = \left[\frac{\sum_{i=1}^{N} (P_i n_i^2)}{P} \right]^{1/2} \quad (3.24)$$

where P_i = length of ith perimeter, n_i = roughness of ith perimeter and P = the length of the total wetted perimeter used in the calculations (see French, 1986);

(b) the friction gradients of the main channel and floodplains are unknown, so it is not possible to obtain directly the discharges. This problem is overcome by first calculating the conveyances of the subsections at two different discharges (15 m^3/s and 12.5 m^3/s) and then equating the total conveyance to the gauged discharge to obtain the gradients. The discharge in the subsections can then be calculated.

LEFT FLOODPLAIN

Sectional area $A_1 = 0.73 \times 10 = 7.30\,\text{m}^2$
Wetted perimeter, $P_1 = 10.73\,\text{m}$
Hydraulic radius, $R_1 = A_1/P_1 = 7.30/10.73 = 0.68\,\text{m}$

From equation 3.24, $n_{E1} = \left(\frac{10 \times 0.035^2 + 0.73 \times 0.15^2}{10.73} \right)^{1/2} = 0.052\,\text{s/m}^{1/3}$

From equation 3.6, conveyance $K_1 = A_1R_1^{2/3}/n_{E1} = 7.30 \times 0.68^{2/3}/0.052 = 109\,m^3/s$

MAIN CHANNEL – PROJECTED BRIDGE OPENING

$A_2 = 6.30\,m^2$, $P_2 = 4.30\,m$ (= width of opening) and $n_{E2} = 0.035\,s/m^{1/3}$
$R_2 = A_2/P_2 = 6.30/4.30 = 1.47\,m$
$K_2 = A_2R_2^{2/3}/n_{E2} = 6.30 \times 1.47^{2/3}/0.035 = 232\,m^3/s$

MAIN CHANNEL – REMAINDER

$A_3 = (2.23 \times 5.5) - 6.30 = 5.97\,m^2$
$P_3 = 2.23 + (5.50 - 4.30) + 2.23 = 5.66\,m$
$R_3 = A_3/P_3 = 5.97/5.66 = 1.05\,m$
Using the same roughness coefficients as for section 1 in Fig. 3.13a:

$$n_{E3} = \left(\frac{2 \times 2.23 \times 0.08^2 + (5.50 - 4.30) \times 0.035^2}{5.66}\right)^{1/2} = 0.073\,s/m^{1/3}$$

$K_3 = A_3R_3^{2/3}/n_{E3} = 5.97 \times 1.05^{2/3}/0.073 = 85\,m^3/s$

RIGHT FLOODPLAIN

$A_4 = ½ \times 0.73 \times 14.98 = 5.47\,m^2$
$P_4 = 15.00 + 0.73 = 15.73\,m$
$R_4 = A_4/P_4 = 5.47/15.73 = 0.35\,m$

$$n_{E4} = \left(\frac{0.73 \times 0.08^2 + 15 \times 0.035^2}{15.73}\right)^{1/2} = 0.038\,s/m^{1/3}$$

$K_4 = A_4R_4^{2/3}/n_{E4} = 5.47 \times 0.35^{2/3}/0.038 = 71\,m^3/s$

TOTAL CONVEYANCE AND RELATIONSHIP TO DISCHARGE

Total conveyance of main channel, $K_M = K_2 + K_3 = 232 + 85 = 317\,m^3/s$
Total conveyance of floodplains, $K_F = K_1 + K_4 = 109 + 71 = 180\,m^3/s$
From equation 3.6, $K_t\, S_{F_t}^{1/2} = Q_t$

In this example, assume that the main channel has a different friction gradient (S_{FM}) from that of the floodplains (S_{FF}), but assume that both floodplains have the same friction gradient. Thus at 15 m^3/s

$$317\, S_{FM}^{1/2} + 180\, S_{FF}^{1/2} = 15.0 \quad (1)$$

If the calculations are repeated for a discharge of $12.5\,m^3/s$ and a corresponding stage of 2.00 m, then another equation is obtained:

$$293\, S_{FM}^{1/2} + 96\, S_{FF}^{1/2} = 12.5 \quad (2)$$

Solving equations (1) and (2) for the two unkowns $S_{FM}^{1/2}$ and $S_{FF}^{1/2}$ then

$S_{FM}^{1/2} = 0.0362$ giving $S_{FM} = 0.00132$ or 1 in 757 and $S_{FF}^{1/2} = 0.0198$ giving $S_{FF} = 0.000376$ or 1 in 2660

As a check, if the conveyances corresponding to 10.8 m^3/s and a stage of 1.85 m are calculated then

$$278\ S_{FM}^{1/2} + 50\ S_{FF}^{1/2} = 10.8 \qquad (3)$$

If the gradients of 1:757 and 1:2660 are substituted into equation (3) the discharge obtained is slightly high at 11.06 m^3/s. A reasonable compromise between all three equations is

$$S_{FM} = 1 \text{ in } 780$$

$$S_{FF} = 1 \text{ in } 2500$$

This gives the three calculated discharges as 14.95 m^3/s, 12.41 m^3/s and 10.95 m^3/s, which is adequate for interpolation between 10.8 m^3/s and 15.0 m^3/s. By 7.00 m^3/s the flow is confined to the main channel so different assumptions are required.

BRIDGE OPENING RATIO AT 15 m^3/s

Conveyance of projected waterway, $K_2 = 232\,\text{m}^3/\text{s}$
Discharge through projected waterway, $Q_2 = K_2 \times S_{FM}^{1/2} = 232 \times (1/780)^{1/2}$
$= 8.31\,\text{m}^3/\text{s}$

Bridge opening ratio, $M = q/Q = 8.31/15.00 = 0.55$

Notes

1. If the calculations are repeated for other stages the results are as in Fig. 3.3b.
2. For 15 m^3/s, if the ratio of the conveyances is used (equation 3.5) then $M = 232/497 = 0.47$, but this erroneously assumes that the floodplain friction gradient equals that for the main channel. If the stored water on the floodplains is not eliminated from the calculations, then the equivalent values of b/B and a/A are about 0.08 and 0.14 respectively, significantly different from 0.55.
3. The hedgebank on the left is similar to situations where levees separate the floodplain from the main channel (see Fig. 8.16). Even though the levees may have been overtopped in one location so that the area behind is inundated, how much water is actually discharged through this region? How this is handled can make a big difference to a hydraulic analysis. Some computer software includes an option to eliminate areas outside the levees.

References

Ackers, P. (1992) Hydraulic design of two stage channels. *Proceedings of the Institution of Civil Engineers, Water Maritime and Energy*, **96**, December, 247–257.

Ackers, P. (1993) Stage–discharge functions for two stage channels: the impact of new research. *Journal of the Institution of Water and Environmental Management*, 7, February, 52–61.

Biery, P.F. and Delleur, J.W. (1962) Hydraulics of single span arch bridge constrictions. *Proceedings of the American Society of Civil Engineers, Journal of the Hydraulics Division*, **88**(HY2), March, 75–108.

Bradley, J.N. (1978) *Hydraulics of Bridge Waterways*, 2nd edn, US Department of Transportation/Federal Highways Administration, Washington DC.

Chow, V.T. (1981) *Open-Channel Hydraulics*, International Student Edition, McGraw-Hill, Tokyo.

French, R.H. (1986) *Open-Channel Hydraulics*, International Student Edition, McGraw-Hill, Singapore.

Hamill, L. (1993) A guide to the hydraulic analysis of single-span arch bridges. *Proceedings of the Institution of Civil Engineers, Municipal Engineer*, **98**, 1–11.

Hamill, L. (1997) Improved flow through bridge waterways by entrance rounding. *Proceedings of the Institution of Civil Engineers, Municipal Engineer*, **121**, 7–21.

Henderson, F.M. (1966) *Open Channel Flow*, Macmillan, New York.

Husain, S.T. and Rao, G.M. (1966) Hydraulics of river flow under arch bridges. *Irrigation and Power*, October, 441–454.

Hydraulics Research. (1988) Assessing the hydraulic performance of environmentally acceptable channels. Report EX 1799, Hydraulics Research Ltd, Wallingford, England.

Hydrologic Engineering Center (HEC) (1990) *HEC-2 Water Surface Profile User's Manual*, US Army Corps of Engineers, Davis, CA.

Kindsvater, C.E. and Carter, R.W. (1955) Tranquil flow through open-channel constrictions. *Transactions of the American Society of Civil Engineers*, **120**, 955–992.

Kindsvater, C.E., Carter, R.W. and Tracy, H.J. (1953) *Computation of peak discharge at contractions*, Circular 284, United States Geological Survey, Washington DC.

Laursen, E.M. (1984) Assessing vulnerability of bridges to floods, in *Transportation Research Record 950*, Second Bridge Engineering Conference, Vol. 2, Transportation Research Board/National Research Council, Washington DC, pp 222–229.

Matthai, H.F. (1967) Measurement of peak discharge at width contractions by indirect methods. *Techniques of Water Resource Investigations of the United States Geological Survey*, Chapter A4, Book 3, *Applications of Hydraulics*, US Government Printing Office, Washington DC.

Tracy, H.J. and Carter, R.W. (1955) Backwater effects of open-channel constrictions. *Transactions of the American Society of Civil Engineers*, **120**, 993–1018.

Yarnell, D.L. (1934) Bridge piers as channel obstructions. Technical Bulletin No. 444, November, US Department of Agriculture, Washington DC.

4 How to calculate discharge and afflux

4.1 Introduction

Usually bridge hydraulics is concerned with determining either the stage–discharge relationship of a structure or the afflux, or possibly both. This chapter shows how to analyse the flow through rectangular and arched bridge waterway openings with either single or multispans, before and after the waterway becomes drowned. The abutments provide the primary contraction, with the piers being a secondary influence: often the gross area of the waterway opening (A_{N2}, which includes the area occupied by any piers) is employed initially in the calculations, and then a correction factor (J) is introduced to allow for the secondary obstruction of the piers. Typically $J = A_p/A_{N2}$ where A_P is the cross-sectional area of the submerged parts of the piers projected onto section 1. On the other hand, Chapter 5 considers long crossings where it is the piers that control the flow while the effect of the abutments is negligible.

The methods described below are relatively simple, most being devised before the computer era. Thus they provide a method of analysis for engineers without access to today's specialist computer software and, for those who do, a means of quickly checking the results. Since there have been relatively few major investigations of bridge hydraulics, some of the principles and equations described in Chapters 4–6 will be included in the software.

Four methods of analysis are considered below; each is introduced, described in more detail, and then its advantages and disadvantages summarised. Only a condensed version of the original publications can be reproduced, so whenever possible readers should refer to the source material. These methods are attributable to:

- US Geological Survey (USGS);
- US Bureau of Public Roads (USBPR);
- Biery and Delleur;
- Hydraulics Research (HR).

The first two listed were developed almost entirely for rectangular openings and 'deck type' bridges, while the last two were mainly concerned with arch bridges. However, all four methods can sometimes be applied to arches, as shown in Section 4.6 (Matthai, 1967; Hamill, 1993). Which method is adopted depends to some extent upon the data available, the ease of application, the type of flow experienced (e.g. subcritical, supercritical, abnormal stage), the waterway geometry (e.g. rectangular, arched, single or multispan), and the purpose of the investigation. The USGS and the USBPR methods are the most frequently employed. Both have some restrictions on their use (Sections 4.2.3 and 4.3.7), which should be understood before undertaking an analysis. With any of the methods large errors may be incurred under adverse conditions or when conditions differ significantly from those of the original research (Skogerboe *et al.*, 1973; Barret and Skogerboe, 1973; Fiuzat and Skogerboe, 1983; Kaatz and James, 1997). Adverse conditions include the following.

- Sites where the flow is non-uniform and abnormal stages exist. Paradoxically, Barret and Skogerboe (1973) found that the USBPR method was more accurate with an abnormal stage, but conjectured that a greater range of data would reveal greater errors. However, it is certainly true that an abnormal profile complicates the calculations, which are much simpler when uniform flow exists and normal depth provides a convenient datum (Figs. 2.3 and 2.4). Generally, with an abnormal stage the bridge exerts a relatively weak control on the flow, reducing the accuracy of an analysis.
- Crossings where the opening ratio is large (e.g. $M = 0.8$) so that the bridge does not exert a strong control on the flow. Under these conditions Fiuzat and Skogerboe (1983) claimed that the USGS method may underestimate the discharge (see Section 4.2.3, items 2 and 3).
- Sites with wide, heavily vegetated floodplains. With wide valleys the flow approaching a bridge has to move from the floodplain into the main channel in order to pass through the constriction. Laursen (1970) termed this a zone of accretion. Similarly, having passed through the opening, the flow expands from the main channel back onto the floodplain. This was termed a zone of abstraction. For flow to expand laterally in the zone of abstraction it follows that the water surface must slope away from the main channel (this is the reverse of the normal situation). If the floodplain is heavily covered with trees and brush, a large transverse gradient is required to overcome the resistance to lateral flow, so backwater can occur downstream of a bridge as well as upstream. Consequently Laursen concluded that in wide valleys the upstream backwater can be much greater than that predicted by either the USGS or USBPR method, so it is not surprising that under these conditions the backwater measured by Kaatz and James (1997) was roughly double that expected (computer models also had significant errors; see Section 4.6). Laursen presented a numerical approach to this problem, but a drawback was that the accretion and abstraction rates had to be assumed.

- Locations where two or more bridges affect each other hydraulically (Figs 4.27 and 4.28).

The comments above regarding inaccuracies arising from the application of the USGS and USBPR techniques do not mean that they should not be used, rather that these are the most commonly employed methods and thus the most critically appraised. The others are used less so there is little or no evidence available from third parties regarding their accuracy. Care needs to be exercised when interpreting articles discussing the (lack of) accuracy of the various methods: sometimes 'errors' arise because the equations are used outside their recommended range or under conditions not included in the original study. As always, it is difficult to obtain accurate field data with which to form an assessment. Generally, models verified using a large body of accurate field data should be the most reliable.

4.2 US Geological Survey (USGS) method

This was one of the largest and most comprehensive investigations of flow through rectangular openings (Kindsvater *et al.*, 1953; Tracy and Carter, 1954; Matthai, 1967). It is based on an extensive laboratory study, which was verified at 30 field sites. It included four different types of opening, eccentricity, skew, entrance rounding, piers, wingwalls and submergence of the opening (but not abnormal stage or supercritical flow). The procedure has been well documented in hydraulics textbooks (Chow, 1981; French, 1986) and is widely used, so it may be the method most familiar to some engineers and hydrologists. The original work was extended by Matthai (1967), who suggested that the Type I opening with vertical abutments and vertical embankments often approximates an arched waterway, but if much of the arch is submerged a reliable answer cannot be obtained. Nevertheless, there is some justification for cautiously applying the technique to arch bridges.

Essentially the bridge was regarded as a form of gauging device, and a procedure was devised for calculating the peak discharge at the contraction from the observed water levels at sections 1 and 3 (in practice this often means using trash marks). The theoretical basis is the energy and continuity equations, which are combined to give a discharge equation (equation 4.5) that can be applied to openings operating in either the channel flow or submerged condition. A method of calculating the afflux was added, but this is not exactly straightforward.

This method needs the water levels at sections 1 and 3. Section 1 is generally located one span (b) upstream. Section 3 is located at the minimum cross-sectional area of flow on a line parallel to the bridge face, generally between the abutments and not necessarily at the downstream face of the opening. Unfortunately during a flood it is not always easy to observe or measure accurately the stage at section 3 since it may be located inside a waterway opening that is flowing almost full. Nor is this stage easy to calculate. There are ways to get around this, which are described in Section 4.2.2, but they tend to be relatively convoluted and tedious. Another difficulty associated with the technique is that it uses the Froude number at section 3, which, as explained in Section 3.3, becomes meaningless as an arched opening closes and the waterway becomes submerged.

4.2.1 *Calculation of discharge (stage–discharge analysis)*

The USGS method assumes that the contracted section formed by the bridge abutments and the channel bed is effectively a discharge meter that can be utilised to calculate flood flows. This is achieved by substituting into the discharge equation the values of a series of experimental coefficients that relate to standard types of opening (see below) and the measured difference in water level between sections 1 and 3.

The discharge equation is derived from the continuity and energy equations. For convenience it has been assumed that the bed is horizontal in Fig. 4.1, and that the vena contracta occurs at section 3. Thus the energy equation can be written as

$$Y_1 + \frac{\alpha_1 V_1^2}{2g} = Y_3 + \frac{\alpha_3 V_3^2}{2g} + h_E + h_F \tag{4.1}$$

where h_E is the head loss (m) due to eddying and turbulence between sections 1 and 3, and h_F represents the head loss (m) due to boundary friction between sections 1 and 3. If $(Y_1 - Y_3) = \Delta h$, then

$$V_3 = \left[\frac{2g}{\alpha_3}\left(\Delta h + \frac{\alpha_1 V_1^2}{2g} - h_E - h_F\right)\right]^{1/2} \tag{4.2}$$

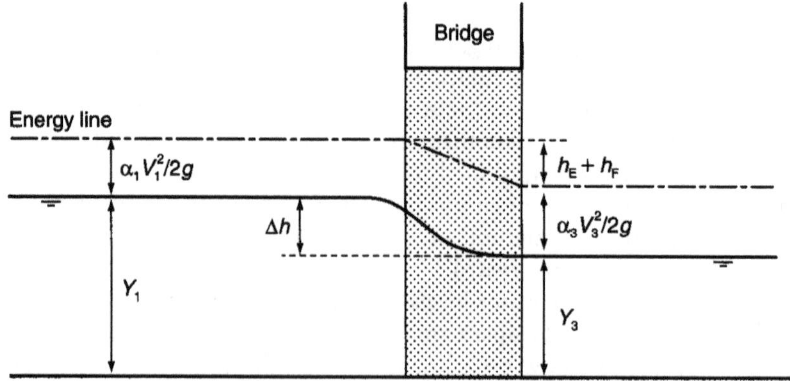

Fig. 4.1 Diagrammatic representation of flow through a bridge opening for the derivation of the USGS discharge equation.

Now from the continuity equation:

$$Q = C_C\, b\, Y_3\, V_3 \tag{4.3}$$

$$Q = C_C\, b\, Y_3 \left[\frac{2g}{a_3}\left(\Delta h + \frac{a_1 V_1^2}{2g} - h_E - h_F\right)\right]^{1/2} \tag{4.4}$$

For a rectangular opening $A_3 = bY_3$ and for an arched waterway A_3 can be taken as the cross-sectional area of the waterway at depth Y_3 so:

$$Q = CA_3 \left[2g\left(\Delta h + \frac{a_1 V_1^2}{2g} - h_F\right)\right]^{1/2} \tag{4.5}$$

where C is a dimensionless coefficient of discharge that is a function of C_C, a_3 and h_E; Q is the discharge (m^3/s); A_3 (m^2) is the gross cross-sectional area of flow (including the area occupied by any piers) at section 3 at the minimum contracted section measured parallel to the bridge face; Δh is the difference in elevation (m) of the water surface between sections 1 and 3; and $a_1V_1^2/2g$ is the weighted average velocity head (m) at section 1. This equation was adopted by Kindsvater *et al.* (1953), Kindsvater and Carter (1955), and Matthai (1967).

If $V_1 = Q/A_1$, then it is apparent that equation 4.5 has to be solved iteratively, since both Q and V_1 are initially unknown. However, in many situations a direct solution is possible since $a_1V_1^2/2g$ and h_F are approximately equal in magnitude and can be cancelled; both are usually small in comparison with Δh anyway (Kindsvater and Carter, 1955). This means that a simplified equation can be obtained without introducing a large error. Equation 4.9 provides an alternative.

The friction loss

The friction loss, h_F, is defined as the total loss of head due to friction between sections 1 and 3. This comprises two parts: the loss in the approach reach between section 1 and section 2 (upstream face), and the loss in the opening (section 2 to 3). The lengths of these two reaches are $L_{1\text{-}2}$ and L respectively. Assuming uniform flow with $S_F = S_O$ the corresponding bed slopes in the two reaches are $S_{O1\text{-}2}$ and $S_{O2\text{-}3}$, so

$$h_F = L_{1\text{-}2}\, S_{O1\text{-}2} + L\, S_{O2\text{-}3} \tag{4.6}$$

From equation 3.6, $S_{Oi} = (Q_i/K_i)^2$. This allows equation 4.6 to be written in terms of the total discharge, Q, and the total conveyances of the two reaches, $K_{1\text{-}2}$ and $K_{2\text{-}3}$, found by averaging the conveyances of the full cross-sections at the ends of each reach. Thus

$$h_F = L_{1\text{-}2}\left(\frac{Q}{K_{1\text{-}2}}\right)^2 + L\left(\frac{Q}{K_{2\text{-}3}}\right)^2 \tag{4.7}$$

Assuming that in the opening $K_2 = K_3$ and that $K_{1-2} = K_{1-3} = (K_1 K_3)^{1/2}$, then

$$h_F = L_{1-2}\left(\frac{Q^2}{K_1 K_3}\right) + L\left(\frac{Q}{K_3}\right)^2 \tag{4.8}$$

$$\text{where } K_i = \frac{A_i R_i^{2/3}}{n_i} \tag{3.6}$$

The presence of piers or a submerged deck reduces the conveyance of sections 2 and 3 (and the discharge through them) so this must be taken into account, as follows.

1. Compute the gross area, A_3, of section 3 including any obstructions.
2. Split section 3 into subsections using dividing lines at the edge of each pier or pile bent or deck.
3. Calulate the total submerged area, A_{3P}, of the obstructions projected on the plane defined by section 3, which is parallel to the bridge face.
4. The net area of flow is $A_i = (A_3 - A_{3P})$, which can be used in equation 3.6 above.
5. Note that the lengths of the sides of the piers, piles or bridge surfaces in contact with the water should be included when calculating the hydraulic radius, R_i (Kindsvater *et al.*, 1953).
6. In laboratory tests, about one-half of the total fall in the water surface between sections 1 and 3 occurred between section 1 and the upstream side of the bridge, which may help when deciding whether the opening is submerged.

If the approach section is thick with vegetation while the reach under the bridge is relatively clear, then h_F may be a significant part of the total fall of the water surface. Under these conditions satisfactory results cannot be obtained, although the accuracy may be improved by using K_b (equation 3.5) instead of K_3 in the first term of equation 4.8 (Matthai, 1967). However, ideally the method should be used only where $\Delta h > 4h_F$, and heavily vegetated floodplains should be avoided.

When conducting preliminary calculations, it should be remembered that there is the option of assuming that h_F equals the velocity head and omitting both of these terms from equation 4.5.

The explicit discharge equation

By putting $V_1 = Q/A_1$ and using equation 4.8 to replace h_F it is possible to write equation 4.5 as

$$Q = 4.43CA_3\left(\frac{\Delta h}{1 - a_1 C^2\left(\frac{A_3}{A_1}\right)^2 + 2gC^2\left(\frac{A_3}{K_3}\right)^2\left(L + \frac{L_{1-2}K_3}{K_1}\right)}\right)^{1/2} \tag{4.9}$$

where all the terms are in metric units. If working in feet the 4.43 should be replaced by the value of $(2g)^{1/2}$, which is 8.02. This equation has the advantage that it can be solved directly for Q without having to iterate. However, if C is a function of the Froude number, as with type I openings, then an iterative solution is still required.

The standard opening types

There are four types of opening (Kindsvater *et al.*, 1953; Matthai, 1967). When slopes are quoted below they are in the form 1 vertical to x horizontal.

TYPE I

Vertical embankments and vertical abutments, with or without wingwalls (Fig. 4.2). In addition to the opening ratio (M) and length ratio (L/b), the Froude number, the degree of entrance rounding or the wingwall angle, and the skew of the contraction with respect to the direction of flow affect the coefficient of discharge (Fig. 4.3). Matthai suggested that this type of opening often approximates an arched waterway, but if much of the arch is submerged a reliable answer cannot be obtained.

TYPE II

Sloping embankments and vertical abutments (Fig. 4.4). In addition to the opening ratio (M) and length ratio (L/b), the depth of water at the abutments, the skew angle and the embankment slope affect the coefficient of discharge. Charts are presented for embankment slopes of 1 to 1 and 1 to 2 (Figs 4.5 and 4.6).

TYPE III

Sloping embankments and sloping spillthrough abutments (Fig. 4.7). In addition to the opening ratio (M) and the length ratio (L/b), the skew angle and the slope of the embankment and abutments affect the coefficient of discharge. The effect of the slope of the type III abutments is represented by x/b, where x is the horizontal thickness of the upstream side slope measured longitudinally at the projected level of the water surface at section 1, and b is the average width of the opening. Note that with all sloping embankments x is included in the waterway length, L. Three embankment slopes were investigated: 1 to 1, 1 to 1.5, and 1 to 2, each of which must be evaluated using its own charts (Figs 4.8–4.10).

TYPE IV

Sloping embankments, vertical abutments with wingwalls (Fig. 4.11). In addition to the opening ratio (M) and length ratio (L/b), the wingwall

Type I opening: vertical embankment and abutments

Without wingwalls

With wingwalls

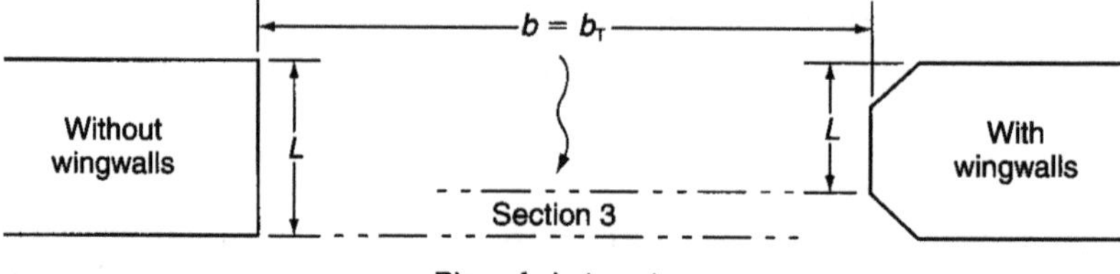

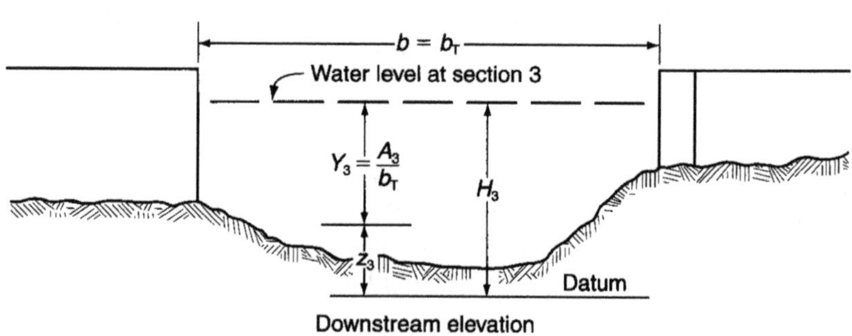

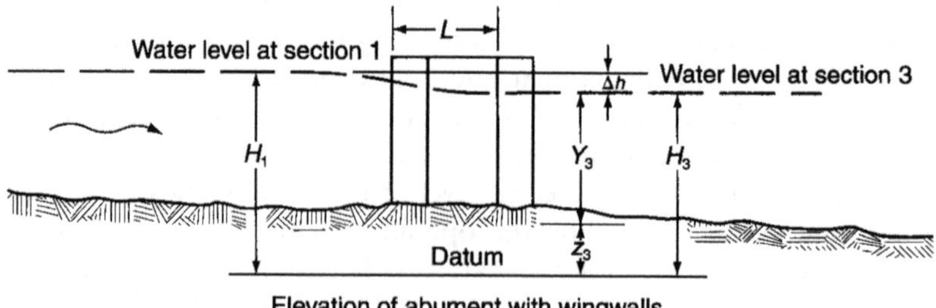

Fig. 4.2 Diagram defining a USGS type I opening with vertical embankments and vertical abutments, with or without wingwalls. (After Matthai, 1967. Reproduced by permission of US Geological Survey, US Dept of the Interior)

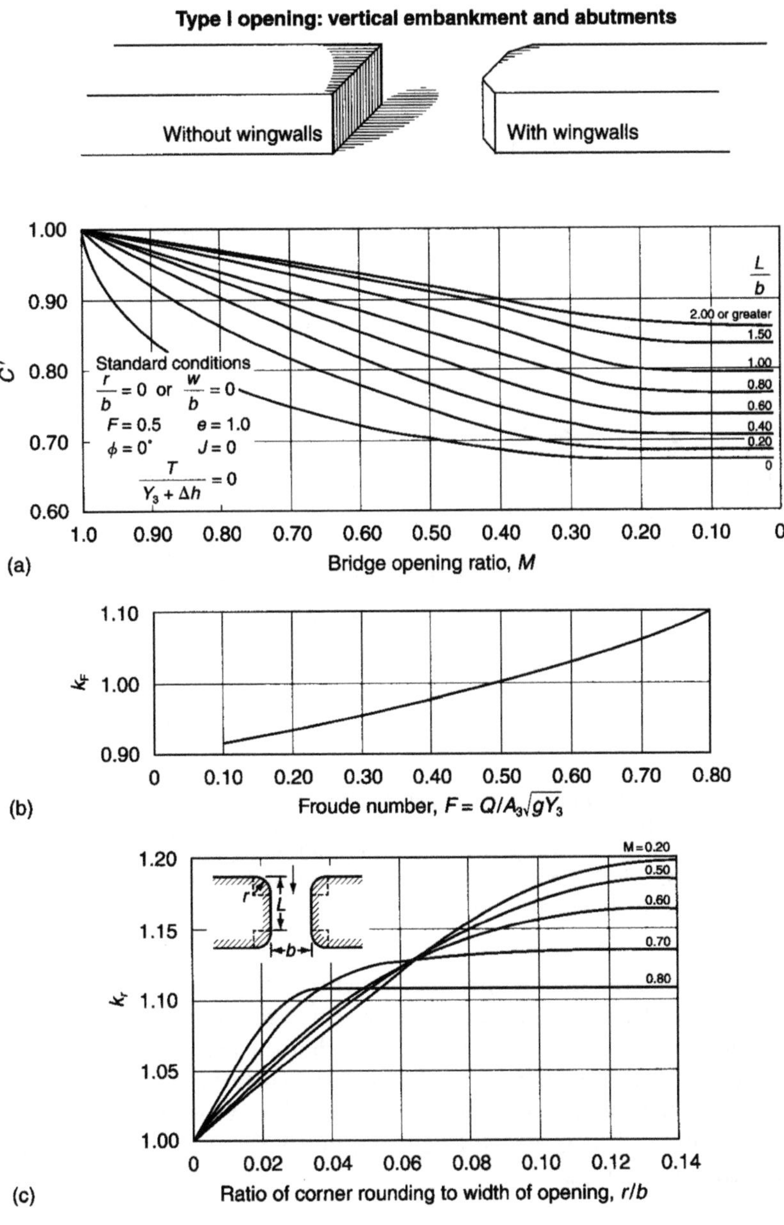

Fig. 4.3 USGS coefficients for type I openings with vertical embankments and vertical abutments. Diagram (a) gives the base coefficient of discharge, C', which is then multiplied by the adjustment factors in diagrams (b)–(g) as appropriate: adjustment factor variation with (b) Froude number; (c) entrance rounding. (After Matthai, 1967). Reproduced by permission of US Geological Survey, US Dept of the Interior)

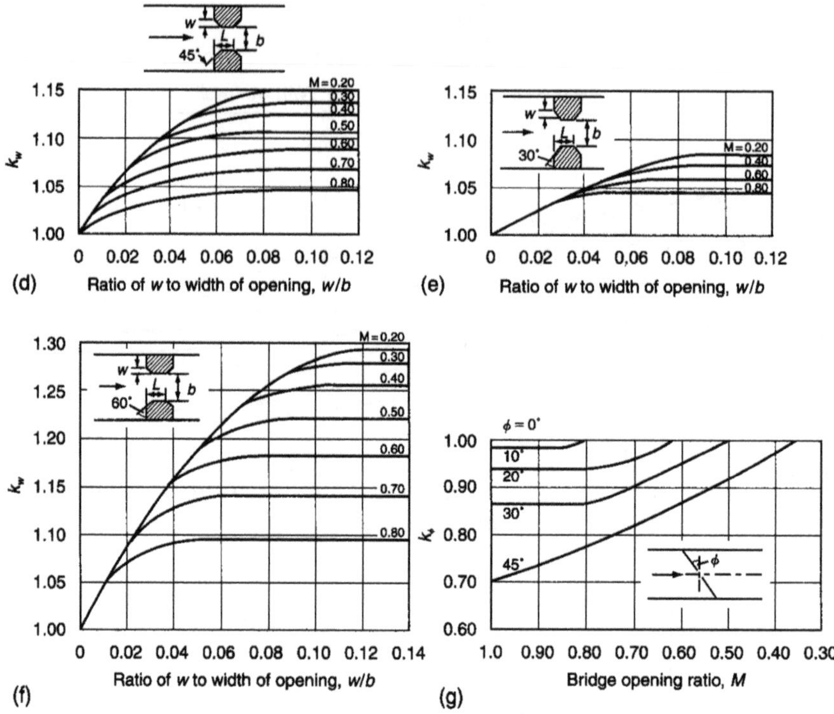

Fig. 4.3 USGS coefficients for type I openings with vertical embankments and vertical abutments (*cont.*). Diagram (d) length of 45° wingwalls or chamfers; (e) length of 30° wingwalls; (f) length of 60° wingwalls; (g) skew, ϕ. (After Matthai, 1967. Reproduced by permission of US Geological Survey, US Dept of the Interior)

angle, skew, embankment slope, and sometimes the Froude number affect the coefficient of discharge. Charts are presented for embankment slopes of 1 to 1 and 1 to 2 (Figs 4.12 and 4.13). Note that the existence of wingwalls does not automatically make a type IV opening, since a type I may also have wingwalls. If the flow passes around a vertical edge at the upstream extremity of the wingwall it is a type I opening, but if the flow passes around a sloping edge it is a type IV.

The coefficient of discharge, C, and the adjustment factors

For any opening type the value of the coefficient of discharge, C, is obtained by means of a 'standard' base coefficient, C', and a series of numerical adjustment factors, k. First, the type of opening is identified (e.g. type I). Next, M and L/b are calculated and used to obtain C' from the appropriate chart (e.g. Fig. 4.3a). This value assumes the standard

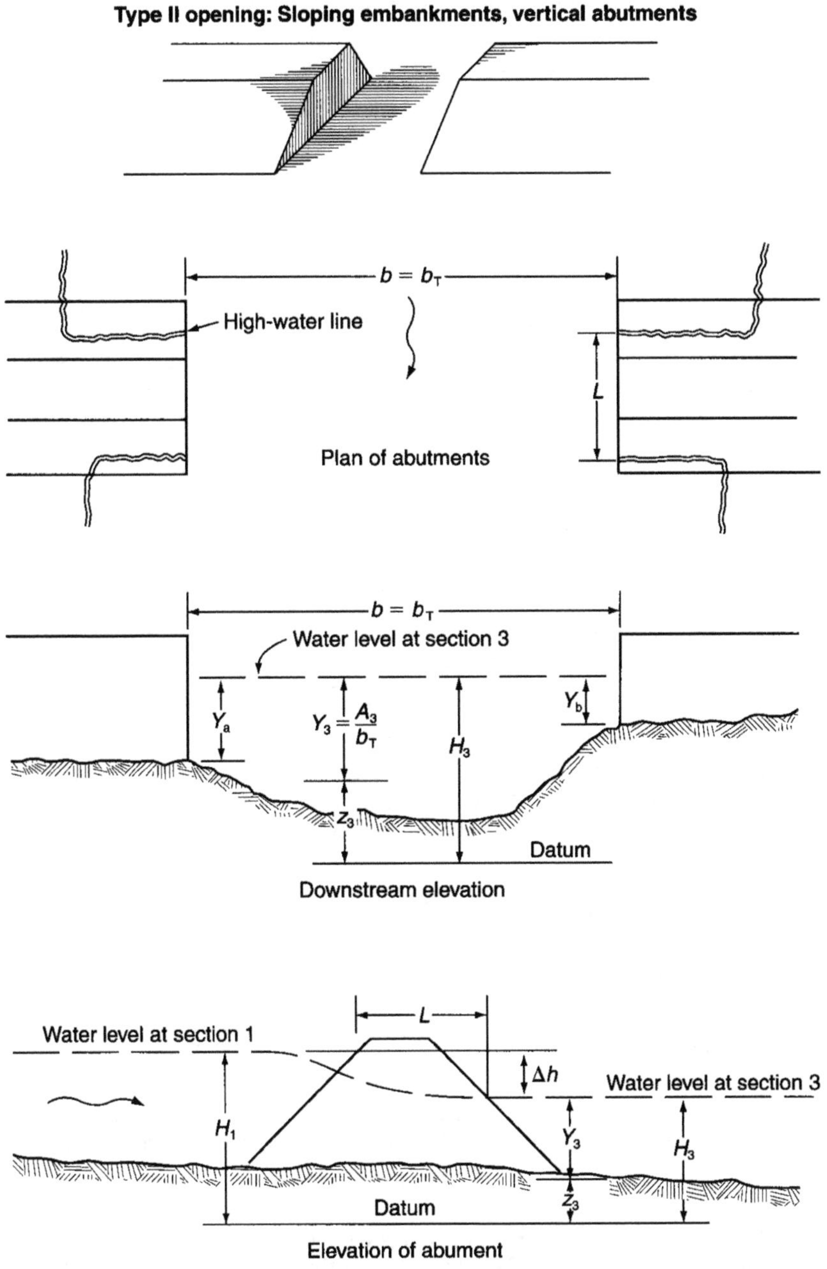

Type II opening: Sloping embankments, vertical abutments
b = b_T
High-water line
L
Plan of abutments
b = b_T
Water level at section 3
Y_a
Y_3 = A_3 / b_T
H_3
Y_b
Z_3
Datum
Downstream elevation
L
Water level at section 1
Δh
Water level at section 3
H_1
Y_3
H_3
Z_3
Datum
Elevation of abument

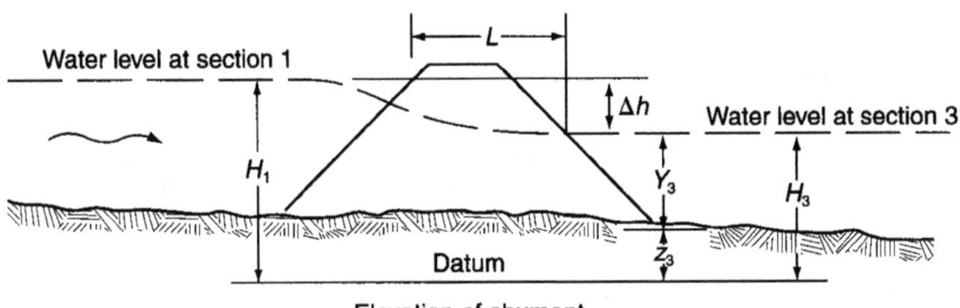

Fig. 4.4 Diagram defining a USGS type II opening with sloping embankments and vertical abutments. (After Matthai, 1967. Reproduced by permission of US Geological Survey, US Dept of the Interior)

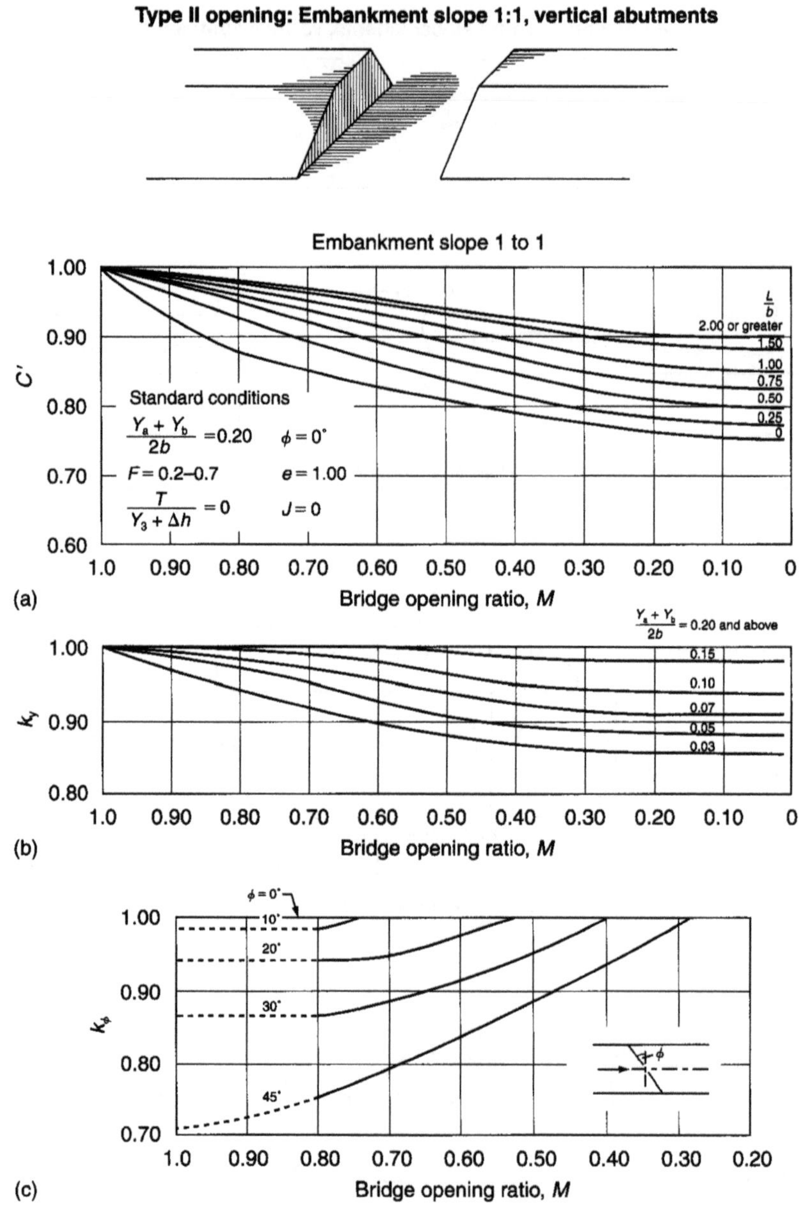

Type II opening: Embankment slope 1:1, vertical abutments

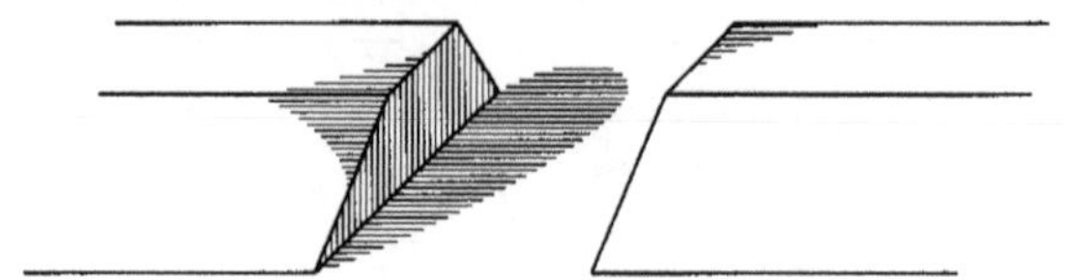

Embankment slope 1 to 1

C': 1.00, 0.90, 0.80, 0.70, 0.60

$\frac{L}{b}$: 2.00 or greater, 1.50, 1.00, 0.75, 0.50, 0.25, 0

Standard conditions

$\frac{Y_a + Y_b}{2b} = 0.20$ $\quad \phi = 0°$

$F = 0.2–0.7$ $\quad e = 1.00$

$\frac{T}{Y_3 + \Delta h} = 0$ $\quad J = 0$

1.0 0.90 0.80 0.70 0.60 0.50 0.40 0.30 0.20 0.10 0

(a) Bridge opening ratio, M

$\frac{Y_a + Y_b}{2b}$ = 0.20 and above, 0.15, 0.10, 0.07, 0.05, 0.03

k_y: 1.00, 0.90, 0.80

1.0 0.90 0.80 0.70 0.60 0.50 0.40 0.30 0.20 0.10 0

(b) Bridge opening ratio, M

$\phi = 0°$, 10°, 20°, 30°, 45°

k_ϕ: 1.00, 0.90, 0.80, 0.70

1.0 0.90 0.80 0.70 0.60 0.50 0.40 0.30 0.20

(c) Bridge opening ratio, M

Fig. 4.5 USGS coefficients for type II openings with an embankment slope of 1 to 1 and vertical abutments. Diagram (a) gives the base coefficient of discharge, C′, which is then multiplied by the adjustment factors in diagrams (b) and (c) as appropriate: adjustment factor variation with (b) $(Y_a + Y_b)/2b$ ratio; (c) skew. (After Matthai, 1967. Reproduced by permission of US Geological Survey, US Dept of the Interior)

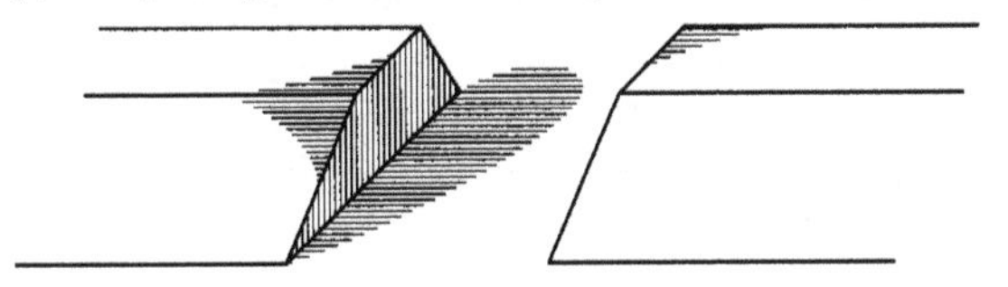

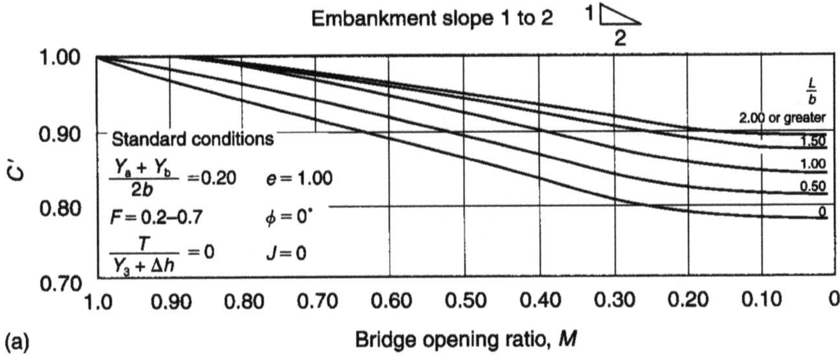

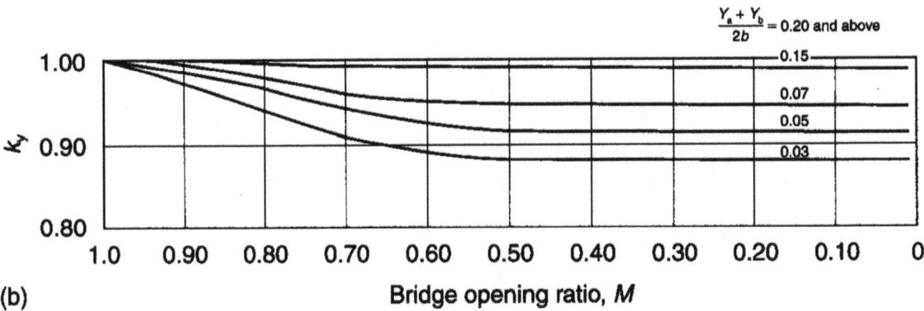

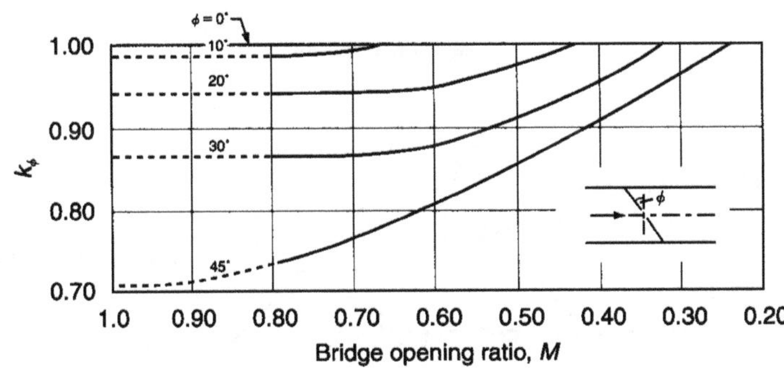

Fig. 4.6 USGS coefficients for type II openings with an embankment slope of 1 to 2 and vertical abutments. Diagram (a) gives the base coefficient of discharge, C′, which is then multiplied by the adjustment factors in diagrams (b) and (c) as appropriate: adjustment factor variation with (b) $(Y_a + Y_b)/2b$ ratio; (c) skew. (After Matthai, 1967. Reproduced by permission of US Geological Survey, US Dept of the Interior)

Type III opening: Sloping embankments and abutments

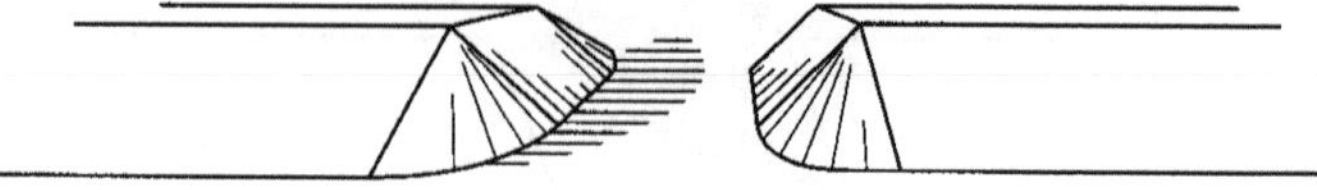

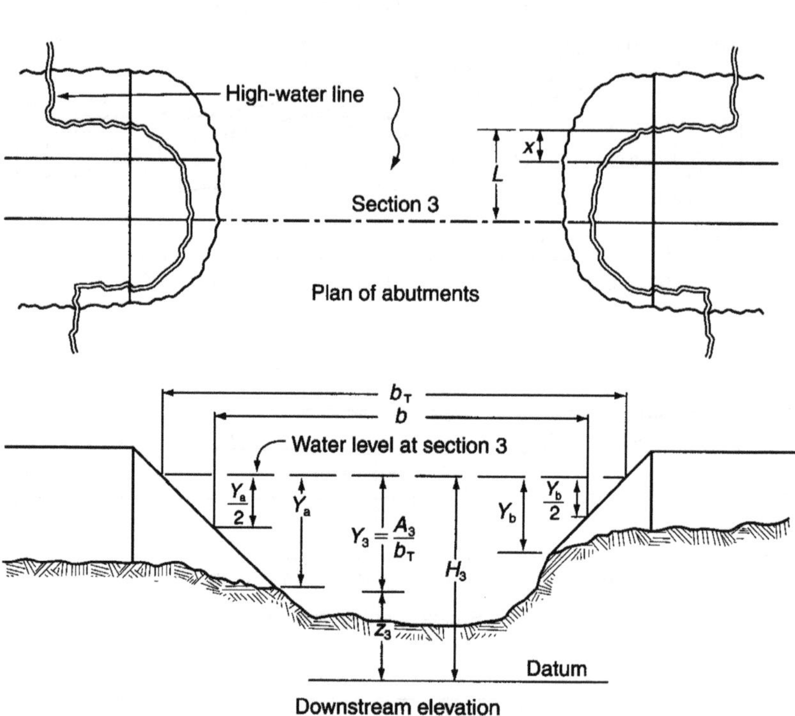

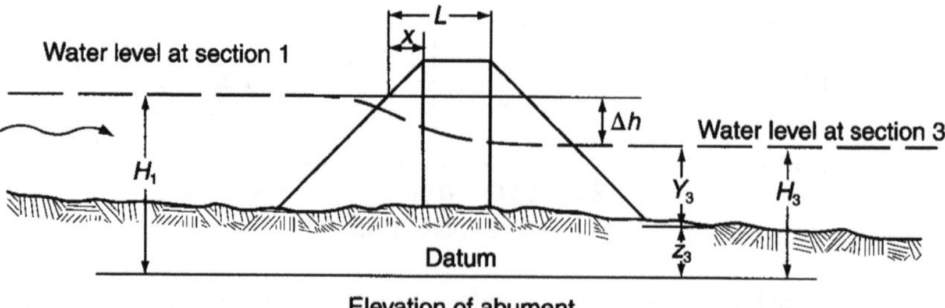

Fig. 4.7 Diagram defining a USGS type III opening with sloping embankments and sloping abutments. (After Matthai, 1967. Reproduced by permission of US Geological Survey, US Dept of the Interior)

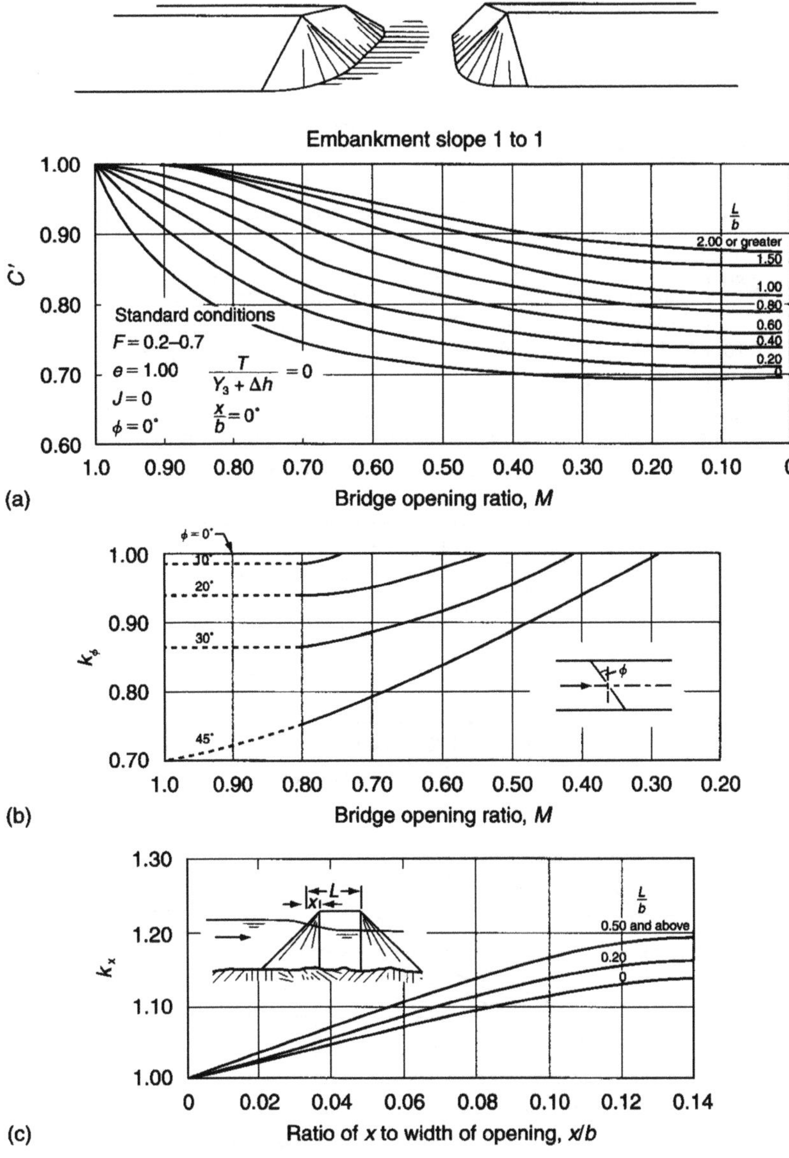

Fig. 4.8 USGS coefficients for type III openings with both embankment and abutment slopes of 1 to 1. Diagram (a) gives the base coefficient of discharge, C′, which is then multiplied by the adjustment factors in diagrams (b) and (c) as appropriate: adjustment factor variation with (b) skew; (c) *x*/*b* ratio. (After Matthai, 1967. Reproduced by permission of US Geological Survey, US Dept of the Interior)

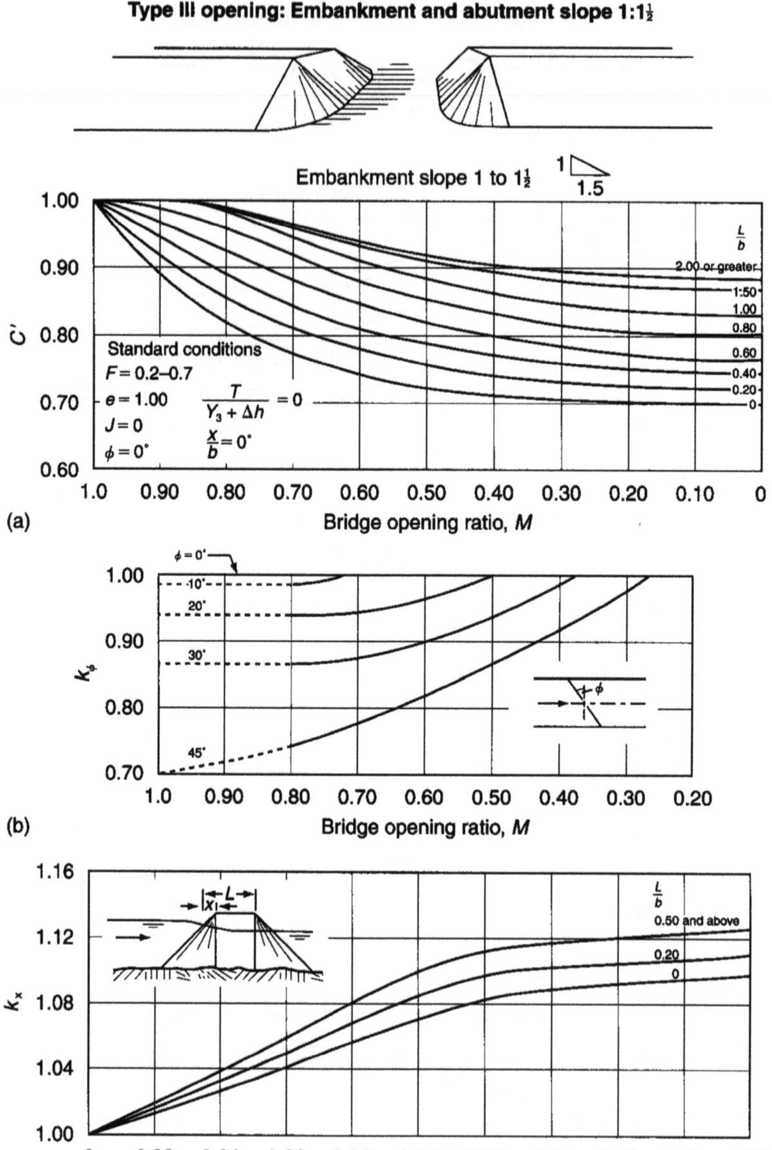

Fig. 4.9 USGS coefficients for type III openings with both embankment and abutment slopes of 1 to 1.5. Diagram (a) gives the base coefficient of discharge, C′, which is then multiplied by the adjustment factors in diagrams (b) and (c) as appropriate: adjustment factor variation with (b) skew; (c) x/b ratio. (After Matthai, 1967. Reproduced by permission of US Geological Survey, US Dept of the Interior)

Type III opening: Embankment and abutment slope 1:2

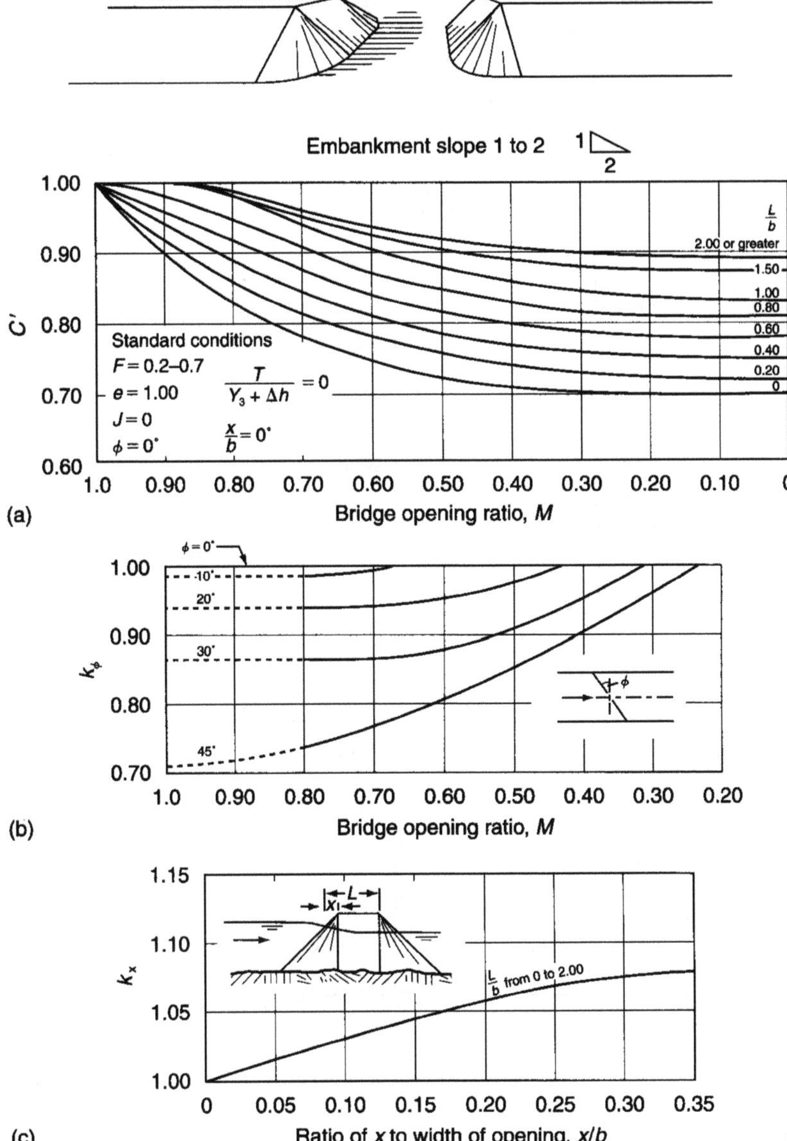

Fig. 4.10 USGS coefficients for type III openings with both embankment and abutment slopes of 1 to 2. Diagram (a) gives the base coefficient of discharge, C′, which is then multiplied by the adjustment factors in diagrams (b) and (c) as appropriate: adjustment factor variation with (b) skew; (c) *x*/*b* ratio. (After Matthai, 1967. Reproduced by permission of US Geological Survey, US Dept of the Interior)

Type IV opening: Sloping embankments, vertical abutments with wingwalls

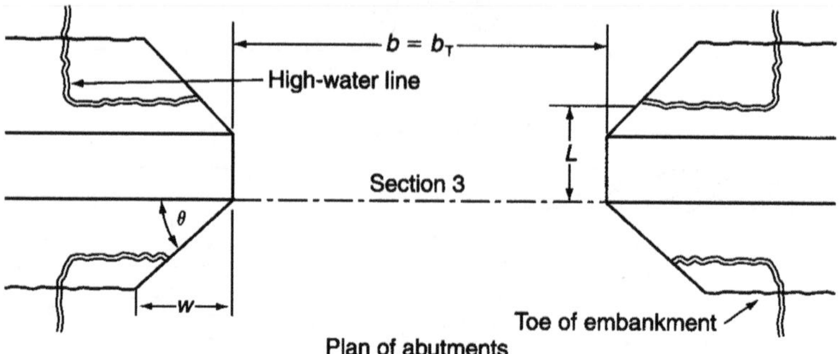

b = b_T
Water level at section 3
$Y_3 = \frac{A_3}{b_T}$
H_3
Z_3
Datum
Downstream elevation

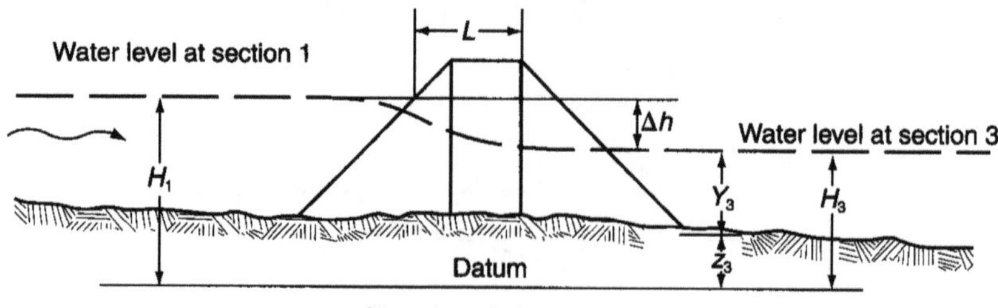

Fig. 4.11 Diagram defining a USGS type IV opening with sloping embankments and vertical abutments with wingwalls. (After Matthai, 1967. Reproduced by permission of US Geological Survey, US Dept of the Interior)

conditions of no entrance rounding ($r/b = 0$), no chamfer ($w/b = 0$), $F = 0.5$, no eccentricity ($e = 1.0$), no skew ($\phi = 0$), no obstruction caused by piers ($J = 0$), and no submergence ($k_T = 1.0$). Then the values of the adjustment factors are obtained from the relevant chart and used to modify C' to allow for

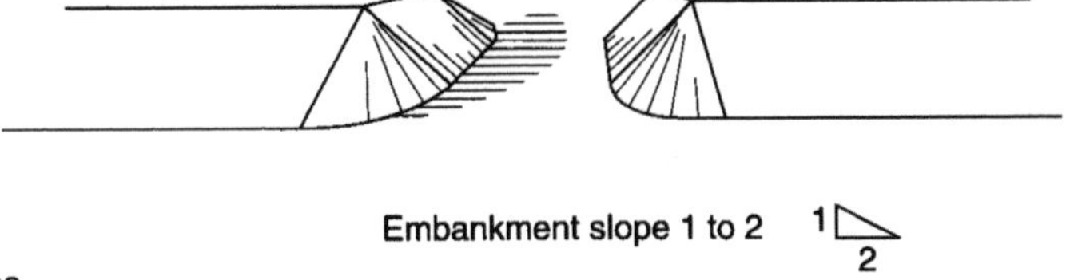

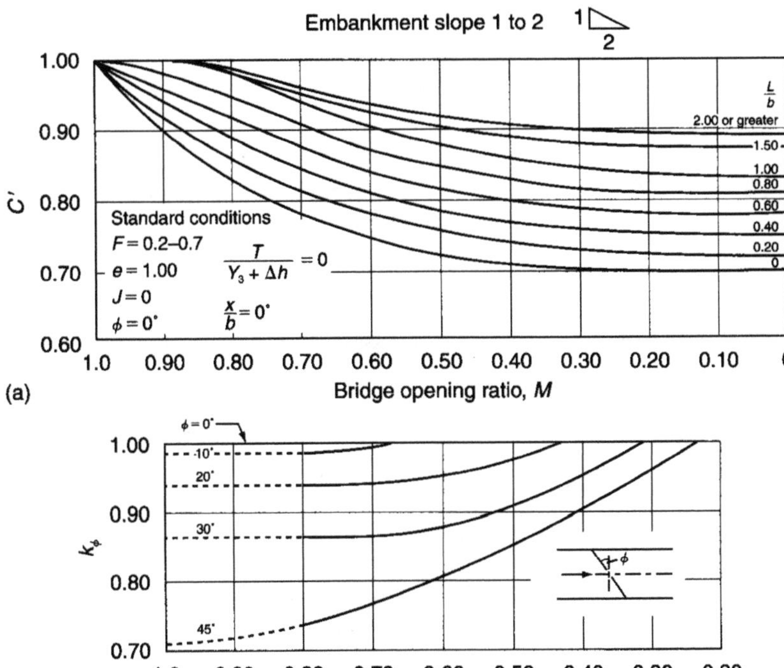

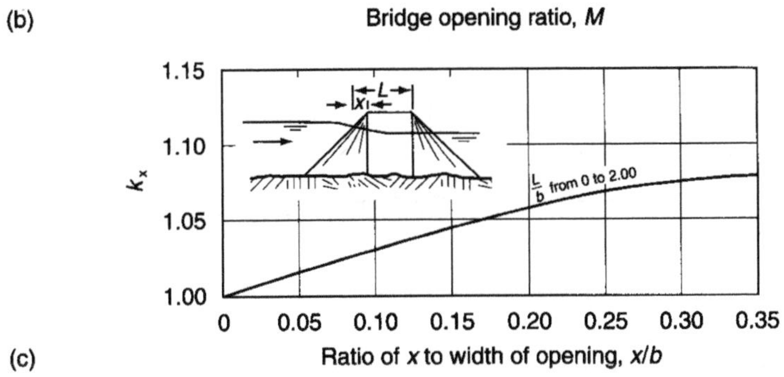

Fig. 4.10 USGS coefficients for type III openings with both embankment and abutment slopes of 1 to 2. Diagram (a) gives the base coefficient of discharge, C', which is then multiplied by the adjustment factors in diagrams (b) and (c) as appropriate: adjustment factor variation with (b) skew; (c) x/b ratio. (After Matthai, 1967. Reproduced by permission of US Geological Survey, US Dept of the Interior)

Type IV opening: Sloping embankments, vertical abutments with wingwalls

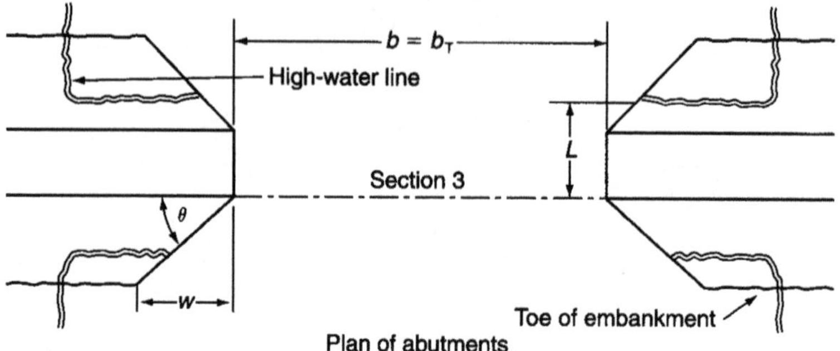

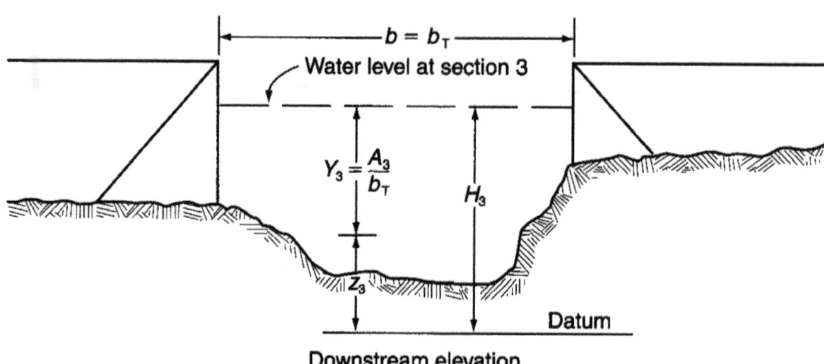

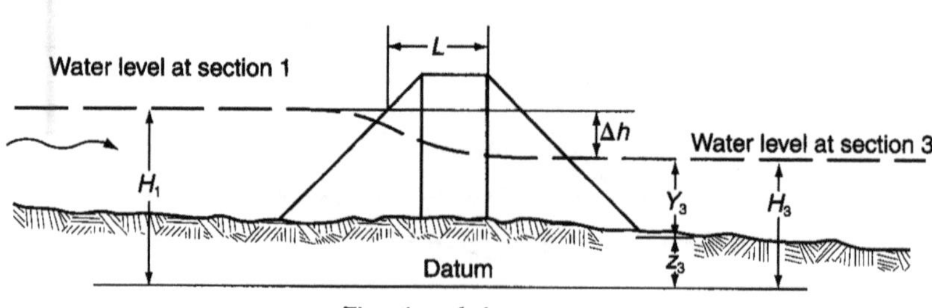

Fig. 4.11 Diagram defining a USGS type IV opening with sloping embankments and vertical abutments with wingwalls. (After Matthai, 1967. Reproduced by permission of US Geological Survey, US Dept of the Interior)

conditions of no entrance rounding ($r/b = 0$), no chamfer ($w/b = 0$), $F = 0.5$, no eccentricity ($e = 1.0$), no skew ($\phi = 0$), no obstruction caused by piers ($J = 0$), and no submergence ($k_T = 1.0$). Then the values of the adjustment factors are obtained from the relevant chart and used to modify C' to allow for

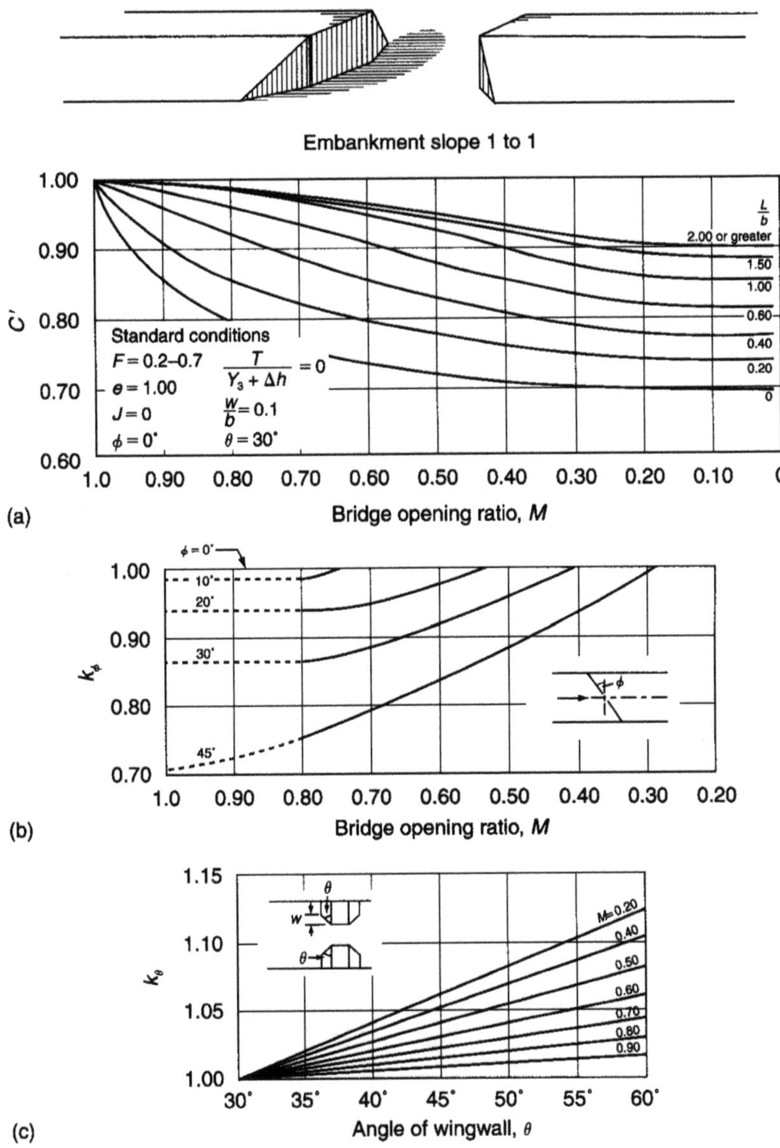

Fig. 4.12 USGS coefficients for type IV openings with an embankment slope of 1 to 1 and vertical abutments with wingwalls. Diagram (a) gives the base coefficient of discharge, C′, which is then multiplied by the adjustment factors in diagrams (b) and (c) as appropriate: adjustment factor variation with (b) skew; (c) wingwall angle. (After Matthai, 1967. Reproduced by permission of US Geological Survey, US Dept of the Interior)

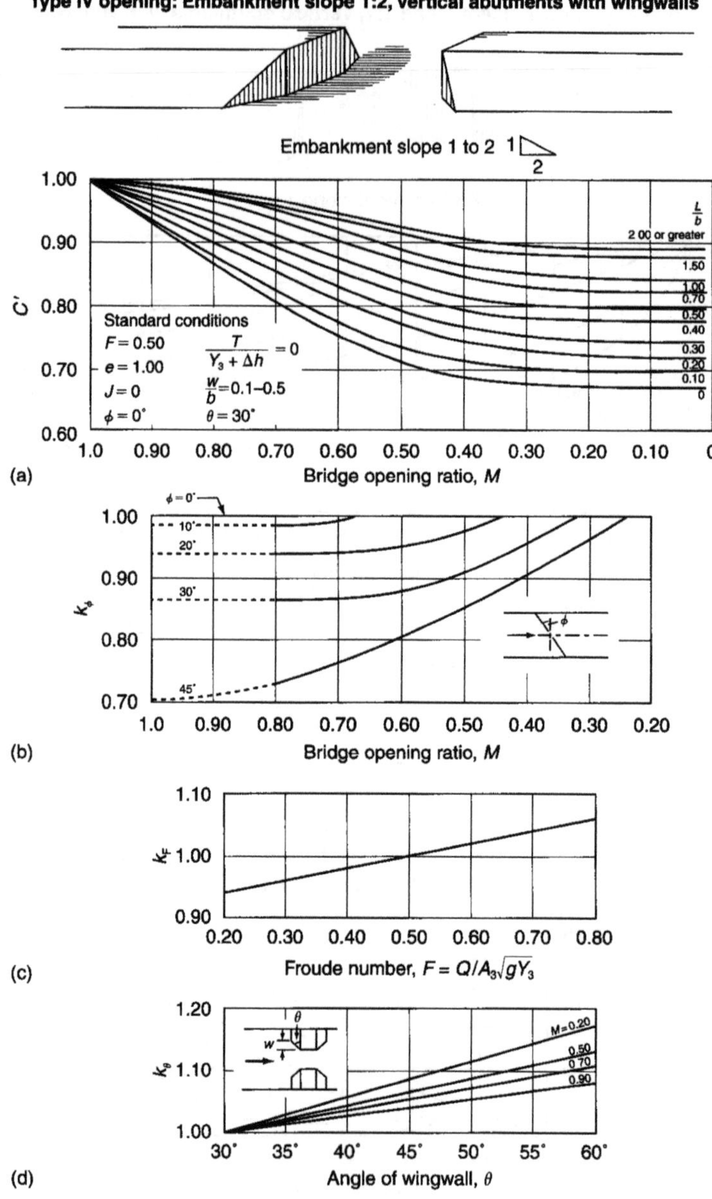

Fig. 4.13 USGS coefficients for type IV openings with an embankment slope of 1 to 2 and vertical abutments with wingwalls. Diagram (a) gives the base coefficient of discharge, C′, which is then multiplied by the adjustment factors in diagrams (b)–(d) as appropriate: adjustment factor variation with (b) skew; (c) Froude number; (d) wingwall angle. (After Matthai, 1967. Reproduced by permission of US Geological Survey, US Dept of the Interior)

the deviation from the standard conditions. Thus the coefficient of discharge, C, which takes into account all of the pertinent variables for the opening type, is given by:

$$C = C' \ (k_F \ k_r \ k_w \ k_\phi \ k_e \ k_y \ k_x \ k_\theta \ k_P \ k_T) \tag{4.10}$$

Of course, not all of the adjustment factors are required and only the relevant ones are used. For instance, the effect of eccentricity is usually small (Table 4.1) so k_e is often omitted, while only one of k_r, k_w and k_θ would be needed to describe the geometry of the waterway. The use of k_P and k_T is described later when piers and submerged openings are considered. Note that the Froude number used is that relating to the contracted section so $F_3 = V_3/(gY_3)^{1/2}$ where $Y_3 = A_3/b_T$ is the mean depth at section 3. As explained in Section 3.3, this causes problems with arches.

Under no circumstances should a value of C greater than 1.0 ever be adopted. Additionally, some adjustment factors are insignificant with certain types of waterway, so only the diagrams associated with a particular type of opening should be used. For embankments of different slope from those shown in the diagrams it is necessary to interpolate, although with type IV openings the coefficients for a 1 in 2 slope can also be used for flatter slopes (i.e. 1 in 3, 1 in 4 etc). If the abutment slope is not equal to the embankment slope, an average value should be used.

If the two abutments are of different type, a value of C should be calculated for each side and a weighted average obtained using the conveyance (K) of the two sides. Using subscripts L and R to denote the left and right sides:

$$C = \frac{C_L K_L + C_R K_R}{K_L + K_R} \tag{4.11}$$

If in doubt, remember that since M and L/b are the most important variables priority should be given to ensuring that they are correct, the other adjustment factors being of lesser importance. For non-standard types of opening some engineering judgement must be used. Matthai suggested that when there is a choice between two types, C should be calculated for each; if the difference is less than 5% use either, if it is greater than 5% use the average value.

Skew (angularity), k_ϕ

If the embankment and face of the bridge are skewed at an angle ϕ to the direction of flow, the abutments may be either parallel to the flow (skew 1 in Fig. 3.7) or perpendicular to the face (skew 2). If $\phi < 20°$ then the adjustment factor for both conditions is the same. For $\phi > 20°$ and skew 2 the adjustment factors in the diagrams are valid (e.g. Fig. 4.5c), but they should not be used for skew 1 with abutments parallel to the flow. The latter condition was not included in the original laboratory investigation so the correction factors are unknown.

Figure 4.14a illustrates how the USGS method treats skew. Unusually the projected width of the opening at section 1 is b, while the distance between the abutments measured along the centreline of the embankment is also b. Thus $b = b_S$; this is not the case in Fig. 3.7 nor in the USBPR method.

Eccentricity, k_e

Originally Kindsvater *et al.* (1953) used the ratio of the length of the two abutments to define eccentricity and the adjustment factor (Fig. 3.1e). Matthai (1967) changed this to the ratio of the conveyances of the floodplains or approach sections (K_a and K_c in Fig. 4.14b) on each side of the projected opening width (b) thus:

$$e = \frac{K_a}{K_c} \leq 1.0 \qquad (4.12)$$

where K_a is always the smaller of the two values. The two conveyances are proportional to the flow that has to deviate from its natural course to enter the bridge opening. No correction is needed for ratios greater than 0.12; otherwise the adjustment factor can be obtained from Table 4.1.

The fully eccentric condition occurs when $X_a = 0$ and the 'abutment' is aligned with the side of the channel. This is analagous to half of a normal crossing, divided along the longitudinal centreline of the channel, which is why the fully eccentric condition is treated as half a normal contraction and why $2b$ is used below. The procedure for coping with a fully eccentric opening is as follows.

1. Locate section 1 at a distance $2b$ upstream of the bridge.
2. Determine the base coefficient C' from the graph of C' against M using a value of $L/2b$. Only the abutment on the contracted side is considered.
3. The elevation of the water surface at section 1 is the average of the values on each bank at the end of the section.
4. The elevation of the water surface at section 3 is measured only at the abutment where the contraction occurs (i.e. $X \neq 0$). This value is used to calculate both A_3 and Δh.

Curved approach

Matthai (1967) pointed out that skew should not be confused with eccentricity, or with curvature of the river. With skew, the bridge is not

Table 4.1 USGS adjustment factors, k_e, for eccentricity

$e = K_a/K_c$	0	0.02	0.04	0.06	0.08	0.10	0.12
k_e	0.953	0.966	0.976	0.984	0.990	0.995	1.000

After Matthai (1967)

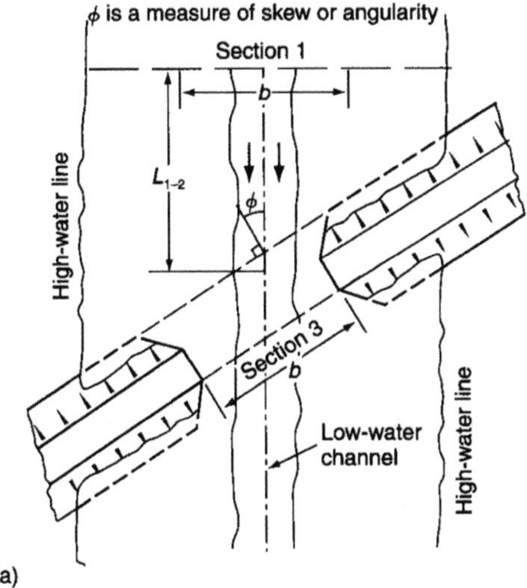

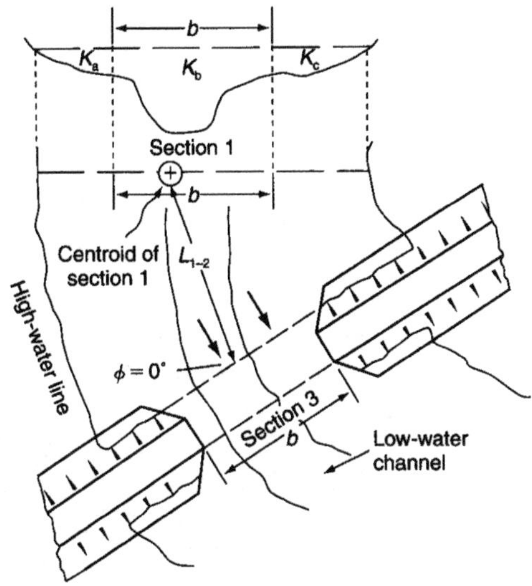

Fig. 4.14 Diagram comparing (a) a skewed crossing with (b) a normal crossing on a bend. In (a) the longitudinal centreline of the highway embankments is not perpendicular to either the approaching flow or the river banks indicating skew. In (b) the highway embankments are perpendicular to the approaching flow and the river banks. Note that with the USGS method the full skewed length of the opening, *b*, is superimposed on the main river channel at section 1, as indicated, with K_a, K_b and K_c being the conveyances of the three channel subsections. (After Matthai, 1967. Reproduced by permission of US Geological Survey, US Dept of the Interior)

perpendicular to the approach flow or the high-water line (Fig. 4.14a), but with a curved approach this is the case (Fig. 4.14b). For a bend where sections 1, 2 and 3 are not parallel, the distance between sections 1 and 2 should be measured from the centroid of section 1 to the centre of section 2. Note that when projecting the bridge opening upstream to section 1 this method uses the full width b, which is centred on the main river channel at section 1.

Piers or piles and multiple openings

The procedure for a bridge with piers or piles is as before, with the exception that another adjustment factor is needed to deal with this new condition. The proportion of the waterway blocked by the piles is

$$J = \frac{A_P}{A_3} \tag{4.13}$$

where A_P is the submerged area of the piers or piles projected onto the plane of section 3, which is the minimum contracted section parallel to the line of the bridge face and approach embankments, and A_3 is the gross area of section 3. The adjustment factor, k_P, is found from Fig. 4.15. Diagram a relates to piers, and its use is quite straightforward. Diagram b is employed for piles: for example, suppose $M = 0.59$, $L/b = 0.69$ and $J = 0.04$. The pro-

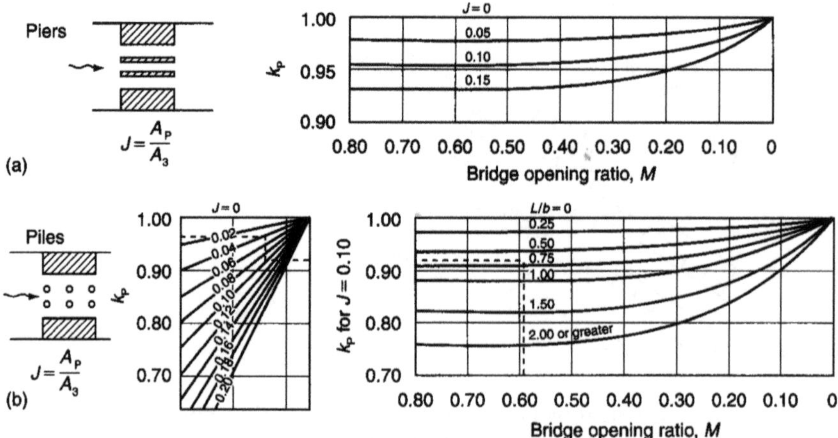

Fig. 4.15 USGS adjustment factors for (a) piers and (b) piles. With (b), the dashed lines illustrate the procedure for $M = 0.59$, $L/b = 0.69$ and $J = 0.04$. Start with the value of M (i.e. 0.59) in the right-hand diagram, move vertically up to the L/b value (0.69), continue across horizontally into the left diagram to the 0.10 diagonal reference line, move up to the $J = 0.04$ line and then horizontally across to obtain $k_P = 0.965$. (After Matthai, 1967. Reproduced by permission of US Geological Survey, US Dept of the Interior)

cedure is to enter the right-hand diagram at the appropriate M (= 0.59 as indicated by the dashed line), move vertically up to the L/b value (0.69), move horizontally to the diagonal line marked $J = 0.10$ on the diagram on the left, move vertically up to the $J = 0.04$ line, move horizontally across to the left-hand axis and read off the value of k_P , which is 0.965 in this particular example. Note that the diagonal $J = 0.10$ line is the reference line from which to move either vertically up or down depending upon whether $J < 0.10$ or $J > 0.10$. The original diagrams presented by Matthai (1967) have the J values above the diagonal 0.10 line shown as 0.8, 0.6, 0.4 and 0.2 when presumably these should be 0.08, 0.06, 0.04 and 0.02.

If a bridge has piles or piers and the openings submerge so that the flow is in contact with the deck then the obstructive effects of the piers and deck are considered separately, the latter being considered under the submergence heading below. How the piers are treated when there is submergence depends upon whether sluice gate or drowned orifice flow is experienced (Fig. 2.6 and Section 2.4). If only the upstream face of the openings is submerged (type 3 sluice gate flow) then again $J = A_P / A_3$ where A_P is only the submerged area of the piers or piles projected on the plane of section 3 (the area of the deck is not included) while A_3 is the gross area of section 3. If the upstream and downstream faces of the openings are submerged (type 1 drowned orifice flow) then

$$J = \frac{A_P}{a_{W3}} \tag{4.14}$$

where A_P is only the area of piers or piles projected onto section 3 (the area of the deck is not included), and a_{W3} is the gross area below the soffit of the deck, i.e. the gross area of the openings at section 3.

The USGS method assumes a single opening; it may have several spans supported by piers or piles, but it is one opening nevertheless with only one pair of abutments. In contrast, a multiple-opening contraction (Fig. 4.16) typically has more than one pair of abutments, which are connected by an embankment within the river channel (an 'interior' embankment). Thus a multiple-opening embankment was defined by Matthai (1967) as 'a series of independent single-opening contractions, all of which freely conduct water from a common approach channel'.

The procedure for determining the discharge through a multiple-opening contraction is basically to split the channel into separate subchannels or subsections (each with its own single-opening contraction), calculate the discharge through each opening using the techniques already described, and then add the discharges to obtain the total. When splitting the channel, the larger openings are assigned a longer length of embankment in direct proportion to the gross area (A_3) of their waterway openings. The dividing lines are then projected upstream. As usual section 1 is located one opening width upstream but, if the multiple openings have different spans, this means that each subchannel has section 1 in a different place so the

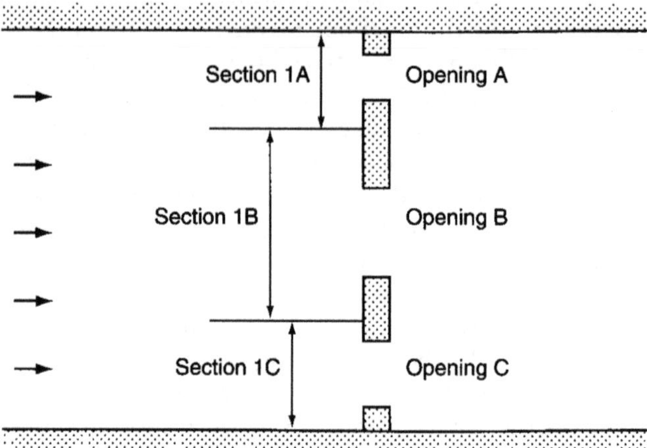

Fig. 4.16 Diagram illustrating a multiple-opening constriction. The length of interior embankment within the subchannels is in proportion to the gross area of the openings. (After Matthai, 1967. Reproduced by permission of US Geological Survey, US Dept of the Interior)

upstream water level (H_1) has to be obtained at several locations within the channel. On the outside of the channel this often means using observed or high-water (trash) marks. Interior values have to be estimated from the corresponding observed or high-water marks on the upstream face of the interior embankments (H_S) as follows:

$$H_1 = H_S - \frac{a_1 Q_1^2}{2gA_1^2} + \frac{Q_1^2 L_{1-2}}{3K_1^2} \tag{4.15}$$

where all the quantities are for a psuedo single-opening channel at section 1: a_1 is the dimensionless velocity coefficient; Q_1, A_1 and K_1 are the discharge (m^3/s), cross-sectional area of flow (m^2) and conveyance (m^3/s); g is the acceleration due to gravity (9.81 m/s^2); and L_{1-2} is the distance (m) between sections 1 and 2. Unfortunately the discharge has to be assumed and verified by later calculations. The downstream level (H_3) has to be measured at each opening to determine Δh for that particular waterway.

Submergence

This condition occurs when the water level rises above the top of the opening(s) and the bridge deck causes an obstruction to the flow. It is assumed that there is no flow over the deck, approach embankments or around the bridge. The vertical distance between the water level at section 1 and the lowest horizontal member of the submerged part of the bridge deck is the distance T in Fig. 4.17. The degree of submergence is defined by

$$\text{bridge submergence ratio} = \frac{T}{Y_3 + \Delta h} \tag{4.16}$$

This ratio is used to obtain the value of the submergence adjustment factor, k_T, from Fig. 4.17, which is then used with equations 4.10 and 4.5 to calculate Q. Note that submergence usually does not occur until $Y_u/Z > 1.1$ or, more specifically here, when the bridge submergence ratio > 0.08. If $k_T <$ 1.00 then submergence is assumed; when estimating the friction losses in the approach reach and through the bridge add the full length of the deck to the wetted perimeter, even with type 3 sluice gate flow (Fig. 2.6) where only part of the opening is running full. This is said to give a better estimate of the friction losses that occur at high discharges.

With type 2 flow where all of the deck beams are in contact with the water, the discharge coefficient (C) obtained from equation 4.10 using all of the adjustment factors (including k_T) may be nearly 1.00. Matthai (1967) advised that after all adjustment factors have been applied the maximum value of C should be restricted to the value of k_T obtained from Fig. 4.17.

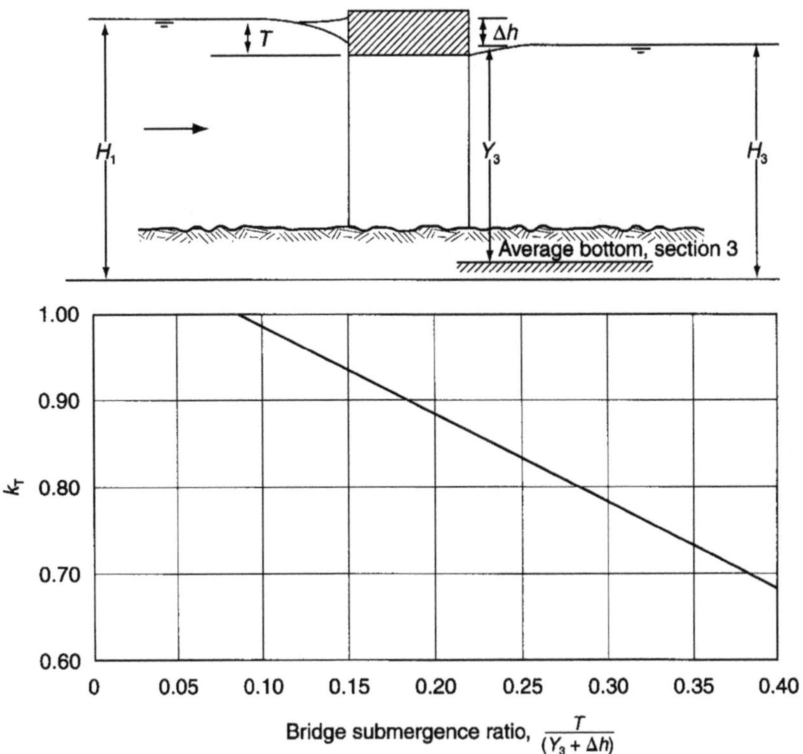

Fig. 4.17 USGS adjustment factor, k_T, for the bridge submergence ratio. (After Matthai, 1967. Reproduced by permission of US Geological Survey, US Dept of the Interior)

Additionally with drowned orifice flow where both the upstream and downstream faces of the opening are submerged and the deck is fully in contact with the water use the following: $k_T = 1.0$ and $A_3 = (bY_3 - A_D)$ where A_D is the cross-sectional area of the submerged deck.

It appears that the USGS technique struggles to find a consistent procedure for dealing with submerged flows, necessitating several minor adjustments according to circumstance. The USBPR technique is simpler in this respect and uses equations 2.8 and 2.9.

Spur dykes, k_a, k_b, k_d

Spur dykes (guide walls) are rare in Britain but frequently used elsewhere with large rivers. When dealing with wide floodplains they can help to funnel the flow through the opening. Although not included in the original study, Matthai presented adjustment factors for use with the USGS method. These are included in Section 7.4 and Figs 7.13 and 7.14.

Summary and example

Table 4.2 provides a summary of the steps involved when applying the USGS method to calculate the discharge at a contraction. Example 4.1 is a numerical application of the technique.

4.2.2 Backwater analysis

This approach to the estimation of backwater was described by Tracy and Carter (1955). The difference in water level between sections 1 and 3 is given by

$$\Delta h = H_1^* + H_3^* + S_O L_{1-3} \qquad (4.17)$$

where Δh, H_1^* and H_3^* have been defined previously and are shown in Fig. 2.3. The fall in the bed of the channel is given by the product of its gradient (S_O) and the distance between sections 1 and 3 (L_{1-3}). Solving equation 4.17 for H_1^*, the true afflux, and writing the resulting expression in dimensionless form gives:

$$\frac{H_1^*}{\Delta h} = 1 - \frac{H_3^*}{\Delta h} - \frac{S_O L_{1-3}}{\Delta h} \qquad (4.18)$$

To solve the equation for the maximum afflux, H_1^*, the following steps are employed.

1. For a given value of total discharge (Q) calculate the bridge opening ratio, M.
2. Estimate the average roughness of the river channel in terms of Manning's n. To be compatible with the laboratory tests, n must repre-

Table 4.2 Summary of the steps in the application of the USGS method to calculate peak discharge

This is an iterative procedure since (for some types of opening) the Froude number has to be estimated in step 7 and the velocity head evaluated before the discharge can be calculated.

1. For the stage under consideration, draw a plan of the site showing water levels, including a longitudinal section along each bank and across the upstream and downstream faces of the constriction. Draw cross-sections of the abutments and sections 1 and 3.
2. Determine the average water surface elevations at sections 1 and 3 and determine the fall, Δh.
3. For short bridges, measure the span (b) in the field; otherwise measure from the plan. Superimpose b on section 1, generally keeping the centre of the low water channel in the same relative position but incorporating the thalweg within b. The full width, b, is used regardless of skew. Measure the length of the waterway (L) and calculate the value of L/b.
4. Split section 1 into subsections using vertical dividing lines at the edges of the opening and where abrupt changes in hydraulic radius occur. Calculate the areas and wetted perimeters of the subsections, plus the total area.
5. For section 1, calculate the conveyances of the sections corresponding to b (i.e. K_b) and to either side (K_a and K_c). For the calculation of eccentricity, K_a should be the smaller. Calculate $M = K_b/K$.
6. Determine the type of abutments (I–IV) and their dimensions and slope.
7. For section 3, calculate the areas and wetted perimeters, if necessary splitting the channel at the edge of each pier or pile bent or at abrupt changes of hydraulic radius. Compute the conveyance of each subsection. Assuming $\alpha_3 =$ 1.0, calculate the Froude number, $F_3 = Q/A_3(gY_3)^{1/2}$.
8. From the M and L/b values obtain C' from the base curve appropriate to the type of opening.
9. Calculate the values of the appropriate adjustment factors from the secondary curves.
10. Obtain the value of C from steps 8 and 9 and equation 4.10.
11. Calculate Q from equation 4.5.
12. Calculate the mean velocities at sections 1 and 3; check they are reasonable. Calculate the Froude numbers, F_1 and F_3; they must be less than 0.8 for a valid analysis.
13. Compare Q and F_3 from steps 11 and 12 with the values in step 7: perform additional iterations as necessary.

sent the average value in the channel reaches both upstream and downstream of the constriction.

3. Obtain the value of the backwater ratio $H_1^*/\Delta h$ from Fig. 4.18.
4. Obtain the value of the coefficient of discharge of the opening (i.e. C for types I–IV) and also the equivalent value (C_{basic}) for the basic condition of an opening with a vertical face and square-edged abutments (i.e. type I with no rounding or wingwalls). Calculate C/C_{basic} and obtain the value of k_c from Fig. 4.19.
5. Adjust the value of $H_1^*/\Delta h$ (from step 3) for constriction geometry by multiplying it by the value of k_c.

6. Calculate the value of Δh corresponding to Q from equation 4.5 or 4.9.
7. Multiply the adjusted backwater ratio $k_c H_1^*/\Delta h$ from step 5 by Δh to obtain H_1^*.
8. If the stage–discharge relationship upstream of the bridge is required, add H_1^* to the normal depth at section 1 without the bridge. Normal depth can be calculated from the Manning equation.
9. H_3^* can be obtained from equation 4.17, if required, having calculated or measured $S_O L_{1-3}$.

The inclusion of Manning's n as a variable in Fig. 4.18 acknowledges the fact that the difference in water level across a bridge increases with increasing roughness. However, the backwater ratio, $H_1^*/\Delta h$, is primarily a function of M, so some error in the estimation of n is tolerable. Constriction geometry is also of secondary importance, but the coefficient k_c is included to relate the basic opening type used in the model tests (vertical face, square-edged abutments) to the other types of opening. Variables that are characteristic of the flow, such as the Froude number and depth, are considered to have an insignificant effect on the backwater ratio and are excluded from the calculations. Before using this technique the limitations described in Section 4.2.3 below should be studied.

The term $H_1^*/\Delta h$ is often called the backwater ratio or backwater function. While this non-dimensional term may be a convenient and frequently used method of expressing the afflux at a particular site, it can be misleading: the value of the backwater ratio may be constant over a wide range of

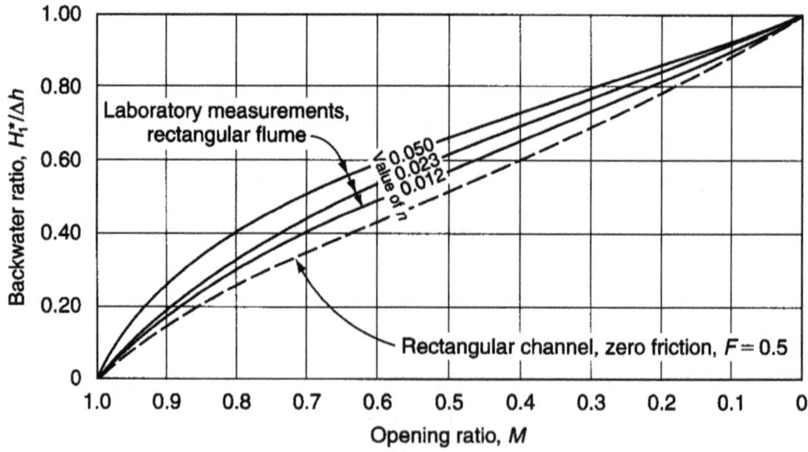

Fig. 4.18 USGS backwater ratio ($H_1^*/\Delta h$) for the basic condition of a constriction with a vertical face and square edged abutments (After Tracy and Carter, 1955; Backwater effects of open-channel constrictions, *Transactions of the ASCE*. Reproduced by permission of ASCE)

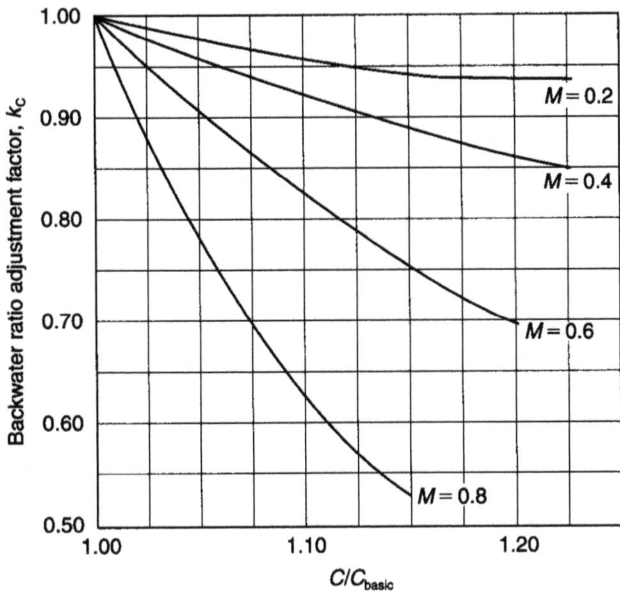

Fig. 4.19 USGS backwater ratio adjustment factor, k_c, which allows for any deviation from the basic constriction geometry of Fig. 4.18. Here C is the coefficient of discharge of the opening type under investigation and C_{basic} is the coefficient of a type I opening with a vertical face, square-edged abutments, and the same opening ratio, *M*. If the ratio C/C_{basic} is known, k_c can be obtained from the diagram. The adjusted backwater ratio is $k_c H_1^*/\Delta h$. (After Tracy and Carter, 1955; Backwater effects of open-channel constrictions, *Transactions of the ASCE*. Reproduced by permission of ASCE)

stages (e.g. 0.02/0.2 and 0.2/2.0) but the afflux is increasing significantly (i.e. from 0.02 m to 0.2 m).

4.2.3 Limitations on the use of the USGS method

There are a number of situations where this method of analysis should not be used, or used with caution. Some of the restrictions are as follows.

1. Only sites with a relatively stable bed should be analysed. It is recognised that some scour may take place in the waterway, so diagrams such as Figs. 4.2 and 4.4 show the river bed below the level of the abutments with $Y_3 = A_3/b_T$ where b_T is the top width of the water surface. However, the method assumes that the minimum flow area occurs at the contracted section (section 3) and that this controls the flow. If a large scour hole develops, the upstream lip of the hole instead of the bridge geometry

may determine the stage–discharge relationship. The coefficients and adjustments factors then cannot be applied (see also Section 3.11).

2. The method should not be applied where the difference in the elevation of the water surface between sections 1 and 3 (i.e. Δh) is less than 150 mm.
3. As mentioned earlier, Δh should be greater than $4h_F$. Problems are encountered with heavily vegetated floodplains, which give rise to large values of h_F.
4. The bridge geometry should be close to the standard types, and the flow conditions should be similar. The flow must be subcritical throughout.
5. In many situations, such as when designing a new bridge, the water levels at sections 1 and 3 (i.e. H_1 and H_3), and hence Δh, are all initially unknown (Fig. 2.3). One approach is to assume a value for Q, calculate Δh from equation 4.5 or 4.9, obtain H_1^* using the procedure in Section 4.2.2, evaluate $S_O L_{1\text{–}3}$ and then use equation 4.17 to obtain H_3^*.
6. It is apparent from Section 4.2.2 and the above that the USGS method is rather clumsy when it is used to evaluate afflux or when the water levels at sections 1 and 3 are unknown. The fact that the calculations do not easily relate to normal depth conditions can be a disadvantage.

The USGS method is best at what it was designed for, estimating the flow from observations of the water level at sections 1 and 3 (the USBPR method provides a more direct means of calculating afflux). Despite the limitations above, this is one of the most reliable and comprehensive investigations of flow through a contraction (see also Section 4.6).

4.3 US Bureau of Public Roads (USBPR) method

This may also be known as either the BPR or DT (Department of Transport) method. It was principally concerned with the evaluation of bridge backwater, and is based on a comprehensive laboratory study (Liu, *et al.*, 1957), which was verified by field measurements. Bradley (1978) produced a detailed report that described how to calculate the afflux (backwater) under normal depth conditions. The report included many useful examples. It also considered skew, eccentricity, the abnormal stage condition, dual bridges, situations where the flow passes through critical depth, the difference in water level across the embankments, and submergence of the opening. Consequently it can be applied to a wide range of problems. Important limitations are that it is not very suitable for irregular channels, there is no discharge equation for a bridge experiencing channel flow (but there is for the submerged condition), and arch bridges were not considered.

Essentially the USBPR method assumes normal depth in a uniform channel so the water surface is parallel to the bed, and then calculates the additional afflux caused by introducing a bridge into the flow (in the laboratory the afflux was measured). Basically all that is required to calculate the afflux is the bridge opening ratio (M), the normal depth velocity head at section 2 (i.e. $\alpha_2 V_{N2}^2/2g$, where the subscript N denotes normal depth), and the value of some empirical coefficients found from charts. It does not require the calculation of the Froude number, which can be very advantageous. The simple equations and discharge coefficients for the sluice gate and drowned orifice flow condition (equations 2.8 and 2.9) are also quick and easy to use, and do not require the calculation of M, which can save much effort at complex sites.

The theoretical basis of the work is the energy and continuity equations applied between sections 1 and 4. Bradley (1978) located section 1 at a variable distance upstream, but in what is sometimes called the 'modified Bradley method' it is located one span (b) upstream (Kaatz and James, 1997). It is assumed that the additional energy represented by the afflux at section 1 is exactly expended as the flow reaches section 4. It is also assumed that the channel is reasonably straight, the cross-sections are reasonably uniform, the gradient of the bed is constant between sections 1 and 4, and that the cross-sections at 1 and 4 are essentially the same so that $\alpha_1 = a_4$ and $A_1V_1 = A_4V_4 = A_{N2}V_{N2}$. In other words, it is assumed that the prototype river and bridge are similar to the laboratory channel and models, which may not be the case.

It is important to appreciate that A_{N2} is not the actual cross-sectional area of flow that occurs in the opening at section 2; it is a theoretical area obtained by calculating the cross-sectional area of flow beneath the normal depth line at section 2 in the opening (Fig. 2.3). Thus for a rectangular opening $A_{N2} = bY_{N2,}$ where b is the opening width and Y_{N2} is the vertical distance between the bed and the normal depth line at section 2. This is consistent with the idea of projecting the width b upstream to section 1 to calculate M under normal depth conditions, as described in detail in Section 3.2. The reference velocity, V_{N2}, can be easily calculated ($= Q/A_{N2}$) but it is not a real, measurable velocity since it corresponds to a situation that does not exist. Note that doubling the opening width to $2b$ would have the effect of halving V_{N2} but would not change Y_{N2}.

With irregular beds or oddly shaped waterways the mean value of Y_{N2} should be used: that is, Y_{N2m}. The effective width of the opening is obtained by considering the equivalent trapezium with $b = A_{N2}/Y_{N2m}$.

4.3.1 *Backwater analysis (normal depth condition)*

In an unconstricted channel of bed slope S_O, for conservation of energy between sections 1 and 4, which are a distance $L_{1\text{–}4}$ apart:

$$S_O L_{1-4} + Y_1 + \frac{a_1 V_1^2}{2g} = Y_4 + \frac{a_4 V_4^2}{2g} + h_L \tag{4.19}$$

where h_L is the total loss of energy between the two sections. The laboratory test procedure was to use uniform flow with normal depth as the datum level, so $S_O L_{1-4} = h_L$ and cancels. However, the introduction of a bridge into the uniform flow creates an additional energy loss, h_b, thus:

$$Y_1 - Y_4 = \frac{\alpha_4 V_4^2}{2g} - \frac{\alpha_1 V_1^2}{2g} + h_b \tag{4.20}$$

The energy loss due to the bridge can be expressed in terms of a coefficient and the reference velocity head:

$$h_b = k^* \left(\frac{\alpha_2 V_{N2}^2}{2g} \right) \tag{4.21}$$

where V_{N2} is defined above, and k^* is the total backwater coefficient, which is explained below. The afflux at section 1 is $H_1^* = Y_1 - Y_4$ as in Fig. 2.3. Thus equation 4.20 becomes

$$H_1^* = k^* \frac{\alpha_2 V_{N2}^2}{2g} + \frac{\alpha_4 V_4^2}{2g} - \frac{\alpha_1 V_1^2}{2g} \tag{4.22}$$

If the cross-sectional areas of flow at sections 1 and 4 are basically the same then α_4 can be replaced by α_1. Similarly, $A_1 V_1 = A_4 V_4 = A_{N2} V_{N2}$ so a velocity can be expressed in terms of the reference velocity and an area ratio. Thus the maximum afflux at section 1 is given by

$$H_1^* = k^* \frac{\alpha_2 V_{N2}^2}{2g} + \alpha_1 \left[\left(\frac{A_{N2}}{A_4} \right)^2 - \left(\frac{A_{N2}}{A_1} \right)^2 \right] \frac{V_{N2}^2}{2g} \tag{4.23}$$

where α_1 and α_2 are the dimensionless velocity distribution coefficients at sections 1 and 2, V_{N2} is the average velocity (m/s) in the constriction at normal depth, A_{N2} is the gross cross-sectional area of water (m^2) in the constriction calculated below normal depth and including the area occupied by any piers, A_4 is the cross-sectional area of water (m^2) at section 4 where normal depth is re-established, and A_1 is the total cross-sectional area of water (m^2) at section 1, including the afflux. The equation has to be solved iteratively by evaluating the first term $k^* \alpha_2 V_{N2}^2/2g$ and then using this approximate value of H_1^* to determine A_1 and hence the value of the second term. This can be used to obtain a better estimate of H_1^* and A_1, and so on.

Figure 4.20 is an aid to determining the value of α_2 from a known value of α_1, and was based on velocity meter traverses of existing bridges. This is intended as a guide for use only when there is no better value available. It should be appreciated that local factors such as asymmetry of the flow and variations in cross-section can significantly affect the value. Bradley recommended that a generous value of α_2 should be adopted (but see Section 3.10).

The total backwater coefficient, k^*

The total backwater coefficient, k^*, is obtained by first determining the base coefficient, k_b^*, whose value depends upon M and the geometry of the

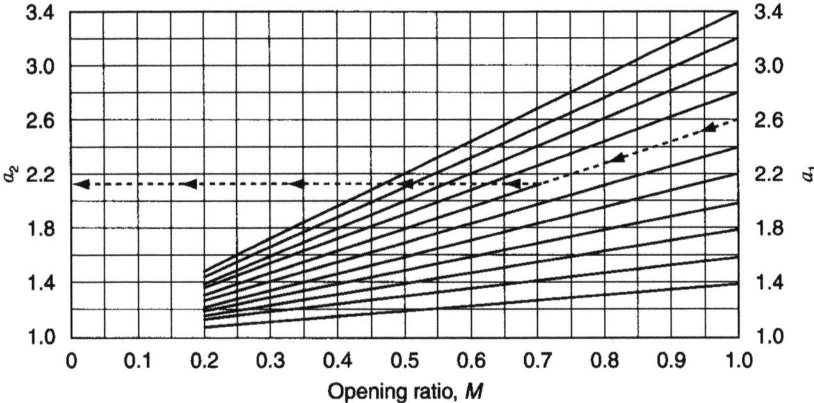

Fig. 4.20 USBPR nomogram for estimating the value of the velocity distribution coefficient at section 2 (i.e. α_2) from α_1 and M (after Bradley, 1978). For example, if $\alpha_1 = 2.6$ and $M = 0.7$, then α_2 has a value of about 2.1. The diagram is only a rough guide and large variations are possible.

abutments (Fig. 4.21). Incremental coefficients are then added to allow for eccentricity (Δk_e^*), skew (Δk_ϕ^*), and bridge piers (Δk_P^*). Thus

$$k^* = k_b^* + \Delta k_e^* + \Delta k_\phi^* + \Delta k_P^* \qquad (4.24)$$

Of course, the last three coefficients are used only if the geometry of the site warrants their inclusion. The most important term is k_b^* since the value of k^* is mainly dependent upon M.

The base coefficient, k_b^*

Figure 4.21 shows the variation of k_b^* with M and abutment geometry. Since k_b^* represents the energy loss it increases as the opening gets narrower and M reduces. The lower curve applies to low energy loss 45° and 60° wingwall abutments and all spillthrough types. The middle curve applies to 30° wingwall abutments, and the top curve to high energy loss 90° vertical-wall abutments. The top two curves are used for waterway openings up to 60 m in width; for widths in excess of 60 m the bottom curve should be used regardless of abutment shape (the shape becomes less important as the opening width increases).

Eccentricity, Δk_e^*

The degree of eccentricity, e, is represented by

$$e = Q_a/Q_c \qquad (4.25)$$

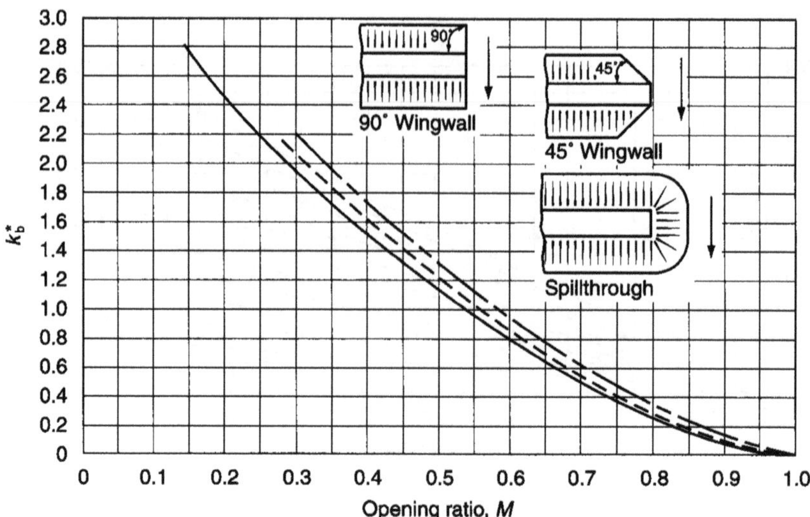

Fig. 4.21 Variation of the USBPR base coefficient, k_b^*, with opening ratio, M, and abutment geometry (after Bradley, 1978). The curve appropriate to the abutment type and span, b, should be used as follows:

top curve	90° wingwall abutments with $b < 60$ m (i.e. high energy loss)
middle curve	30° wingwall abutments with $b < 60$ m
lower curve	all 45° and 60° wingwall abutments (i.e. low energy loss) all spillthrough abutments all bridges with $b > 60$ m.

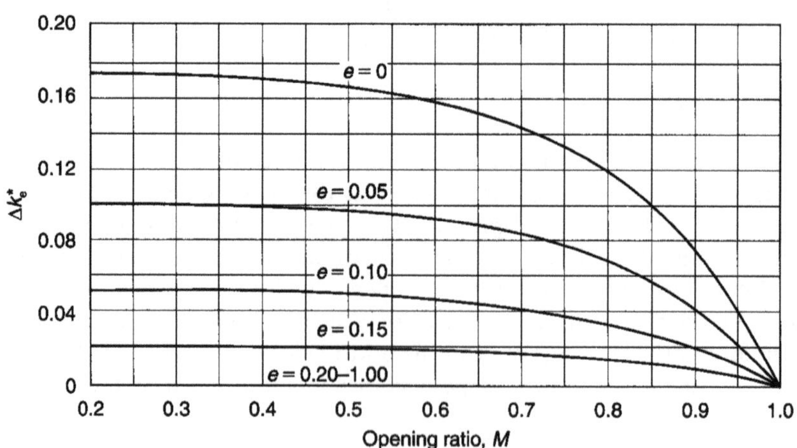

Fig. 4.22 USBPR variation of the incremental backwater coefficient for eccentricity, Δk_e^*, with M (after Bradley, 1978). The coefficient should be added to the base coefficient, k_b^*.

where Q_a and Q_c represent the discharge obstructed by the bridge outside the projected width b, the smaller of the two values always being the numerator (e.g. Fig. 3.1e). If $e > 0.20$ no correction is required, but if $e < 0.20$ then the incremental coefficient, Δk_e^*, can be obtained from Fig. 4.22, which shows the variation of Δk_e^* with M. The greatest energy loss, and hence largest Δk_e^* value, is experienced where the contraction occurs only on one side of the channel: that is, with the fully eccentric condition where $X_a = 0$ in Fig. 3.1e, $Q_a = 0$ and $e = 0$.

Skew (angularity), Δk_ϕ^*

With the USBPR method the opening width is projected upstream onto section 1, which is perpendicular to the general direction of flow (Figs 3.8 and 4.23). It is this projected width, $b_S \cos \phi$, that is used to calculate M and A_{N2}. Similarly, the blockage caused by any piers is calculated using the projected area on a plane perpendicular to the flow (see below).

The value of $b_S \cos \phi$ has a different meaning depending upon whether the abutments are parallel to the flow (skew 1) or perpendicular to the face of the embankments (skew 2). In the former case, $b_S \cos \phi$ represents the actual unobstructed distance between the abutments, and in the latter case it does not (see Fig. 3.7). Hence Fig. 4.23 is in two parts: part (a) is for abutments parallel to the flow, part (b) is for abutments perpendicular to the embankment. In both parts the variation of Δk_ϕ^* with M for various skew angles is shown. Note that some values are negative; the sign arises only from the means of computation and is not indicative of increasing hydraulic efficiency. The incremental backwater coefficient for skew (Δk_ϕ^*) should be added to the base coefficient as indicated by equation 4.24.

Figure 4.24 represents the data in Fig. 4.23a replotted so that it is possible to determine from M and the angle of skew (ϕ) the width of skewed opening (b_S) needed to give the same backwater as a normal opening of width b. For example, if $b = 60\,\text{m}$, $M = 0.6$ and $\phi = 40°$, from the diagram $b_S \cos \phi/b \approx 0.932$ so $b_S = (0.932 \times 60)/\cos 40° = 73.00\,\text{m}$.

Bradley pointed out that crossings skewed at up to 20° produced no objectionable result regardless of abutment geometry, but when there was a significant constriction of the channel skews over 20° produced large eddies, reduced efficiency, and increased the possibility of scour (see Section 3.7).

Piers and pile bents, Δk_P^*

With a crossing that is perpendicular to the flow direction the backwater arising from the obstruction due to piers or piles is allowed for using the incremental pier coefficient, Δk_P^*. The value of the coefficient depends upon the type of pier and the proportion of the opening they occupy, which is

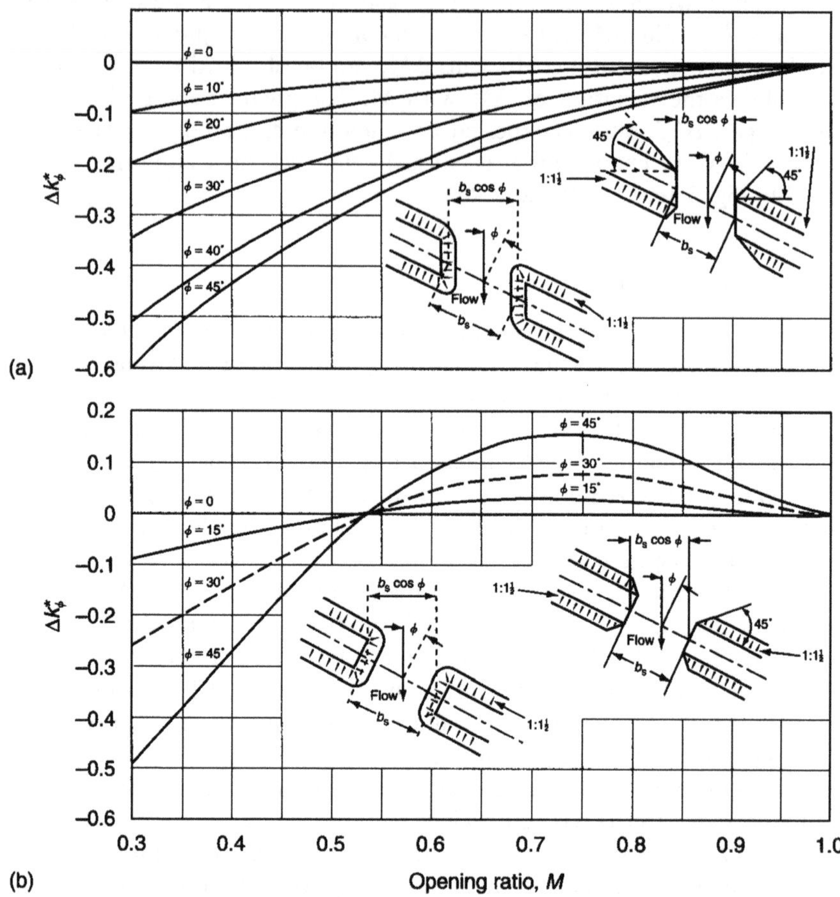

Fig. 4.23 USBPR variation of the incremental backwater coefficient for skew (angularity), Δk_ϕ^*, with M (after Bradley, 1978): (a) for abutments parallel to the flow; (b) for abutments perpendicular to the embankment. Δk_ϕ^* should be added to the base coefficient, k_b^*. Note that the skewed width of the opening or span is b_S, while the width perpendicular to the flow direction is $b = b_S \cos \phi$, where ϕ is the skew angle.

$$J = \frac{A_P}{A_{N2}} \tag{4.26}$$

where A_P is the total submerged area occupied by the piers and A_{N2} is the gross area of the opening (including the piers), both measured below the normal depth at section 2. The value of Δk_P^* is obtained from Fig. 4.25. First, enter part (a) with the appropriate value of J, move upwards to the $M = 1.0$ line corresponding to the pier type, then read off the value of Δk.

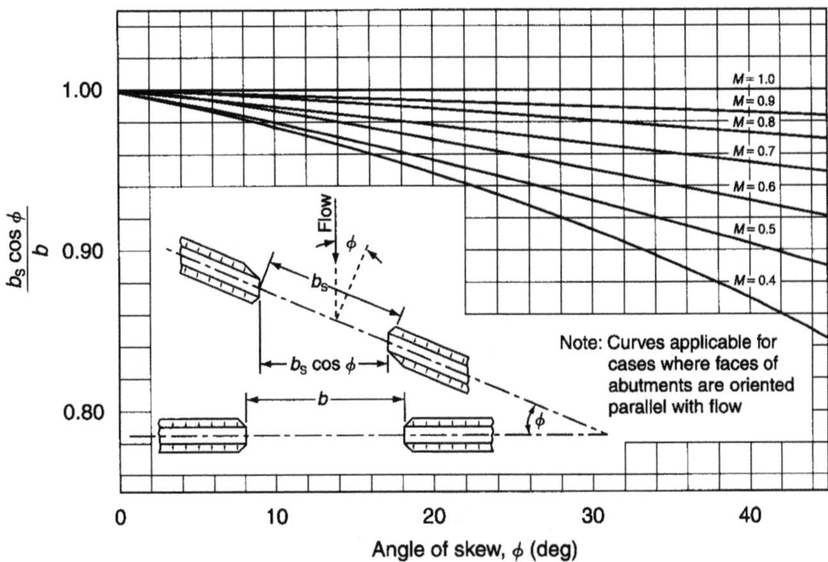

Fig. 4.24 USBPR diagram to determine the skewed span (b_S) required to give the same backwater as an opening of width b perpendicular to the direction of the approaching flow (after Bradley, 1978). The graph is applicable only to openings with abutments parallel to the flow as in Fig. 4.23a.

Second, correct for the fact that $M \neq 1.0$ by entering part (b) with the correct M value, moving down to the line representing the pier type, then across to obtain the correction factor, σ. Then

$$\Delta k_P^* = \sigma\, \Delta k \tag{4.27}$$

Bradley advised that the backwater coefficients for pile bents could effectively be considered independent of the diameter, width or spacing of the piles but should be increased when a bent has more than five piles. A ten-pile bent should have a value of Δk_P^* about 20% higher than the standard five-pile bent in the diagrams. Any sway bracing should be included in the width of pile bents. To allow for trash collecting on the piers use a larger value of J.

With skewed crossings the procedure is basically the same as above but the width of the piers and piles is calculated on a plane that is perpendicular to the general flow direction (Fig 4.25). The total area of the piers (A_P) is the sum of the individual areas.

Spur dykes

These are considered in detail in Section 7.4.

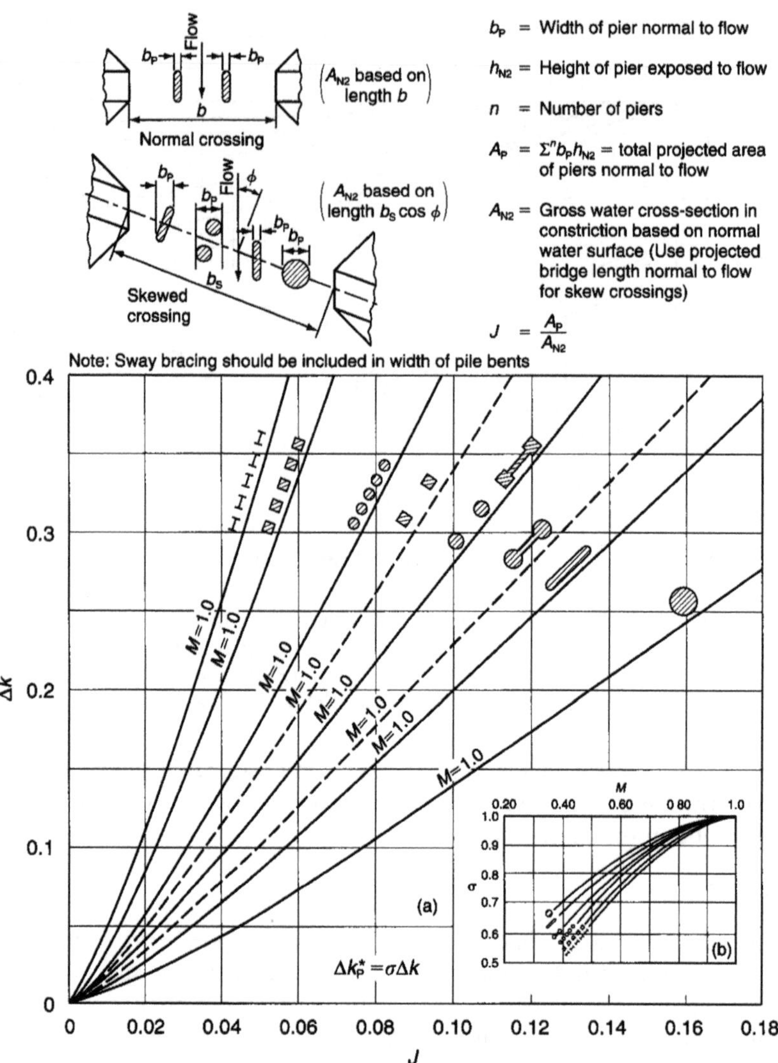

Fig. 4.25 USBPR incremental backwater coefficient for piers, Δk_P^* (after Bradley, 1978). First, the value of Δk for the appropriate pier type is found from the main diagram (a), which assumes $M = 1.0$. Second, the value of σ is obtained from inset diagram (b) using the true value of M and the pier type. Third, the value of Δk_P^* is obtained from $\Delta k_P^* = \sigma \Delta k$. The value of Δk_P^* is added to the base coefficient, k_b^*.

Summary and example

Table 4.3 presents a summary of the USBPR procedure for calculating the afflux arising from flow at the normal depth. Example 4.2 illustrates its use.

Table 4.3 Summary of the USBPR procedure for estimating the maximum bridge afflux with normal depth

The last steps are iterative because the full solution requires in step 13 the value of A_1 (which includes H_1^*) before the afflux, H_1^*, is known.

1. Determine the magnitude of the flood for which the bridge is being designed, Q.
2. Determine the river stage at the design discharge. This would be the normal depth, Y_N, unless an abnormal stage is experienced.
3. Plot the natural channel cross-section under the proposed bridge and superimpose the design river stage from step 2. At section 2, perpendicular to the flow direction, calculate the gross area of the opening (including any piers) between the normal water surface and the river bed, A_{N2}. Thus $V_{N2} = Q/A_{N2}$.
4. Project the opening area (A_{N2}) upstream to section 1, which must be perpendicular to the general flow direction; centre A_{N2} on the main channel if necessary. Check that skew is dealt with correctly.
5. At section 1, plot a representative cross-section and subdivide it where there is a marked change in depth of flow and/or roughness and at the edges of the projected opening. Calculate the conveyance and discharge in each subsection, and the total conveyance and discharge.
6. Calculate the bridge opening ratio, M, from equation 3.1 or 3.5.
7. Determine the value of α_1 (equations 3.17–3.19). Estimate α_2 from Fig. 4.20 and allow for anything that may alter the value, such as vegetation.
8. Obtain the value of the base coefficient, k_b^*, for normal symmetrical crossings from Fig. 4.21.
9. If the bridge is eccentric, calculate the value of e from equation 4.25 and determine the value of the incremental coefficient Δk_e^* from Fig. 4.22.
10. If the bridge is skewed, determine the value of the incremental coefficient, Δk_ϕ^*, using Fig. 4.23 as described in the text.
11. If the bridge has piers, project them onto a plane perpendicular to the flow direction and calculate J from equation 4.26. Determine the value of the incremental coefficient, Δk_p^*, from Fig. 4.25 and equation 4.27.
12. Determine the total backwater coefficient, k^*, from equation 4.24.
13. For the first iteration use $H_1^* = k^* \alpha_2 V_{N2}^2/2g$, otherwise calculate the maximum afflux, H_1^*, from equation 4.23 (or equation 4.28 if the stage is abnormal).
14. For the normal depth situation, obtain the total stage from $Y_N + H_1^*$.
15. Repeat the calculations, if necessary, using the new stage to obtain revised estimates of A_1 (in equation 4.23) and H_1^*.

4.3.2 *Abnormal stage*

If the site of a bridge is affected by a backwater from further downstream the resulting stages are referred to as abnormal, since the depth of flow is greater than the normal depth predicted by the Manning equation (see Section 2.3.2 and Fig. 2.4). This is quite a common occurrence, but it is a more difficult problem to analyse because the profile of the water surface without the bridge has to be determined by a backwater curve analysis.

If the stage is abnormal then the asumptions inherent in the USBPR method (e.g. uniform flow, constant cross-section) are no longer valid, so theoretically the procedure above is no longer applicable. Therefore, Bradley (1978) presented the results of a limited model study (scale 1:40, $S_O = 0.0012$, $n = 0.024$), which allowed an approximate solution for afflux with abnormal flow. In fact, this study suggested that the procedure adopted for normal depth flow could also be applied to abnormal flow, with two modifications. One is that the reference velocity head used in the afflux equation must relate to the abnormal stage condition (subscript A). The second is that the last term of equation 4.23 is no longer appropriate, so the new equation for the maximum backwater is

$$H_{1A}^{*} = k^{*} \frac{\alpha_2 V_{2A}^2}{2g} \tag{4.28}$$

where H_{1A}^{*} is the afflux measured above the abnormal stage at section 1 without the bridge, k^{*} is still the total backwater coefficient as defined by equation 4.24, the mean waterway velocity is $V_{2A} = Q/A_{2A}$, and A_{2A} (= bY_{2A} for a rectangular opening) is the gross area of flow in the constriction at the abnormal stage (Y_{2A}) at section 2. Figures 4.21–4.23 and 4.25 are used to obtain k^{*}, as before.

Abnormal stages result in a larger cross-sectional area of flow (than normal depth) for a given discharge, and consequently a reduced velocity. Since the energy loss due to flow through a bridge is proportional to the square of the velocity, this means that there will also be a reduced energy loss and a smaller afflux. These are the same characteristics that might be expected at a site with a small bed or friction gradient that results in a large normal depth, low velocity and small Froude number.

4.3.3 Difference in water level across the approach embankments

One of the problems with afflux is that it cannot be identified visually during a flood, or measured directly, because it cannot be separated from the normal depth. On the other hand it is relatively easy to measure the difference in water level across a relatively long embankment: the flood level on each side can be marked with a few pegs and then the difference in level (Δh) surveyed at leisure.

Being visible, Δh is often mistaken for the afflux. Of course, the afflux (H_1^{*}) is the maximum backwater measured above normal depth. Similarly, H_3^{*} is the amount by which the water surface falls below normal depth at the downstream face of the bridge (Fig. 2.3). The exception to this is if the flow cannot return to the downstream floodplain because it is covered in dense vegetation or if there is an abnormal stage caused by an obstruction further downstream. However, for the normal situation the relationship between these variables is shown by equation 4.17.

Before a bridge is constructed H_1^{*} can be calculated using the method described above. It may also be desirable to calculate the water level on the downstream side of a proposed structure (Y_3). To facilitate this a model study was undertaken of bridges that were perpendicular to the flow and without piers, eccentricity or skew so H_1^* can be represented by H_b^*, which is evaluated from

$$H_b^* = k_b^* \frac{\alpha_2 V_{N2}^2}{2g} \tag{4.29}$$

where H_b^* is the bridge backwater (exclusive of piers) calculated from the base coefficient (k_b^*) in Fig. 4.21. The variation of H_b^* with the difference in water level across the model embankments was expressed in terms of the differential level ratio, D_b. If the values of H_b^* and M have been evaluated, the value of D_b can be obtained from Fig. 4.26 where

$$D_b = \frac{H_b^*}{(H_b^* + H_3^*)} \tag{4.30}$$

The lower curve in Fig. 4.26 should be used for openings with a relatively low energy loss, such as 45° and 60° wingwall abutments and all spillthrough abutments, regardless of waterway width. The upper, broken line should be used for relatively high energy loss bridges with opening widths up to 60 m or having 90° vertical-wall or other abutment shapes that severely constrict the flow. Rearranging equation 4.30 gives

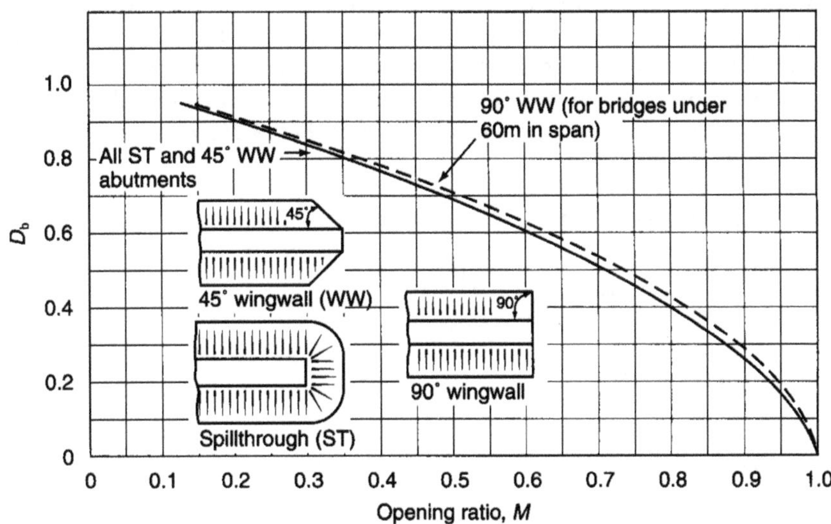

Fig. 4.26 Variation of the USBPR differential level ratio (D_b) with M (after Bradley, 1978). The ratio D_b is used with equations 4.30–4.37 to calculate the difference in water level across the bridge or embankments. Use the top curve for high energy loss 90° wingwall abutments and bridges less than 60 m span; use the lower curve for low energy loss spillthrough and 45° wingwall abutments, and all bridges over 60 m span.

$$H_3^* = H_b^*\left(\frac{1}{D_b} - 1\right) \tag{4.31}$$

$$\text{and } Y_3 = Y_N - H_3^* \tag{4.32}$$

where Y_N is the normal depth and Y_3 is defined as the average depth at section 3 measured along the downstream face of the highway embankments. If the floodplains are wide and the embankments are long, Y_3 is the average depth over a distance of not more than two opening widths (i.e. $2b$) from the banks. If required Δh can be calculated from equation 4.17. The effects of piers, eccentricity (angularity) and skew can be allowed for as follows.

Piers

The effect of piers is to increase the backwater without significantly affecting H_3^*, regardless of the blockage afforded by the piers. Thus the equation for H_3^* above is still valid.

Eccentricity

For severe eccentricity the difference in level calculated below applies only to the side of the river with the largest floodplain discharge (i.e. Q_c in equation 4.25); for mild eccentricity it applies to both. While eccentricity alters the values of H_b^* and H_3^* compared with a symmetrical crossing, for any given value of M the ratio of the two values stays the same. Thus the afflux (H_b^*) for a normal crossing should be calculated; then for an eccentric crossing (with or without piers) the value of D_b corresponding to M can be obtained from the appropriate curve in Fig. 4.26, where now

$$D_b = \frac{H_b^* + \Delta H_e^*}{H_b^* + \Delta H_e^* + H_3^*} \tag{4.33}$$

$$\text{so } H_3^* = (H_b^* + \Delta H_e^*)\left(\frac{1}{D_b} - 1\right) \tag{4.34}$$

$$\text{where } \Delta H_e^* = \Delta k_e^* \frac{\alpha_2 V_{N2}^2}{2g} \tag{4.35}$$

Skew (angularity)

With skewed crossings the difference in level across the bridge or embankment will be different on the two floodplains and dependent upon the geometry of the particular site. Although the absolute values of H_1^* and H_3^* are different from those for a symmetrical crossing, the value of D_b across either end of the embankment can be considered to be the same as for a normal crossing for any given value of M (remember that the projected width is used to calculate M). This means that Fig. 4.26 can be used as before with the normal bridge afflux (H_b^*), so for bridges with or without piers:

$$H_3^* = (H_b^* + \Delta H_\phi^*)\left(\frac{1}{D_b} - 1\right) \tag{4.36}$$

$$\text{where } \Delta H_\phi^* = \Delta k_\phi^* \frac{a_2 V_{N2}^2}{2g} \tag{4.37}$$

Abnormal stage, no eccentricity or skew

If an abnormal stage exists (Fig. 2.4) then equation 4.17 is no longer valid because of the higher profile of the water surface, but equations 4.29–4.31 can be used (with a suffix A to denote abnormal stage).

4.3.4 *Dual bridges*

There are many locations where two bridges for a dual carriageway have been built in close proximity, or where a new crossing has been built alongside an existing one (Fig. 4.27). This obviously increases the difficulties of

Fig. 4.27 The problem of three bridges in close proximity causing hydraulic interference. The most modern upstream bridge has a large single span, through which can be seen a four-span masonry arch bridge, through which can be seen the low spans of a brick bridge carrying the Trent and Mersey Canal over the River Dove. The openings of the latter regularly submerge during flood.

conducting a hydraulic analysis, and requires the exercise of judgement. The USBPR study considered only identical parallel crossings arranged perpendicular to the flow, there being too many possible combinations to do otherwise. Most of the tests were conducted on 45° wingwall abutments with a few spillthrough types being included, all with embankment slopes of 1 vertical to 1.5 horizontal. However, there are some limited guidelines, which are described below.

Backwater

With dual carriageways it is quite common to have virtually identical bridges a short distance apart (Fig. 4.28). This results in a larger afflux than for a single bridge, but less than that obtained by considering the two bridges separately. This rule of thumb should be remembered when dealing with bridges that are not identical.

When the two bridges are near together the flow pattern is similar to that for a single opening, but slightly elongated. However, as the distance between the bridges increases the expanding jet from the first opening encounters the embankment of the second bridge and has to contract and then expand again. This results in additional turbulence and loss of energy, so the afflux upstream of the two bridges is larger than for a single structure. The water level between the two bridges is usually above the normal stage, and is higher than it would be with just the upstream bridge. The water level downstream of the second bridge is lower than it would be if only the upstream bridge existed.

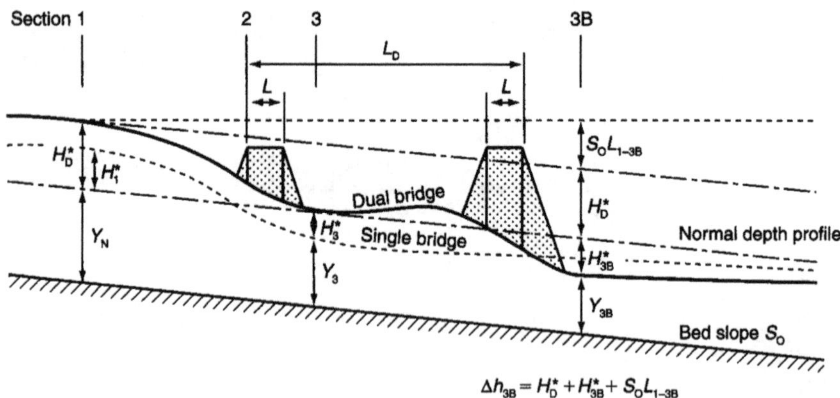

Fig. 4.28 Longitudinal section through a site with dual bridges (after Bradley, 1978). If normal depth conditions existed initially, the water surface profile resulting from a single bridge (the one furthest upstream) is shown by the dashed line and dual bridges by the solid line. Note that the second (downstream) bridge increases the water level everywhere upstream of it, but lowers the level immediately downstream.

The maximum afflux due to the upstream bridge by itself is H_1^*, and is defined as before. H_D^* is the maximum combined afflux due to the dual bridges, also measured upstream of the first bridge. Having established H_1^* using the method described previously, the combined afflux for the dual crossing, measured upstream of the first bridge, is

$$H_D^* = H_1^* \left(\frac{H_D^*}{H_1^*}\right) \tag{4.38}$$

The value of the ratio in the brackets is obtained from Fig. 4.29 according to the distance between the bridges as defined by L_D/L, where L_D is the distance from the upstream face of the first bridge to the downstream face of the second (excluding the slope length of embankments), and L is the length of each waterway in the direction of flow (i.e. the road width, excluding the slope length of any embankments or wingwalls). The ratio H_D^*/H_1^* at first increases as the distance between the bridges increases, then reaches a limit, then decreases as the distance is increased further and the influence of the second bridge diminishes. Figure 4.29 relates only to the first part of the curve, and was obtained from tests made with and without piers.

Downstream water level

The first step towards calculating the water level downstream of the second bridge is to consider the upstream bridge by itself and to calculate H_1^* and H_3^* using the standard procedures described above. The equivalent values for the dual bridge situation are H_D^* and H_{3B}^*, the suffix 3B referring to cross-section 3B, which is located at the minimum depth downstream of the

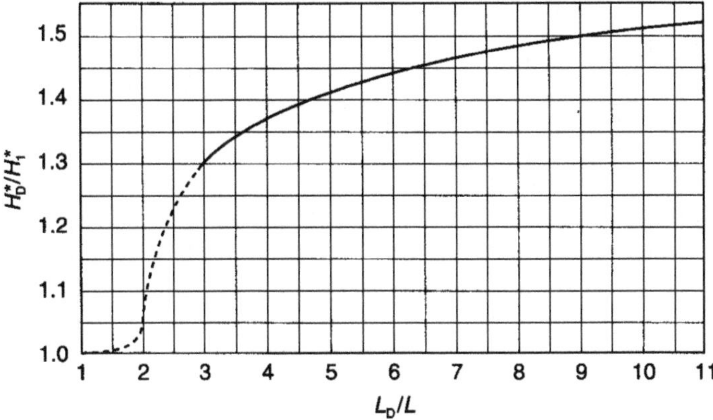

Fig. 4.29 Variation of the USBPR afflux multiplication factor (H_D^*/H_1^*) with the distance between dual bridges expressed as L_D/L (see Fig. 4.28). The dashed part of the curve denotes uncertainty. The factor is used in equation 4.38 to calculate the combined afflux from two bridges in close proximity. (After Bradley, 1978)

second opening (Fig. 4.28). The difference in the water level between the upstream side of the first bridge and the downstream side of the second is

$$\Delta h_{3B} = (H_D^* + H_{3B}^*) + S_O L_{1-3B} \tag{4.39}$$

$$\text{where } (H_D^* + H_{3B}^*) = (H_1^* + H_3^*) \left[\frac{(H_D^* + H_{3B}^*)}{(H_1^* + H_3^*)} \right] \tag{4.40}$$

The value of the term in the square brackets is obtained from Fig. 4.30 for the particular value of L_D/L; H_D^* can be obtained from equation 4.38, so H_{3B}^* can be calculated, if desired.

4.3.5 *Flow passes through critical depth*

Bradley (1978) pointed out that if the backwater calculated for a bridge on a river with a fairly steep gradient appears to be unrealistic, then this may indicate that the flow has passed through critical depth and is of either type 5 or type 6 (Fig. 2.6). In this case an alternative method of analysis to that described above has to be adopted. Bradley outlined such a method based on a laboratory study that covered only a limited range of M and which did not take into account the effect of piers, eccentricity or skew. The tentative result of this simplified analysis is summarised in Fig. 4.31, which shows the variation of the total critical depth backwater coefficient, k_C^*, with M. There are no incremental coefficients. Thus the afflux for type 5 or 6 flow is

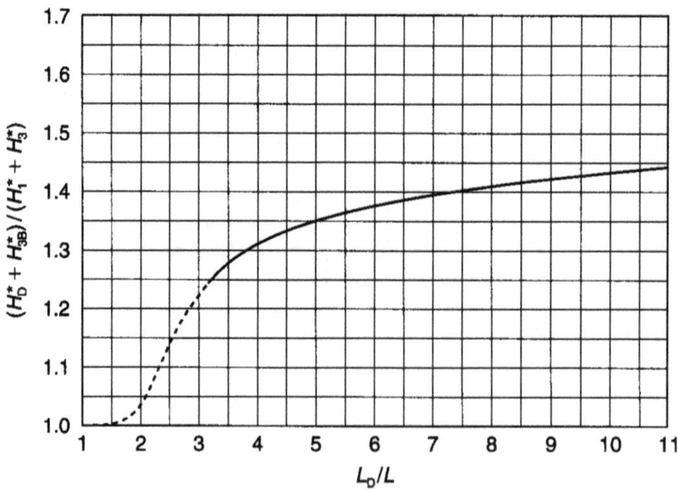

Fig. 4.30 Variation of the USBPR differential level multiplication factor for dual bridges with the distance between the bridges expressed as L_D/L (after Bradley, 1978). The dashed part of the curve denotes uncertainty. The factor is used with equation 4.40 to calculate the water level downstream of the second bridge.

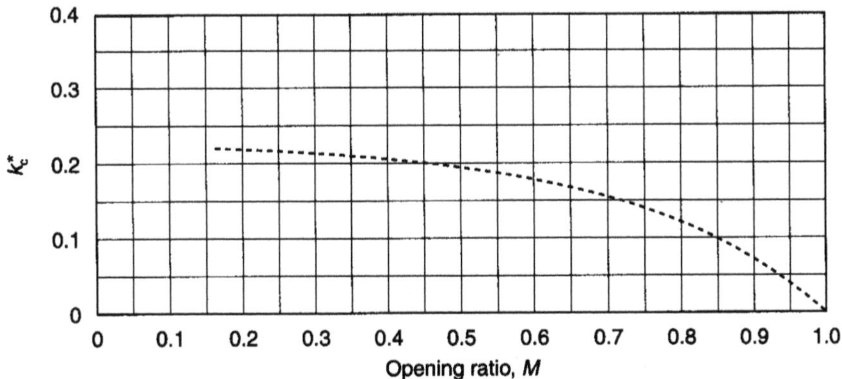

Fig. 4.31 Tentative USBPR curve for the variation of the total critical depth backwater coefficient (k_C^*) with M. (After Bradley, 1978)

$$H_1^* = (k_C^* + 1)\frac{\alpha_2 V_{2C}^2}{2g} - \frac{\alpha_1 V_1^2}{2g} + Y_{2C} - Y_{N2} \tag{4.41}$$

where V_{2C} is the critical velocity (m/s) in the constriction = Q/A_{2C}; A_{2C} is the net area of flow (m^2) between the bed and the water surface with critical flow (= bY_{2C} for a rectangular waterway of width b); Y_{2C} is the critical depth (m) in the constriction, i.e. $(Q^2/gb^2)^{1/3}$ for a rectangular waterway; Y_{N2} is the depth (m) in the waterway assuming normal depth at section 2 (= A_{N2}/b for a rectangular waterway); α_2 and α_1 are the velocity distribution coefficients (dimensionless) for the constriction and the approach section; and the other variables are as before. Note that in the absence of an incremental coefficient for bridge piers, the net cross-sectional area at section 2 should be used instead of the gross area.

Equation 4.41 is derived by equating the energy at section 1 to the energy at the section in the constriction at which the water surface passes through the critical depth, Y_{2C}. This recognises that the backwater is no longer influenced by the conditions downstream (see Section 3.3).

With type 7 flow (Fig. 2.6) there is no backwater in theory. Additionally, it is questionable whether any obstruction should be placed in a channel where the flow is supercritical throughout. This condition is rarely met in practice.

4.3.6 *Calculation of discharge (stage–discharge analysis)*

The USBPR method does not include a discharge equation as such for open channel flow (although equations are given for submerged openings: see equations 2.8–2.10). However, the relationship between the upstream stage (Y_1) and discharge can be obtained easily because the afflux (H_1^*) is calculated above normal depth (Y_N) and $Y_1 = Y_N + H_1^*$. The procedure is to assume a discharge, Q, and then calculate the corresponding Y_N using the Manning equation; calculate the afflux, H_1^*, from equation 4.23 and thus $Y_1 = Y_N + H_1^*$. This can be repeated for other discharges to obtain the stage–discharge relationship.

A similar procedure can be devised for the abnormal stage condition, but at any discharge it would be necessary to use the abnormal profile without the bridge instead of normal depth.

4.3.7 Limitations on the use of the USBPR method

One of the strengths of this method is that it uses normal depth, which is often the only thing that can be calculated easily (although not necessarily accurately) at the start of an analysis. Consequently the absence of a discharge equation for the channel flow condition is not a problem, since it is easy to calculate the stage–discharge relationship upstream of the bridge, afflux included. However, for a valid analysis the conditions on site must resemble those in the laboratory, where there was a straight, uniform channel between sections 1 and 4. Most natural channels do not meet these criteria, and at some sites the bed level can be so variable that depth becomes meaningless. Under these conditions it may be difficult to apply the USBPR method, and if it is used there must be some question as to the accuracy and veracity of the result. Similarly, with overbank flow on wide heavily vegetated floodplains the calculated backwater may have to be multiplied by about 2 to obtain the true value (Kaatz and James, 1997). This was one of the adverse conditions described at the start of the chapter.

Despite its limitations the USBPR method provides a relatively quick and easy way to calculate the afflux due to a bridge, which is exactly what it was designed for. The fact that dual bridges, abnormal stages and flows that pass through critical depth are considered is an advantage, although these aspects of the study were very limited so the results have to be used cautiously. On the other hand, the USGS method was intended mainly to calculate discharges from observed water levels, and the process for estimating afflux was rather laborious. Thus the two methods complement each other. The accuracy that can be achieved is discussed further in Section 4.6.

4.4 Biery and Delleur

This study by Biery and Delleur (1962) is one of the few specifically concerned with the hydraulics of arch bridges, and consequently provides some useful but limited information about their hydraulic performance. Unfortunately skew, eccentricity, entrance rounding, piers and abnormal stages were not considered. The effect of skew on the flow through arch bridges was subsequently investigated by Husain and Rao (1966), but they did not produce any usable results.

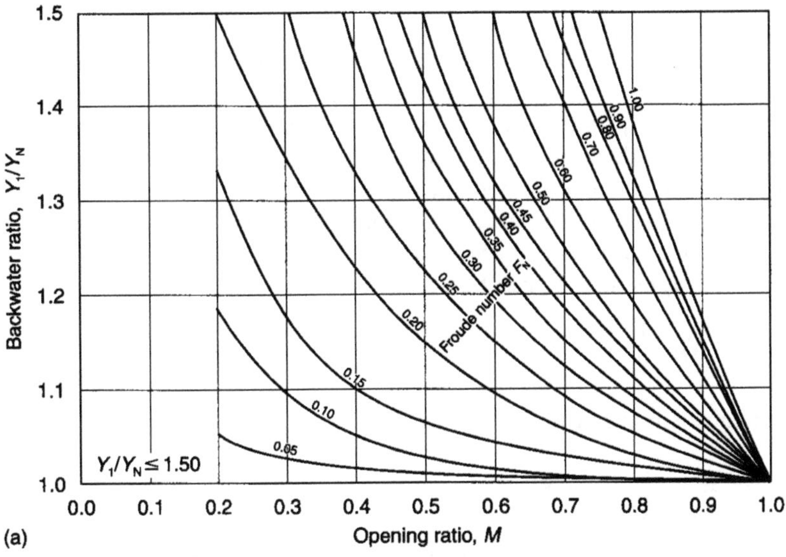

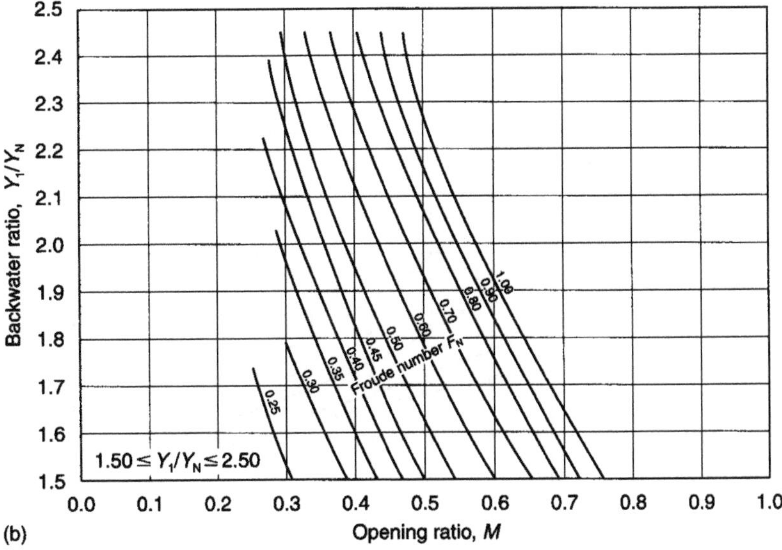

Fig. 4.32 Biery and Delleur's (1962) curves showing the variation of the backwater ratio (Y_1/Y_N) with M and normal depth Froude number (F_N). Use diagram (a) with $Y_1/Y_N < 1.50$ and diagram (b) with $Y_1/Y_N = 1.50$–2.50. (From Hydraulics of single span arch bridge constrictions, *Proceedings of the ASCE*. Reproduced by permission of ASCE)

The research paper by Biery and Delleur is confusing at times and lacks clear examples, so the method is not easy to apply. The laboratory study is not supported by much in the way of comparison with prototype data. Another disadvantage is that the calculations use the normal depth Froude number (F_N or F_4) so that all the reservations expressed previously about the validity of Froude numbers in complex channels apply here.

Biery and Delleur presented equations for the calculation of the backwater ratio $[Y_1/Y_N]$, but the graphs in Fig. 4.32 appear to be a better alternative. Thus if Y_N, M and F_N have been calculated and $[Y_1/Y_N]$ is evaluated from the graphs then Y_1 can be obtained:

$$Y_1 = Y_N\left(\frac{Y_1}{Y_N}\right) \tag{4.42}$$

The equivalent afflux is

$$H_1^* = Y_N\left(\frac{Y_1}{Y_N}\right) - Y_N \tag{4.43}$$

Hamill (1993) suggested if the technique had to be applied to a site with an abnormal stage (Fig. 2.4) the equivalent expression for the afflux, H_{1A}^*, would be

$$H_{1A}^* = Y_{1A}\left(\frac{Y_1}{Y_{1A}}\right) - Y_{1A} \tag{4.44}$$

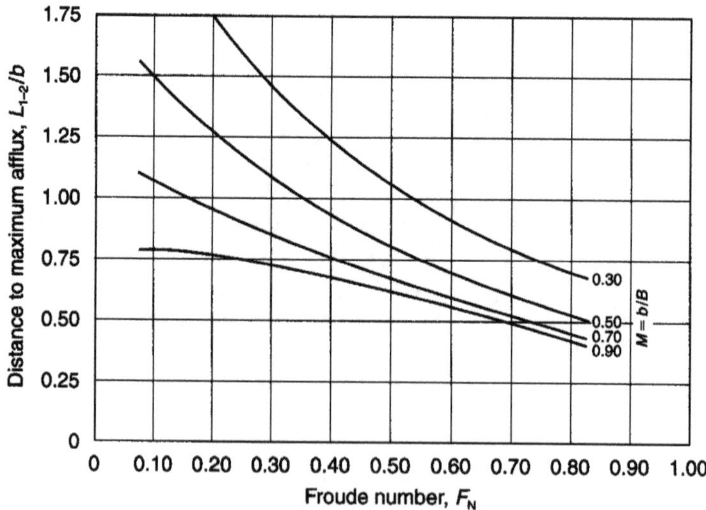

Fig. 4.33 Distance from the bridge face to the section of maximum afflux (L_{1-2}/b) as a function of normal depth Froude number F_N and M. (After Biery and Delleur, 1962; Hydraulics of single span arch bridge constrictions, *Proceedings of the ASCE*. Reproduced by permission of ASCE)

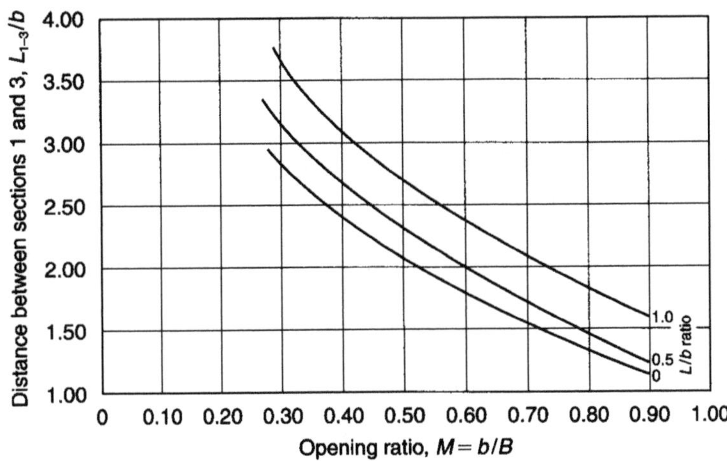

Fig. 4.34 Distance between the sections of maximum (1) and minimum (3) water level (L_{1-3}/b) as a function of M and waterway length (L/b). (After Biery and Delleur, 1962; Hydraulics of single span arch bridge constrictions, *Proceedings of the ASCE*. Reproduced by permission of ASCE)

where Y_{1A} is the abnormal stage without the bridge and Y_1 is the depth including the afflux.

Figure 4.33 shows the distance from the section of maximum afflux to the bridge face (L_{1-2}), while Fig. 4.34 indicates the distance between the section of maximum afflux and the section with the minimum water surface elevation (L_{1-3}). In the former case similar graphs are available from other publications (e.g. Bradley, 1978), but not from measurements involving arched waterways, although there is no obvious reason why the waterway shape should be significant.

The discharge ($Q\ \mathrm{m^3/s}$) through a two-dimensional semicircular arch in a rectangular channel can be calculated from the upstream depth, thus:

$$Q = 0.7083\ C_D\ (2g)^{1/2}\ Y_1^{\ 3/2}\ b \left[1 - 0.1294\left(\frac{Y_1}{r}\right)^2 - 0.0177\left(\frac{Y_1}{r}\right)^4\right] \tag{4.45}$$

where C_D is the Biery and Delleur coefficient of discharge (dimensionless), g is the acceleration due to gravity (9.81 m/s^2), Y_1 is the depth of flow (m) including the afflux at the section of maximum afflux, b is opening width (m) at the springline of the arch, and r is the radius of curvature (m) of the arch. The value of C_D can be obtained from Fig. 4.35, where F_N is the normal depth Froude number in the rough rectangular channel calculated from equation 3.10 and M is the bridge opening ratio. When applied to Canns Mill, where there was an abnormal stage, Hamill (1993) obtained more accurate results using Y_{1A} instead of Y_1 in equation 4.45, although this may not always be the case (see Section 4.6 and Example 4.3).

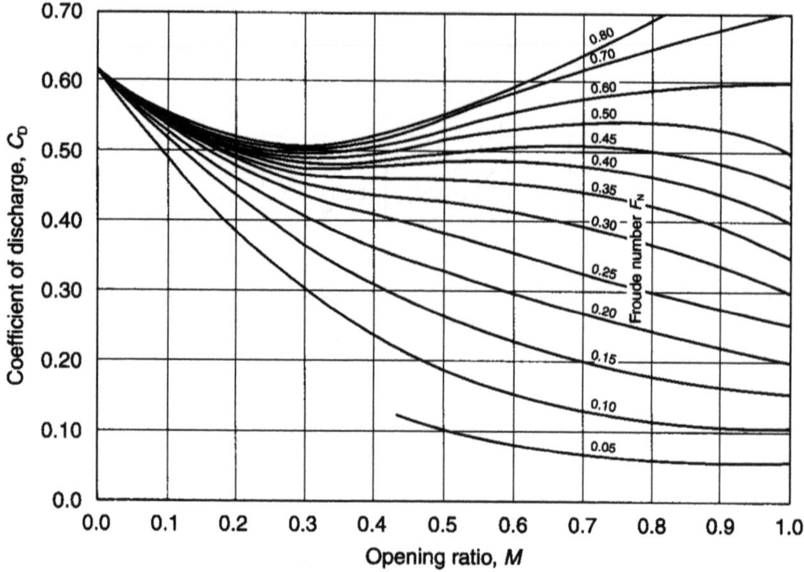

Fig. 4.35 Variation of the Biery and Delleur (1962) coefficient of discharge (C_D) for two-dimensional semicircular arch models with M and normal depth Froude number (F_N). (From Hydraulics of single span arch bridge constrictions, *Proceedings of the ASCE*. Reproduced by permission of ASCE)

Limitations of this method include the ones mentioned earlier: it is applicable only to normal, centrally located, square-edged single-span openings in a channel flowing at the normal depth. The values of both M and F_N must be calculated under all conditions, which is a disadvantage since these variables can be difficult to estimate accurately. The method is not very user friendly, since the original research paper is relatively long at 34 pages and can be confusing. However, the discharge equation and graphs for the backwater ratio have been applied to Canns Mill with some success, and despite the numerous limitations under some circumstances this method appears to work quite well.

4.5 Hydraulics Research (HR) method

This investigation by Hydraulics Research, Wallingford, England, included a laboratory study of single and multispan arched openings with the objective of evaluating the afflux (Brown, 1985, 1989). The theoretical background stems from an analogy between the blockage effect of channel flow past smooth circular cylinders and afflux at bridge obstructions (Ranga Raju *et al.*, 1983; Brown, 1989). A laboratory investigation was undertaken that included the performance of single semicircular arches, single elliptical arches, multiple semicircular arches and multiple semicircular arches with different soffit levels. These structures were tested either at normal depth or by introducing a low flow down the channel and then (at the same discharge) incrementally increasing the height of the tailgate to vary the depth of flow. The laboratory results were compared with field measurements from selected bridges, but the accuracy of some of this data is perhaps questionable.

Details of the investigation and its application are not as comprehensive or helpful as those relating to the USBPR or USGS methods. In the 1985 report afflux is defined rather vaguely as 'the difference in river level either side of the bridge' and as 'the difference between the upstream and gauged heads measured furthest from the bridge'. Adopting the notation of prevous chapters, the investigation produced charts showing the relationship between H_1^*/Y_4, F_4 ($= F_N$), and either the upstream or downstream blockage ratio (J_1 or J_4 respectively), which is vaguely defined as 'the ratio between the area of structural blockage to flow and the total flow area'. Thus the blockage ratio is an alternative to the opening ratio, M. The structure can be analysed in terms of either the upstream or downstream blockage ratio, the downstream providing the easier and more direct solution; the upstream ratio has to be obtained by iteration since it involves the unknown afflux.

The magnitude of the afflux (H_1^*) for single or multiple arches at any particular discharge can be determined from the appropriate chart (Figs 4.36–4.38) once the values of Y_4, F_4 and the blockage ratio have been calculated. The polynomials corresponding to the lines on the chart were also presented, to facilitate inclusion in a computer model (FLUCOMP). The percentage standard deviation of the laboratory and field data from the computed curves is shown in Table 4.4. See also Section 4.6 for an indication of the accuracy of the method.

The advantages of this method are that it is quick and easy to apply, and that the charts appear to show the relationship between the three variables quite well. However, there are some significant disadvantages. There is no

Table 4.4 Percentage standard deviation from calculated HR curves

	Blockage ratio	
Bridge type	*Upstream*	*Downstream*
Prototype single-arched bridges	13.60	12.48
Model elliptical arched bridges	10.45	12.00
Model semicircular arched bridges	9.56	8.43
All model single-arched bridges	9.96	9.78
All model and prototype single-arched bridges	10.15	9.97
Prototype multiple arched bridges	36.75	45.42
All model multiple arched bridges	10.0	8.80

After Brown (1989)

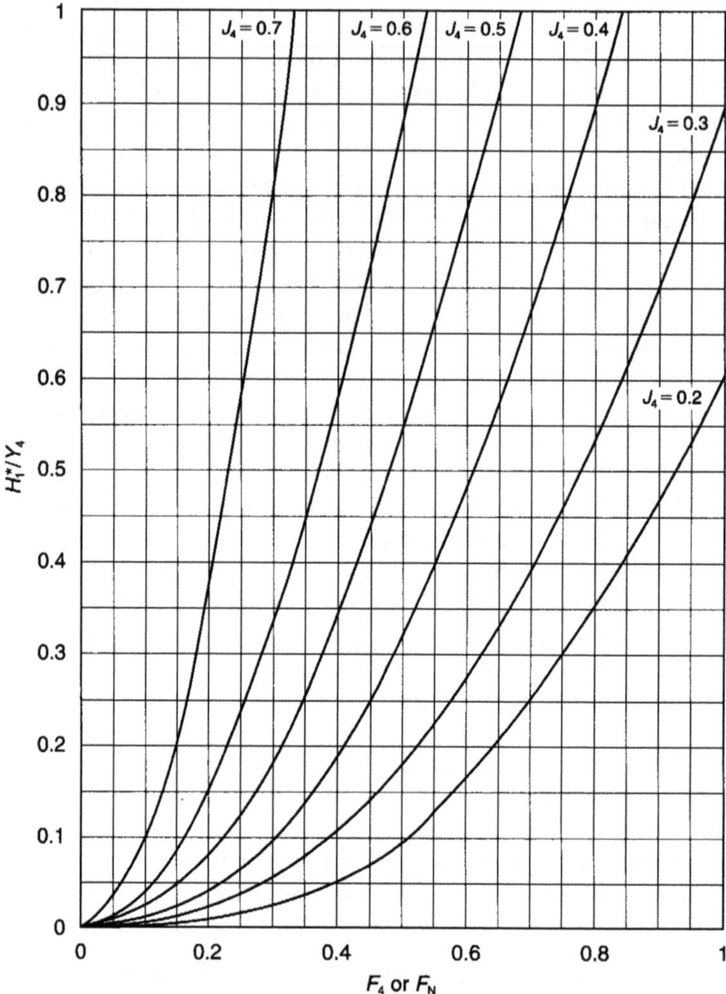

Fig. 4.36 HR method afflux ratio curves applicable to all bridges, showing the variation of H_1^*/Y_4 with F_4 and downstream blockage ratio, J_4. (After Brown, 1989. Reproduced by permission of the Ministry of Agriculture Fisheries and Food)

means of calculating the discharge from observed water levels, unlike most of the other methods, and the arches of a multispan bridge must be 'separated only by typical pier widths'. It is not clear how this method should be applied to sites with compound channels and extensive floodplains, since the laboratory investigation and the worked examples relate to a rectangular section. At Canns Mill the values obtained for the blockage ratio during flood were unreasonable and off the scale of the charts, basically for the

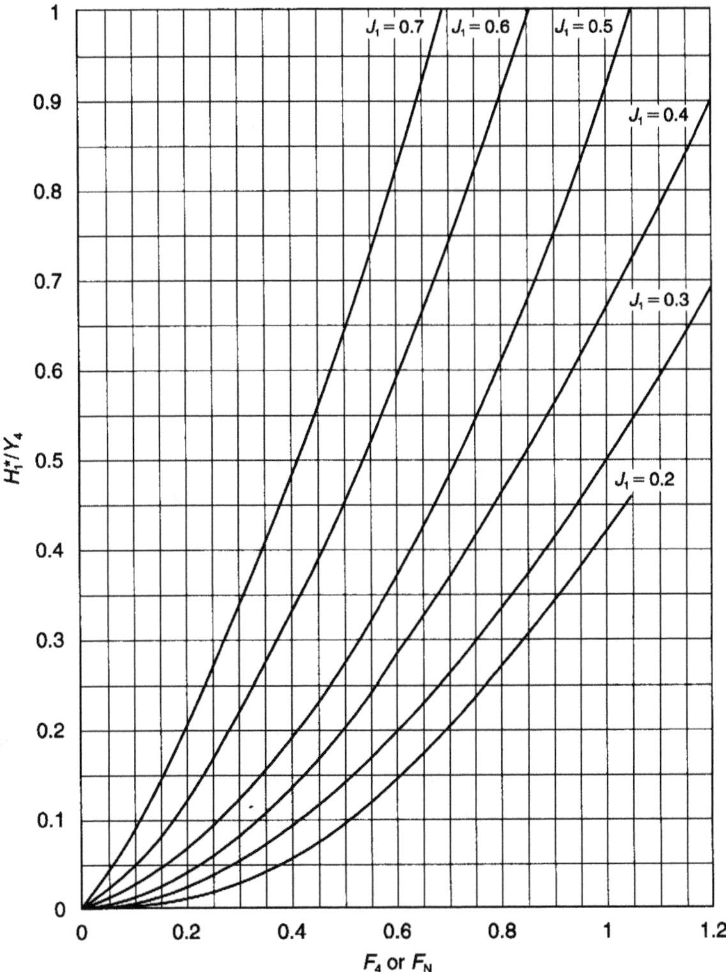

Fig. 4.37 HR curves applicable to all single arches, showing the variation of H_1^*/Y_4 with F_4 and upstream blockage ratio, J_1. (After Brown, 1989. Reproduced by permission of the Ministry of Agriculture Fisheries and Food)

reasons described in Section 3.2 regarding the calculation of M values. To obtain an accurate solution, a rectangular main channel without the floodplains was assumed, while retaining the stage–discharge relationship recorded in the field. Thus the compound channel was effectively turned into a stylized laboratory channel (see Example 4.4). This must be regarded as a dubious practice, particularly if there is a large flow over the floodplains, which may not work everywhere. At Canns Mill the answers were already known, of course, a luxury not usually available. Consequently

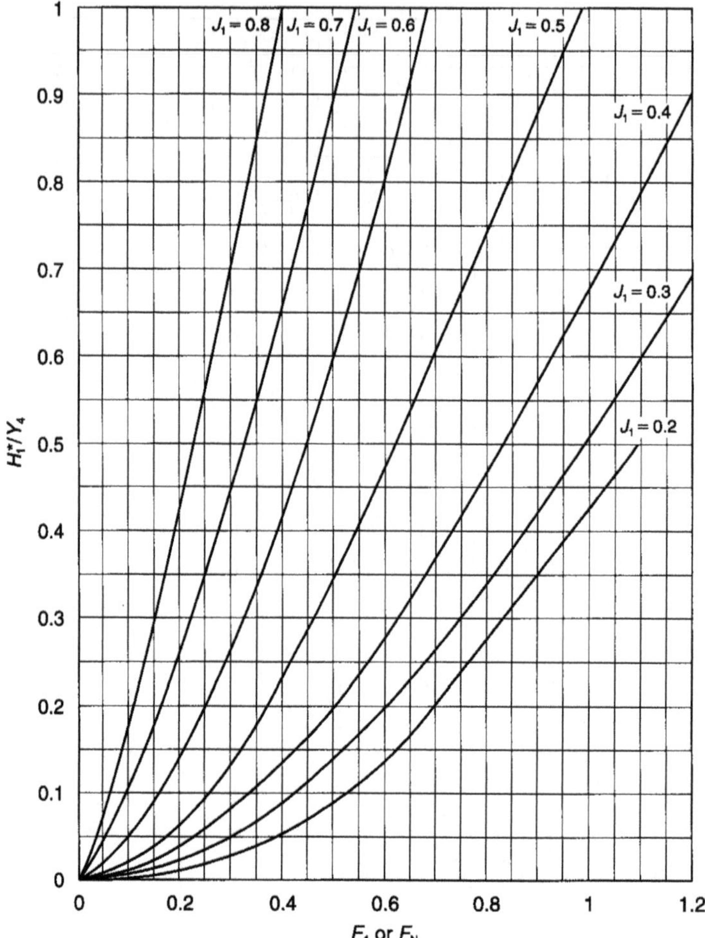

Fig. 4.38 HR curves applicable to all multiple arches, showing the variation of H_1^*/Y_4 with F_4 and upstream blockage ratio, J_1. (After Brown, 1989. Reproduced by permission of the Ministry of Agriculture Fisheries and Food)

when comparing the results and relative accuracy of the four methods in Section 4.6, this must be taken into account.

4.6 The accuracy of a hydraulic analysis and numerical models

If the best possible estimate of discharge or afflux is required, then the average of several methods should be used. Although an individual technique may be unreliable over part of the range of flow, when averaged a reasonably accurate result may be obtained.

Regardless of how the hydraulic analysis of a bridge is undertaken, it is important that the input data are accurate and that the flow characteristics of the site during flood are closely observed. Without well-positioned water level recording equipment backed up by personal observation during flood, it may be the input data rather than the analytical method itself that causes inaccurate results.

When considering the accuracy of the various methods, the potential sources of error in any analysis should be remembered.

- Errors in the field data. An inspection of the discharge equations shows that Q is directly proportional to the cross-sectional area of flow, A. Hence a 5% error in A results in a 5% error in discharge. Scouring of the bed during flood could change the area. Although the error depends upon the equation used and the flow condition, typically a 5% error in measuring the difference in water level across the bridge results in a 2% error in the calculated discharge and a 5% error in afflux.
- Errors in calculated variables such as F and M. These have a less direct effect on accuracy than some of the field data. The error again depends upon the equation used and the flow condition, but typically a 5% error in F produces an error of 1% in Q. At the maximum discharge a 5% error in M can result in a 1% error in the calculated Q and a 4% error in afflux (sometimes much more).
- Errors inherent in the theoretical equations. The equations only approximate the flow phenomena concerned, and are unlikely to reflect accurately the prototype performance over the whole range of flow. In other words, even if the field data are totally accurate an error may still be incurred. This will be most significant when an equation or model is used outside its recommended range, or under conditions for which the model has not been verified using field data.
- Errors in the empirical coefficients (used with the equations) and charts, and the difficulty of drawing, printing and reading such charts precisely.
- Differences between the laboratory and prototype conditions, and scale effects. All the methods of analysis are based on laboratory studies, which may not truly reflect the site or conditions being analysed. All hydraulic models are compromises, so scale effects are inherent.

It is difficult to obtain reliable field data, as opposed to laboratory data, against which to test various methods of hydraulic analysis. In this respect Canns Mill Bridge is relatively unique since its hydraulic performance has been closely observed and recorded for many years, providing an ideal opportunity to test some of the standard methods (Hamill, 1993). The results indicated that it is possible to undertake a reasonably successful analysis using only relatively limited survey data and simple calculations, without having to rely on sophisticated computer programs (see Examples 4.1–4.4). Indeed, some of these calculations can be used to provide a quick check on the output from such programs.

When applying the methods of analysis described above to Canns Mill, all had to be modified to some extent, to make them applicable either to arched openings or to abnormal flow conditions, or both. In general, these modifications appear to have been successful, but it should be remembered that Canns Mill is a segmental arch bridge which experiences an abnormal stage and relatively low Froude numbers (in the region of 0.3 or less). If the same techniques are applied to a bridge or site that is significantly different, the same accuracy may not be obtained.

It is not easy to define what constitutes a successful or accurate hydraulic analysis of a bridge during flood. Arguably, anything that estimates the discharge to within about 10% could be considered very good. Gauging a flood from a bridge with a velocity meter is unlikely to be more accurate (if this is attempted a careful study of the isovels must be undertaken first, particularly with arch bridges; such a study was conducted for Canns Mill). Similarly, any method that can estimate the afflux during flood to within 25 mm is probably performing better than could reasonably be expected. After all, the water surface is likely to be fluctuating by much more than this anyway (see Fig. 2.7). Of course, because Canns Mill Bridge has been so closely observed it is much easier to conduct an accurate analysis of this structure than of any other. Frequently sites have to be analysed without the benefit of a detailed knowledge of their hydraulic performance, so less accurate answers than those described below should be expected.

The results from all the methods of analysis are shown in Table 4.5. As an indication of the relative accuracy achieved, the root mean squared (rms) deviations of the calculated results from the values in Table 4.5 are shown in Tables 4.6a (afflux) and 4.6b (discharge) for the three flow conditions: channel flow, sluice gate flow, and orifice flow. It should be noted that these figures are not always based on the same number of results, and that the rms deviation obtained is always positive regardless of whether the results are larger or smaller than the true values. The rms deviation for the whole range of flow has also been calculated, as has the average value for all of the methods.

Considering first the calculated discharges, it is apparent from Tables 4.5 and 4.6b that all of the methods produced answers within 10% over at least part of the range of flows. The worst results were obtained from the USBPR drowned orifice equation, but this is often the case since it is difficult to obtain accurate values of the coefficient of discharge (see Fig. 2.12) because small differences in fluctuating water levels have to be measured. The best results were obtained from the USGS technique. This is not too surprising, since the principal purpose of this investigation was to enable the peak discharge to be obtained by using the bridge constriction as a discharge meter.

Table 4.5 Comparison of observed and calculated results

Observed values		USBPR			USGS		Biery and Delleur		HR		Average (all methods)	
Discharge (m^3/s)	*Afflux (mm)*	*Afflux eqn 4.23 (mm)*	*Discharge eqn 2.8 (m^3/s)*	*Discharge eqn 2.9 (m^3/s)*	*Afflux eqn 4.18 (mm)*	*Discharge eqn 4.5 (m^3/s)*	*Afflux (graph) (mm)*	*Discharge eqn 4.45 (m^3/s)*	*Afflux upstream ratio (mm)*	*Afflux downstream ratio (mm)*	*Discharge (m^3/s)*	*Afflux (mm)*
2.4	3	4	–	–	1[a]	1.84[a]	28	2.07	–	–	1.96	11
4.5	5	7	–	–	2[a]	3.88[a]	31	3.81	–	–	3.85	13
5.8	11	11	–	–	3[a]	4.82[a]	27	5.11	–	14	4.97	14
7.0	17	17	–	–	8[a]	6.29[a]	42	6.08	30	19	6.19	25
8.3	45	27	–	6.15[a]	17[a]	8.02[a]	100	10.54	51	52	8.24	49
9.5	72	40	8.21	7.67[a]	34[a]	10.01[a]	114	11.48	83	77	9.34	70
10.8	115	64	11.37	9.26[a]	68	11.44	130	12.22	112	107	11.07	96
12.5	175	119	13.50[a]	10.85	137	12.71	183	13.09	164	160	12.54	153
13.8	220	195	14.83[a]	12.45	211	13.7	220	13.82	209	202	13.70	207
15	270	269	16.02[a]	14.01	288	14.7	282	14.41	255	260	14.79	271

[a] Equation used outside recommended range of application

Hamill (1993); reproduced with permission, Institution of Civil Engineers

Table 4.6 Root mean square (rms) deviations of the results in Table 4.5

$$\text{Root mean square deviation} = \sqrt{\frac{\sum_{i=1}^{n}(x_o - x_c)^2}{n}}$$

x_o = observed value (from Table 4.5)
x_c = calculated value
n = number of configurations (results)

(a) Root mean square deviations of afflux (mm)

Type of flow	*USBPR eqn 4.23*	*USGS eqn 4.18*	*Biery and Delleur*	*HR upstream ratio*	*HR downstream ratio*	*Average (all methods)*
Channel	1	6	23	13	3	7
Sluice gate	36	39	41	7	7	11
Orifice	35	25	8	13	15	15
All types	28	25	27	11	10	11

(b) Root mean square deviations of discharge (m^3/s)

Type of flow	*USBPR eqn 2.8*	*USBPR eqn 2.9*	*USGS eqn 4.5*	*Biery and Delleur eqn 4.45*	*Average (all methods)*
Channel	–	–	0.74	0.69	0.70
Sluice gate	1.00	1.86	0.50	1.91	0.18
Orifice	1.02	1.36	0.22	0.48	0.14
All types	1.01	1.63	0.55	1.16	0.46

Hamill (1993); reproduced with permission, Institution of Civil Engineers

The reduced accuracy at the lower flows is to be expected, since Matthai (1967) warned that the method should not be used when the head loss between sections 1 and 3 is less than 150 mm, which occurs at about 9.5 m^3/s at Canns Mill. Below this discharge the bridge does not exert a sufficient control on the flow for it to act as an efficient meter. These values are marked with the letter 'a' in Table 4.5, as are the equivalent values of the afflux for which a similar argument applies. With the Biery and Delleur method the results are somewhat flattering in that the upstream abnormal stage without afflux (Y_{1A}) was used instead of the recommended value Y_1, simply because this gave the best results (see Example 4.3). Thus the comparison is unfair, and it is not certain whether the same improvement in accuracy would be obtained at other sites and without abnormal stages. Consequently it has to be concluded that the USGS method is the better option, possibly even with arch bridges.

With respect to afflux, there was little to choose between the USBPR, USGS, and Biery and Delleur methods in terms of overall accuracy. All three appeared to be good in some parts of the range and poor in others.

This being so, the USBPR technique has the merit of being relatively simple and easy to apply. Whether or not to apply the USGS technique may depend upon the availability of accurate values for the Froude number and the difference in water level between sections 1 and 3. The Biery and Delleur method was not easy to apply because of the use of the Froude number and the fact that by using slightly different approaches within the same general guidelines it was possible to obtain a range of results. As for the HR method, this technique was applied in such a way as to obtain results in agreement with the field data. The first attempts at using the method were totally unsuccessful due to the difficulty of defining the blockage ratio; the wide compound channel had to be treated like a simple channel equal in width to the bridge structure without its approach embankments (see Example 4.4). However, at Canns Mill it did reproduce the afflux very well, so if experience can be built up regarding how to apply the technique to complex sites and confidence in the method established, then it could provide a very quick and easy method of calculating afflux.

Kaatz and James (1997) analysed the backwater at nine different bridge sites during 13 flood events using four one-dimensional flow models: the modified Bradley method and three computer models (the HEC-2 normal bridge method, the HEC-2 special bridge method, and WSPRO). All of the sites had wide, flat heavily vegetated floodplains, where subcritical channel flow occurred. These are the adverse conditions described at the start of the chapter that are outside the range of the Bradley (USBPR) method, so it is not surprising that this technique resulted in the average computed backwater for all events being 51% less than the measured value. Individual values ranged from about 0.1 to 0.8 of the actual values. However, it was observed that the accuracy improved as the velocity in the constriction increased. Of course, this raises the question as to whether normal or abnormal stages were experienced at the sites and how they were analysed. However, perhaps the tentative conclusion is that under the conditions described above the backwater obtained from the USBPR method should be doubled, with possibly a larger factor being used if the velocities are very low and the conditions extreme. To put this in perspective, WSPRO produced an average backwater 31% greater than the true values; the range of individual results was about 0.65–2.65. With the HEC-2 special bridge method the average was 26% less with a range of about 0.30–1.25, and with the normal method an average 2% less with a range of about 0.40–1.75. In the latter case, this would have been 36% higher had the recommended 1:4 downstream expansion been adopted (Hydrologic Engineering Center, 1990). Thus all of the methods produced a wide range of results, so under these conditions all of the models can be quite inaccurate. It should not be assumed that similar accuracy would be obtained at sites with completely different characteristics.

As stated at the begining of this section, if accuracy is required then use more than one method of analysis. Engineering judgement and common sense must also be applied.

4.7 Examples

An example is given below of each of the four methods of analysis described in this chapter. For brevity, they build on previous examples and only illustrate the general principle of how the calculations proceed; there are too many possible combinations to include every aspect.

Example 4.1: the USGS method

A bridge with an opening 10 m wide (= b) spans the main channel analysed in Example 3.2 and shown in Fig. 3.12. The bridge has vertical embankments and vertical abutments with no entrance rounding or skew. The opening is 5 m high. The observed depth at section 1 (10 m upstream from the face of the bridge) is 4.175 m while that at section 3 at the waterway exit is 3.807 m. The slope of the channel is 1: 1000 and the normal depth Y_N is 4 m. (a) Calculate the discharge corresponding to the observed water levels. (b) Check the validity of the observed water levels by calculating the afflux (H_1^*) and the value of H_3^* (Fig. 2.3).

(a) The general procedure shown in Table 4.2 is the starting point for the calculations.

1. A plan and other relevant details are shown in Fig. 4.39.
2. Water depths are as follows: $Y_N = 4.000$ m, $Y_1 = 4.175$ m and $Y_3 =$ 3.807 m. Thus:

$$H_1^* = 4.175 - 4.000 = 0.175\,\text{m}$$

$$H_3^* = 4.000 - 3.807 = 0.193\,\text{m}$$

It must be remembered that these depths are measured relative to a bed that has a slope of S_O and that in the distance $L_{1\text{-}3}$ between sections 1 and 3 the bed falls by an amount $S_O L_{1\text{-}3}$ so that the difference in the elevation of the water surface between 1 and 3 (Δh) is as shown in Fig. 4.39c and equation 4.17:

$$\Delta h = H_1^* + H_3^* + S_O L_{1\text{-}3}$$

$\Delta h = 0.175 + 0.193 + (0.001 \times 20.000) = 0.388$ m (> 0.150 m, as recommended)

3. The span, $b = 10$ m and the waterway length, $L = 10$ m so $L/b = 1.000$.
4. The channel is split into subsections in Fig. 3.12 and some variables are listed in Example 3.2.

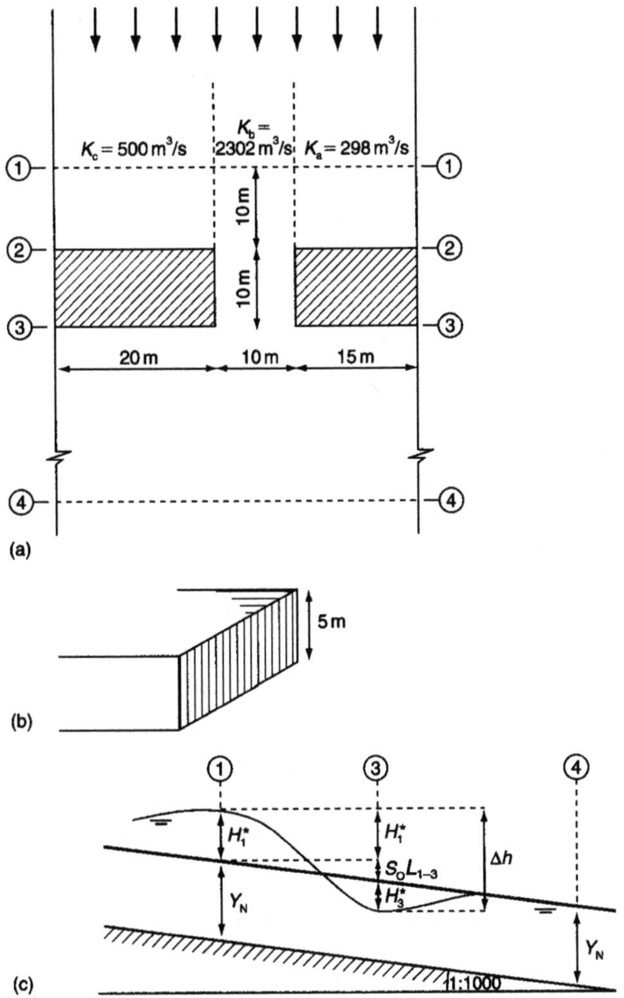

Fig. 4.39 Example 4.1: (a) plan of channel and crossing; (b) details of abutments; (c) longitudinal section.

5. From Example 3.2, $K_a = 298\,m^3/s$, $K_b = 2302\,m^3/s$ and $K_c = 500\,m^3/s$. The total conveyance of the channel is $K = (298 + 2302 + 500) = 3100\,m^3/s$ so from equation 3.5 the opening ratio, M is

$$M = \frac{K_b}{K} = \frac{2302}{3100} = 0.743$$

The discharge ratio in equation 3.1 is often a better basis for the calculations, but has the same value in this instance since the main channel and the floodplains have the same longitudinal gradient.

6. The bridge corresponds to type I abutments and embankments.
7. In this simple example, section 3 is a rectangle 10.000 m wide by 3.807 m deep. With opening types where the coefficient of discharge is a function of the Froude number (F) the calculations below generally have to be repeated several times, since it is necessary to guess a discharge in order to calculate the Froude number. Say the discharge (Q) is 98.030 m^3/s, then

$$F = \frac{Q}{A_3\sqrt{gY_3}} = \frac{98.030}{10.000 \times 3.807\sqrt{9.81 \times 3.807}} = 0.421$$

8. From Fig. 4.3a if $M = 0.743$ and $L/b = 1.000$ then $C' = 0.950$.
9. With $F = 0.421$, from Fig. 4.3b the adjustment factor for the Froude number is $k_F = 0.982$.
   The eccentricity of the opening is $e = K_a/K_c = 298/500 = 0.596$. From Table 4.1 it is apparent that no adjustment factor is needed. There is no entrance rounding, skew or submergence (4.175 m < 5.000 m) so these factors are not needed.
10. From equation 4.10, $C = C' \times k_F = 0.950 \times 0.982 = 0.933$
11. Assuming that $\alpha_1 V_1^2/2g = h_F$ in equation 4.5 so $Q = CA_3\ (2g\Delta h)^{1/2}$, then

$$Q = 0.933 \times 10.000 \times 3.807\ (2 \times 9.81 \times 0.388)^{1/2} = 98.000\,\text{m}^3/\text{s}$$

12. Checks: $V_1 = 98.000/(10.000 \times 4.175) = 2.347$ m/s, which is reasonable.
    $F_1 = 2.347/(9.81 \times 4.175)^{1/2} = 0.367$
    $V_3 = 98.000/(10.000 \times 3.807) = 2.574$ m/s, which is reasonable.
    $F_3 = 0.421 < 0.80$ so flow is subcritical throughout, as required for a valid analysis.
13. Q and F_3 are practically the same as the values in step 7, so another iteration is not needed. However, in most real applications several rounds of calculations may be necessary.

(b) The calculations follow the general procedure given in Section 4.2.2.
1. $Q = 98.000$ m^3/s and $M = 0.743$ as above.
2. The average or effective roughness of the channel (n_E) can be estimated from equation 3.24 using the data in Example 3.2.

$$n_E = \left[\frac{\Sigma(P_i n_i^2)}{P}\right]^{1/2} = \left[\frac{(20.100\times0.040^2)+(14.000\times0.035^2)+(15.130\times0.050^2)}{49.230}\right]^{1/2}$$

$$n_E = 0.042\,\text{s/m}^{1/3}$$

3. With $M = 0.743$ and $n_E = 0.042$ s/m$^{1/3}$ from Fig. 4.18, $H_1^*/\Delta h = 0.45$.
4. The opening is a type I with the basic configuration of vertical abutments so $C/C_{basic} = 1.0$ and hence from Fig. 4.19 it is apparent that $k_c = 1.0$.
5. $k_c\ H_1^*/\Delta h = 0.45$
6. Assuming that $\alpha_1 V_1^2/2g = h_F$ in equation 4.5, so $Q = CA_3\ (2g\Delta h)^{1/2}$, then

$98.000 = 0.933 \times 10.000 \times 3.807\ (2 \times 9.81 \times \Delta h)^{1/2}$

$\Delta h = 0.388\,\text{m}$

7. The afflux $H_1^* = 0.45 \times 0.388 = 0.175\,\text{m}$.
8. Depth at section 1, $Y_1 = Y_N + H_1^* = 4.000 + 0.388 = 4.175\,\text{m}$.
9. From equation 4.17:
   $\Delta h = H_1^* + H_3^* + S_O L_{1-3}$

   $0.388 = 0.175 + H_3^* + (0.001 \times 20.000)$

   $H_3^* = 0.193\,\text{m}$

The depth at section 3, $Y_3 = Y_N - H_3^* = 4.000 - 0.193 = 3.807\,\text{m}$. Thus the depths in part (a) are correct.

Example 4.2: the USBPR method

A bridge is being designed to cross the channel shown in Fig. 3.12 and described in Example 3.2. The crossing will have a skew (ϕ) of 30° with vertical wingwall abutments parallel to the flow, as shown in Fig. 4.40. As in the previous example, assume that the design flood is 98.030 m^3/s, Y_N = 4 m, and the height of the opening is 5 m above bed level. Two alternate designs are being considered: (a) a single span with a skewed width of 11.547 m, which leaves the main channel free of obstruction, and (b) a two-span structure with a skewed width of 11.547 m between the abutments including a round-nosed pier with a skewed width of 1.000 m in the centre of the main channel. There is some concern as to whether the bridge will cause flooding upstream, so it is necessary to calculate the afflux arising from the two designs and the effect of the central pier.

The calculations proceed following the general procedure outlined in Table 4.3.

(a) No pier

1. $Q = 98.030\,\text{m}^3/\text{s}$.
2. The normal depth is already known to be $Y_N = 4.000\,\text{m}$.
3. The channel is shown in cross-section in Fig. 4.40b. The area of the waterway opening below normal depth at section 2 measures 4.000 m deep by 10.000 m wide in a direction perpendicular to the flow. Thus $A_{N2} = 40.000\,\text{m}^2$. The corresponding waterway velocity $V_{N2} = 98.030/40.000 = 2.451\,\text{m/s}$.
4. The opening is shown projected on section 1 in Fig. 4.40a. The skewed width $b_S = 11.547\,\text{m}$ so the width perpendicular to the flow is $b = 11.547\cos 30° = 10.000\,\text{m}$.
5. The channel at section 1 is shown in cross-section in Fig. 4.40b and is subdivided where the changes in depth and roughness occur. From Example 3.2, $Q_a = 9.420\,\text{m}^3/\text{s}$, $Q_b = 72.800\,\text{m}^3/\text{s}$ and $Q_c = 15.810\,\text{m}^3/\text{s}$. As already known, this gives the total discharge $= 98.030\,\text{m}^3/\text{s}$.

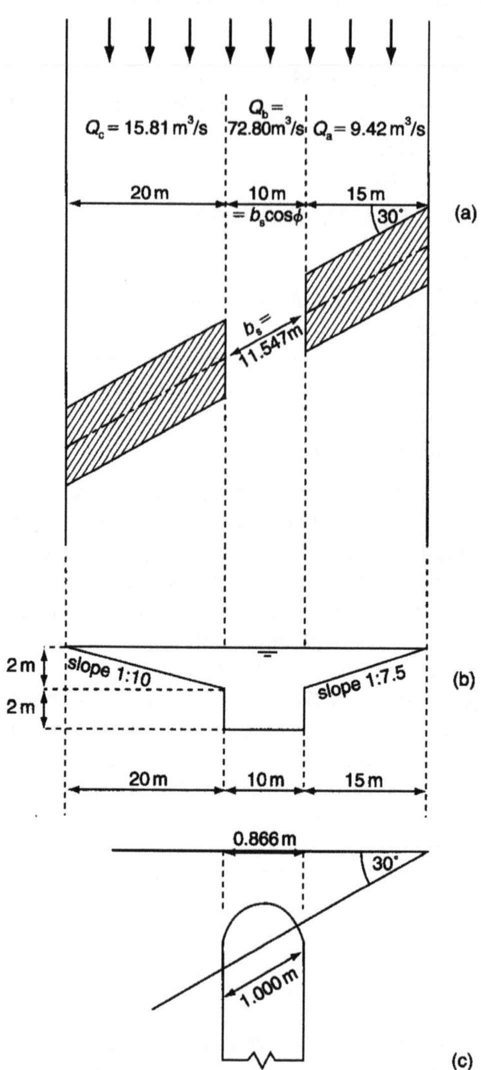

Fig. 4.40 Example 4.2: (a) plan of channel and crossing; (b) details of channel cross-section; (c) details of central pier.

6. From equation 3.1, the opening ratio $M = 72.800/98.030 = 0.743$.
7. From Example 3.2, a_1 is 1.51. From Fig. 4.20 it is estimated that a_2 = 1.38.
8. From the top curve of Fig. 4.21 (wingwall abutments, $b < 60\,\text{m}$) the base coefficient for a symmetrical crossing is $k_b^* = 0.52$.
9. The bridge is eccentric with $e = Q_a/Q_c = 9.420/15.810 = 0.60$. This is greater than 0.20 so no correction is needed and $\Delta k_e^* = 0$.

10. The correction for skew is obtained from Fig. 4.23a (abutments parallel to flow). If $M = 0.743$ and $\phi = 30°$ then $\Delta k^*_\phi = -0.07$.
11. In this part of the example there are no piers so $\Delta k^*_p = 0$.
12. From equation 4.24 the total backwater coefficient $k^* = k^*_b + \Delta k^*_e + \Delta k^*_\phi + \Delta k^*_p$.
    Thus $k^* = 0.52 + 0 + (-0.07) + 0 = 0.45$.
13. The afflux has to be found iteratively by first evaluating the first term of equation 4.23, namely:

$$H^*_1 = \frac{k^* a_2 V_{N2}^2}{2g} = \frac{0.45 \times 1.38 \times 2.451^2}{19.62} = 0.190\,\text{m}$$

Thus the first estimate is that $Y_1 = Y_N + H^*_1 = 4.000 + 0.190 = 4.190\,\text{m}$.

Using this depth, from Fig. 4.40b the cross-sectional areas of flow can be calculated as

$$A_{1a} = {}^1/_2 \times 16.425 \times 2.190 = 17.985\,\text{m}^2$$

$$A_{1b} = 10.000 \times 4.190 = 41.900\,\text{m}^2$$

$$A_{1c} = {}^1/_2 \times 21.900 \times 2.190 = 23.981\,\text{m}^2$$

Thus $A_1 = 17.985 + 41.900 + 23.981 = 83.866\,\text{m}^2$

From Example 3.2, $A_4 = 75.000\,\text{m}^2$. Using the complete equation 4.23:

$$H^*_1 = k^* \frac{a_2 V_{N2}^2}{2g} + a_1 \left[\left(\frac{A_{N2}}{A_4}\right)^2 - \left(\frac{A_{N2}}{A_1}\right)^2\right] \frac{V_{N2}^2}{2g} \tag{4.23}$$

$$= 0.190 + 1.51 \left[\left(\frac{40.000}{75.000}\right)^2 - \left(\frac{40.000}{83.866}\right)^2\right] \frac{2.451^2}{2 \times 9.81}$$

$$H^*_1 = 0.216\,\text{m}$$

14. $Y_1 = Y_N + H^*_1 = 4.000 + 0.216 = 4.216\,\text{m}$.

Repeating steps 13 and 14 starting with $Y_1 = 4.216\,\text{m}$ gives a new value for A_1, which can be substituted into equation 4.23 with the result that $H^*_1 = 0.219$ m. Another iteration produces the same result: hence $H^*_1 = 0.219\,\text{m}$ and $Y_1 = 4.219\,\text{m}$.

Note that Fig. 4.24 suggests that the skewed width of opening (b_S) required to give the same backwater as a normal opening with a 10 m span is $b_S = (0.973 \times 10/\cos 30°) = 11.235\,\text{m}$, slightly less than that adopted. Note also that the afflux calculated above using the USBPR method is broadly consistent with the value of 0.175 m obtained from the USGS method in the previous example.

(b) With the addition of a single round-nosed pier with a skewed width of 1 m, as in Fig. 4.40c, most of the calculations are the same as in part (a) up to step 11. The gross area of the opening at section 2 measured below normal depth is still $A_{N2} = 10.000 \times 4.000 = 40.000\,m^2$.

11. The width of the pier when projected onto section 1 perpendicular to the direction of flow is $1.000 \times \cos 30° = 0.866\,m$. Thus the area of the pier below normal depth is $A_P = 0.866 \times 4.000 = 3.464\,m^2$. From equation 4.26, $J = A_P/A_{N2} = 3.464/40.000 = 0.087$.
From Fig. 4.25a, with $M = 1.0$ a rectangular pier with a round nose has $\Delta k = 0.17$.
Correcting for $M = 0.743$ using Fig. 4.25b, a rectangular pier with a round nose has $\sigma = 0.92$.
Thus from equation 4.27, $\Delta k_P^* = \sigma \Delta k = 0.92 \times 0.17 = 0.16$.
12. From equation 4.24, the total backwater coefficient $k^* = k_b^* + \Delta k_e^* + \Delta k_\phi^* + \Delta k_P^*$.
Thus $k^* = 0.52 + 0 + (-0.07) + 0.16 = 0.61$.
13. The afflux has to be found iteratively by first evaluating the first term of equation 4.23, namely:

$$H_1^* = \frac{k^* \alpha_2 V_{N2}^2}{2g} = \frac{0.61 \times 1.38 \times 2.451^2}{19.62} = 0.258\,m$$

Thus the first estimate is that $Y_1 = Y_N + H_1^* = 4.000 + 0.258 = 4.258\,m$.
Recalculating the cross-sectional areas using the same method as above gives $A_{1a} = 19.120\,m^2$, $A_{1b} = 42.580\,m^2$, $A_{1c} = 25.493\,m^2$ and the total area as $A_1 = 87.193\,m^2$. As before, $A_4 = 75.000\,m^2$. Using the complete equation 4.23 gives

$$H_1^* = 0.258 + 1.51\left[\left(\frac{40.000}{75.000}\right)^2 - \left(\frac{40.000}{87.193}\right)^2\right]\frac{2.451^2}{2 \times 9.81}$$

$$H_1^* = 0.292\,m$$

14. $Y_1 = 4.000 + 0.292 = 4.292\,m$.

Repeating steps 13 and 14 starting with this depth gives $Y_1 = 4.295\,m$. Thus the effect of the pier is to increase the upstream depth by $4.295 - 4.219 = 0.076\,m$.

Example 4.3: the Biery and Delleur method

(a) Calculate the afflux at Canns Mill Bridge when the discharge is $15\,m^3/s$. Assume that the general channel cross-section is as shown in Fig. 3.13, the natural maximum flood stage at section 1 is $Y_{1A} = 1.96\,m$ (i.e. the abnormal stage without the afflux), and that $M = 0.69$. (b) If $Y_1 = 2.23\,m$ (i.e. the stage with the bridge afflux) when $Y_{1A} = 1.96\,m$, $M = 0.69$, $b = 4.28\,m$ and the radius of the arch is $r = 2.7\,m$, calculate the corresponding discharge.

(a) This method calculates the opening ratio (M) assuming normal depth flow exclusive of afflux. From above, when $Y_{1A} = 1.96\,m$ then $M = 0.69$.

At this stage the horizontal top width of the water on the right floodplain (Fig. 3.13) is 9.441 m.

The total top width of the water surface is $B_T = 10.000 + 5.500 + 9.441 = 24.941\,\text{m}$.

The cross-sectional areas of flow can be obtained from the simplified cross-section:

Left floodplain: $A_{1a} = (1.960 - 1.500) \times 10.000 = 4.600\,\text{m}^2$
Main channel: $A_{1b} = 1.960 \times 5.500 = 10.780\,\text{m}^2$
Right floodplain: $A_{1c} = {}^1/_2 \times (1.960 - 1.500) \times 9.441 = 2.171\,\text{m}^2$
Total cross-sectional area of flow $A_{1A} = 4.600 + 10.780 + 2.171 = 17.551\,\text{m}^2$.

Thus $V_{1A} = Q/A_{1A} = 15.000/17.551 = 0.855\,\text{m/s}$.

Usually the normal depth Froude number ($F_N = F_4$) is calculated, but because of the abnormal stage F_{1A} is more appropriate. The mean depth of flow $Y_{M1A} = A_{1A}/B_T = 17.551/24.941 = 0.704\,\text{m}$.

$$F_{1A} = \frac{V_{1A}}{(gY_{M1A})^{1/2}} = \frac{0.855}{(9.81 \times 0.704)^{1/2}} = 0.325$$

There is no consideration of skew or eccentricity.

From Fig. 4.32a with $M = 0.690$ and $F_{1A} = 0.325$ the value of $[Y_1/Y_N] = 1.144$ so for this example involving an abnormal stage assume that $Y_1 = 1.144Y_{1A} = 1.144 \times 1.960 = 2.242\,\text{m}$.

Thus the afflux $H_1^* = 2.242 - 1.960 = 0.282\,\text{m}$.

(The actual recorded value at Canns Mill was 0.270 m.)

(b) The discharge can be calculated from equation 4.45. Although $Y_1 = 2.23$ m (including the afflux), the value of M is calculated from the stage without the afflux which is $Y_{1A} = 1.96\,\text{m}$ when $M = 0.690$ as in part (a). If $F_{1A} = 0.325$ as above, then from Fig. 4.35 this gives $C_D = 0.42$, so

$$Q = 0.7083\, C_D\, (2g)^{1/2}\, Y_1^{3/2}\, b \left[1 - 0.1294\left(\frac{Y_1}{r}\right)^2 - 0.0177\left(\frac{Y_1}{r}\right)^4\right] \qquad (3.45)$$

$$= 0.7083 \times 0.42 \times (19.62)^{1/2} \times 2.23^{3/2} \times 4.28 \left[1 - 0.1294\left(\frac{2.23}{2.70}\right)^2 - 0.0177\left(\frac{2.23}{2.70}\right)^4\right]$$

$$= 17.04\,\text{m}^3/\text{s}$$

The true discharge is, of course, $15.00\,\text{m}^3/\text{s}$ as in part (a). With the abnormal stages at Canns Mill it was found that the accuracy of this method was improved by using Y_{1A} (= 1.96 m) in equation 4.45 instead of $Y_1 = 2.23\,\text{m}$, thus:

$$Q = 0.7083 \times 0.42 \times (19.62)^{1/2} \times 1.96^{3/2} \times 4.28 \left[1 - 0.1294\left(\frac{1.96}{2.70}\right)^2\right.$$

$$- \; 0.0177 \left(\frac{1.96}{2.70}\right)^4\Bigg] = 14.41\,\text{m}^3/\text{s}$$

It is these figures that appear in Table 4.5, which may indicate a higher level of accuracy than is justified. Additionally there is no allowance for skew or eccentricity with this method.

Example 4.4: the HR method

At Canns Mill the downstream depth Y_4 = 2.00 m at the flood discharge of 15.00 m^3/s. Assuming the area of the waterway opening (A_3) is 6.30 m^2 and that the main channel in which the bridge is situated is 5.50 m wide as in Fig. 3.13, calculate the afflux.

This example uses the downstream blockage ratio (J_4) because this is simpler and avoids the need for iteration (the upstream ratio includes the unknown afflux).

$A_4 = 5.50 \times 2.00 = 11.00\,\text{m}^2$
$J_4 = (A_4 - A_3)/A_4 = (11.00 - 6.30)/11.00 = 0.427$
$V_4 = Q/A_4 = 15.00/11.00 = 1.363\,\text{m/s}$
$F_4 = V_4/(gY_4)^{1/2} = 1.363/(9.81 \times 2.00)^{1/2} = 0.308$
If $F_4 = 0.308$ and $J_4 = 0.427$ then Fig. 4.36 gives $H_1^*/Y_4 = 0.13$
Thus $H_1^* = 0.13 \times Y_4 = 0.13 \times 2.00 = 0.260\,\text{m}$

Note that there was considerable overbank flow at section 4 during the flood, but this has been ignored when calculating the blockage ratio. Exactly how to calculate the value of J is one of the principal difficulties in applying this method.

References

Barrett, J.W.H. and Skogerboe, G.V. (1973) Computing backwater at open channel constrictions. *Proceedings of the American Society of Civil Engineers*, **99**(Hy7), July, 1043–1056.

Biery, P.F. and Delleur, J.W. (1962) Hydraulics of single span arch bridge constrictions. *Proceedings of the American Society of Civil Engineers, Journal of the Hydraulics Division*, 88(HY2), March, 75–108.

Bradley, J.N. (1978) *Hydraulics of Bridge Waterways*, 2nd edn, US Department of Transportation/Federal Highways Administration, Washington DC.

Brown, P.M. (1985) Afflux at British bridges. Interim Report, no. SR60, Hydraulics Research, Wallingford, England.

Brown, P.M. (1989) Afflux at arch bridges. Report No. SR 182, Hydraulics Research, Wallingford, England.

Chow, V.T. (1981) *Open-Channel Hydraulics*, International Student Edition, McGraw-Hill, Tokyo.

Fiuzat, A.A. and Skogerboe, G.V. (1983) Comparison of open channel constriction ratings. *American Society of Civil Engineers, Journal of Hydraulic Engineering*, **109**(12), December, 1589–1602.

French, R.H. (1986) *Open-Channel Hydraulics*, International Student Edition, McGraw-Hill, Singapore.

Hamill, L. (1993) A guide to the hydraulic analysis of single-span arch bridges. *Proceedings of the Institution of Civil Engineers*, **98**, March, 1–11.

Husain, S.T. and Rao, G.M. (1966) Hydraulics of river flow under arch bridges. *Irrigation and Power*, October, 441–454.

Hydrologic Engineering Center (HEC) (1990) *HEC-2 Water Surface Profile User's Manual*, US Army Corp of Engineers, Davis, CA.

Kaatz, K.J. and James, W.P. (1997) Analysis of alternatives for computing backwater at bridges. *American Society of Civil Engineers, Journal of Hydraulic Engineering*, **123**(9), September, 784–792.

Kindsvater, C.E. and Carter, R.W. (1955) Tranquil flow through open-channel constrictions. *Transactions of the American Society of Civil Engineers*, **120**, 955–992.

Kindsvater, C.E., Carter, R.W. and Tracy H.J. (1953) *Computation of peak discharge at contractions*. Circular 284, United States Geological Survey, Washington DC.

Laursen, E.M. (1970) Bridge backwater in wide valleys. *Proceedings of the American Society of Civil Engineers*, **96**(Hy4), April, 1019–1038.

Liu, H.K., Bradley, J.N. and Plate, E.J. (1957) Backwater effects of piers and abutments. Report no. CER57HKL10, Colorado State University.

Matthai, H.F. (1967) Measurement of peak discharge at width contractions by indirect methods. *Techniques of Water Resource Investigations of the United States Geological Survey*, Chapter A4, Book 3, *Applications of Hydraulics*, US Government Printing Office, Washington DC.

Ranga Raju, K.G., Asawa, G.L., Rana, O.P.S. and Pillai, A.S.N. (1983) Rational assessment of blockage effect in channel flow past smooth circular cylinders. *Journal of Hydraulics Research*, **21**(4), 289–302.

Skogerboe, G.V., Barrett J.W.H., Walker, W.R. and Austin, L.H. (1973) Comparison of bridge backwater relations. *Proceedings of the American Society of Civil Engineers, Journal of the Hydraulics Division*, **99**(Hy6), June, 921-938.

Tracy, H.J. and Carter, R.W. (1954) Backwater effects of open channel constrictions. *Transactions of the American Society of Civil Engineers*, separate 415.

Tracy, H.J. and Carter, R.W. (1955) Backwater effects of open channel constrictions. *Transactions of the American Society of Civil Engineers*, **120**, 993–1018.

5 How to analyse flow past piers and trestles

5.1 Introduction

Unlike the previous chapter, which dealt with channel constrictions where the abutments formed the primary obstacle to flow and the piers were the secondary consideration, this chapter deals with situations where the piers themselves cause the primary obstruction, while the secondary effect of the abutments is negligible. This may happen either with short bridges where the abutments are not in the flow (just the piers) or when the the bridge is long with a large number of piers in the watercourse. Although the latter may be uncommon in Britain, where rivers and floodplains are relatively small, in some countries the floodplains of large rivers can be several miles across and bridges may have well over 100 spans. Under these conditions it is easy to see how the piers, rather than the abutments, become the primary factor in the hydraulic analysis of the bridge.

The obstruction of bridge piers to the flow of water has been studied for over 150 years. One of the earliest investigations was by d'Aubuisson (1840), followed by Nagler (1918) and Yarnell (1934a). A typical approach is to use a long, laboratory channel in which the flow is initially uniform. By introducing one or more equally spaced piers to the central part of the channel their effect on the flow can be measured. If the uniform flow is supercritical the result will be to split the flow, causing a disturbed wake downstream and spray upstream. The amount of spray depends upon the shape of the piers. There is no other effect upstream, since a characteristic of supercritical flow is that disturbances do not propagate upstream. However, the more usual scenario is that the flow will be initially subcritical. As described in Section 2.2, a backwater occurs upstream, increasing the depth at section 1, while the depth between the piers decreases to a minimum at section 3 as the flow accelerates (Fig. 5.1). At section 4, well downstream of the bridge, the flow is again at normal depth. Between sections 3 and 4 there is a region of turbulence due to the wake created by the piers and the expanding flow.

It is as a result of Yarnell's work that there are fairly reliable coefficients to use with the d'Aubuisson and Nagler equations, and that there is some understanding of which equation is the most suitable under various circumstances. Consequently it is advisable to read all of Sections 5.2–5.4 rather

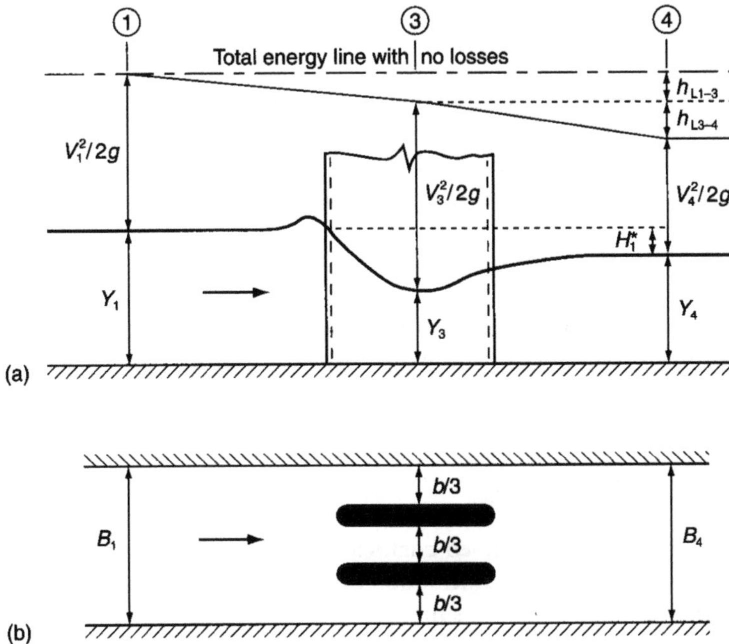

Fig. 5.1 Flow past bridge piers. (a) Longitudinal section, with the velocity heads exaggerated for clarity. The afflux is H_1^*. (b) Plan view, showing the total opening width, b.

than arbitrarily opt for one of the formulae. Some of the equations have been included in computer software (e.g. HEC, 1990).

5.2 The d'Aubuisson equation

The d'Aubuisson equation is obtained by assuming a horizontal bed, and then applying the Bernoulli equation to sections 1 and 3. With reference to Fig. 5.1:

$$Y_1 + \frac{V_1^2}{2g} = Y_3 + \frac{V_3^2}{2g} + h_{L1-3} \tag{5.1}$$

where h_{L1-3} is the head loss between sections 1 and 3 caused by the contraction. The d'Aubuisson method assumes that the relatively high velocity head at section 3 is not converted back into potential energy (i.e. an increased depth of flow) at section 4. Thus $Y_3 = Y_4$, which means that the afflux (H_1^*) in Fig. 5.1 is

$$H_1^* = Y_1 - Y_3 \tag{5.2}$$

so rearranging equation 5.1 gives

$$H_1^* = \frac{V_3^2}{2g} - \frac{V_1^2}{2g} + h_{L1-3} \qquad (5.3)$$

If it is assumed that the head loss during the contraction (h_{L1-3}) is negligibly small, then it is apparent from equation 5.3 that the afflux can be estimated from the difference in the velocity heads at sections 1 and 3.

In order to introduce the discharge, equation 5.3 can be rewritten as

$$V_3 = \left[2g\left(H_1^* + \frac{V_1^2}{2g} - h_{L1-3}\right)\right]^{1/2} \qquad (5.4)$$

If b is the total width of the waterway openings (Fig. 5.1b), then with the assumption above that $Y_3 = Y_4$ the discharge (Q) can be expressed as

$$Q = bY_4 V_3 \qquad (5.5)$$

$$= bY_4\left[2g\left(H_1^* + \frac{V_1^2}{2g} - h_{L1-3}\right)\right]^{1/2} \qquad (5.6)$$

If an experimentally determined coefficient (K_A) is introduced to allow for the head loss caused by the contraction of the flow between the piers (h_{L1-3}) and any error arising from the simplifying assumptions, then

$$Q = K_A\, bY_4 (2gH_1^* + V_1^2)^{1/2} \qquad (5.7)$$

Frequently the value of K_A is obtained from Yarnell's work and typically has a value of between 0.90 and 1.05 (Table 5.1). Thus for a given discharge the normal depth (Y_4) can be calculated and then H_1^* obtained from a trial and error solution of equation 5.7. A trial and error approach is needed because V_1 depends upon H_1^*. Example 5.1 illustrates the use of the d'Aubuisson equation.

The d'Aubuisson equation is said to model completely turbulent flows quite well, but for flows of low to moderate turbulence the Nagler equation is better, although neither is very good at high velocities. The d'Aubuisson equation's assumption that $Y_3 = Y_4$ has to be questionable under some conditions, such as when the piers form a large obstacle to flow and the water depth between the piers (Y_3) falls considerably below the normal depth (Y_4).

5.3 The Nagler equation

Nagler (1918) presented the results of 256 experiments that were conducted on 34 different models of bridge piers, each of which obstructed 23.4% of the total channel area. The objective was to determine the relative obstruction to the flow of water afforded by different pier shapes. He derived his own equation for the discharge (Q) between the piers. This derivation is rarely presented in full, perhaps because the original paper is not always clear in its meaning. However, the basis is simply to consider the flow

Table 5.1 Values of K_A and K_N for bridge piers. For use with equation 5.7 (K_A) or 5.10 (K_N)

Type of pier	Bridge opening ratio, M									
	0.90		0.80		0.70		0.60		0.50	
	K_A	K_N	K_A	K_N	K_A	K_N	K_A	K_N	K_A	K_N
Square nose and tail (Length; Width)	0.96	0.91	1.02	0.87	1.02	0.86	1.00	0.87	0.97	0.89
Semicircular nose and tail	0.99	0.94	1.13	0.92	1.20	0.95	1.26	1.03	1.31	1.11
90° triangular nose and tail	–	0.95	–	0.94	–	0.92				
Twin cylinder pier without diaphragm	–	0.91	–	0.89	–	0.88				
Twin cylinder pier with diaphragm	–	0.91	–	0.89	–	0.88				
Lens-shaped nose and tail	1.00	0.95	1.14	0.94	1.22	0.97				
Angle of attack with respect to approach flow: ϕ degrees										

After Yarnell, 1934a

between the piers (Fig. 5.2) as that through a small orifice operating under a differential head, h. Thus:

$$Q = K_N A_3 \sqrt{2gh} \tag{5.8}$$

where K_N is a coefficient of discharge (Table 5.1) and A_3 is the total cross-sectional area of flow between the piers (equivalent to the area of the orifice). If b is the total opening width between the piers (Fig. 5.1b) and Y_3 is the minimum depth at section 3 then $A_3 = bY_3$, so:

$$Q = K_N bY_3 \sqrt{2gh} \tag{5.9}$$

Nagler chose to present the equation with $Y_3 = (Y_4 - \theta V_4^2/2g)$ where θ is an adjustment factor, and with $h = H_1^* + \beta V_1^2/2g$ where β is another adjustment factor that corrects the velocity of approach. Thus the Nagler equation is

$$Q = K_N b \sqrt{2g} \left(Y_4 - \theta \frac{V_4^2}{2g}\right)\left(H_1^* + \beta \frac{V_1^2}{2g}\right)^{1/2} \tag{5.10}$$

The values of the adjustment factors θ and β were evaluated experimentally. Normally θ can be taken as 0.3, although (logically) it has a value of zero when the piers have little or no effect on the flow and $Y_3 = Y_4$, but is larger

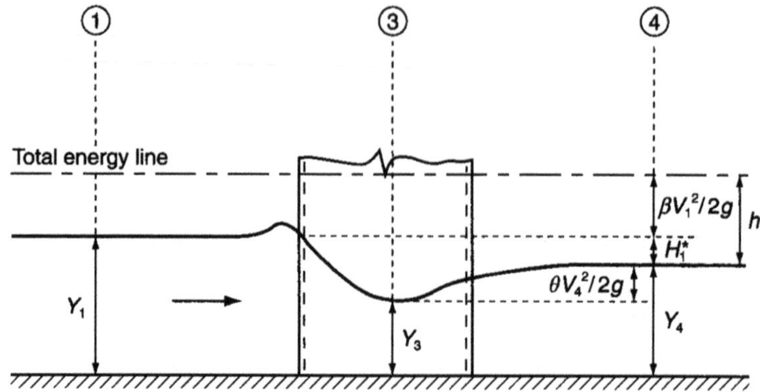

Fig. 5.2 Longitudinal section of flow past bridge piers for the derivation of the Nagler equation.

when the flow is shooting and turbulent. The value of β varies with the opening ratio (M), as shown in Fig. 5.3. Example 5.1 provides an illustration of the application of the equation.

The Nagler equation is rather more complicated than d'Aubuisson's but, because there is some allowance for the recovery of head downstream of section 3 and it is not assumed that $Y_3 = Y_4$, it is more accurate under conditions of low turbulence. Like the d'Aubuisson equation it is not very accurate at high velocities.

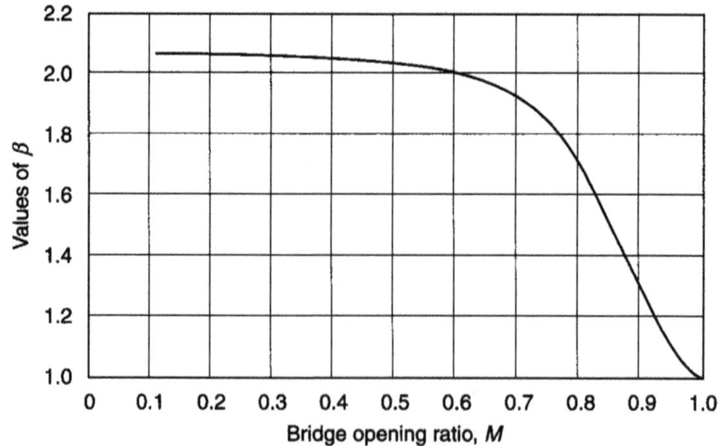

Fig. 5.3 Values of β in the Nagler equation for flow past piers. (After Nagler, 1918; Obstructions of bridge piers to the flow of water, *Transactions of the ASCE*. Reproduced by permission of ASCE)

5.4 The work of Yarnell

5.4.1 Flow between bridge piers

Between 1927 and 1931 Yarnell conducted about 2600 experiments on the obstructive effects of bridge piers to the flow of water. He used larger piers and a more extensive range of conditions than previous investigators. The opening ratios adopted were between 0.88 and 0.50, quite severe by modern standards, but the results can still be applied to more slender piers. Yarnell published his results in 1934, and listed the following as affecting the amount of obstruction caused by a bridge pier, and hence the resulting afflux:

- the shape of the pier nose;
- the shape of the pier tail;
- the channel contraction caused by the pier;
- the length of the pier;
- the angle between the longitudinal centreline of the pier and the approaching stream of water (the angle of attack, ϕ, in Table 5.1);
- the quantity of flow.

Yarnell's aim was to evaluate these factors by investigating the flow past piers of different shape and length to width ratio. He also determined the values of the coefficients (K) in the d'Aubuisson and Nagler equations (as shown in Table 5.1) and investigated the accuracy of the equations. Yarnell made the following observations.

- The d'Aubuisson and Nagler equations are not accurate at high velocities, but for ordinary velocities the coefficients in Table 5.1 are generally satisfactory.
- For low-velocity flows and little turbulence, piers with lens-shaped noses and tails, semi-circular noses and tails, or some combination of these, are the more efficient. Twin-cylinder piers (with or without connecting diaphragms), piers with 90° noses, and piers with recessed webs are less efficient, while piers with square noses and tails are the least hydraulically efficient. A lens-shaped nose or tail is formed from two convex curves that are tangent to the sides of the pier and which have a radius twice the width of the pier.
- The optimum pier length-to-width ratio is probably between 4 and 7 depending upon velocity. Typically increasing the ratio from 4 to 13 will increase the values in Table 5.1 by 3–5%, but this is not guaranteed.
- There is little difference between similar piers that are parallel to the approaching flow and at an angle of less than 10°, but if the angle of attack (ϕ) is 20° then typically the values in Table 5.1 will decrease by 7%. Over 20° the afflux increases significantly, the actual increase depending upon the flow rate and depth, and the severity of the obstruction.

Two basic flow types were identified: class A or subcritical flow, and class B or supercritical flow. Yarnell used equation 3.13 to determine the limiting condition at which the flow becomes choked and supercritical flow may commence. With subcritical conditions the afflux $H_1^* = Y_1 - Y_4$ and can be calculated from

$$H_1^* = K\, Y_4\, F_4^2\, (K + 5F_4^2 - 0.6)(a + 15a^4) \qquad (5.11)$$

where K is Yarnell's pier shape coefficient (Table 5.2), $Y_4 = Y_N$ is the normal depth at section 4, $F_4 = F_N$ is the normal depth Froude number, and a is the channel contraction ratio = $(1 - b/B)$ or $(1 - M)$. Here a can also be defined as the ratio of the obstructed area to the total channel area. Equation 5.11 is conveniently based on normal depth conditions, which the designer can calculate easily using conventional open channel hydraulics (see Example 5.2). It may also be written as

$$H_1^* = 2K\,(K + 10\omega - 0.6)(a + 15a^4)V_4^2/2g \qquad (5.12)$$

where ω is $V_4^2/2gY_4$ so $F_4^2 = 2\omega$. This equation is used in some computer software (e.g. Hydrologic Engineering Center, 1990)

With supercritical flow there will be a hydraulic jump downstream. The afflux can be calculated by using the critical flow section (say section 3) as a control point and working back upstream to section 1 allowing for the energy loss $(E_1 - E_3)$, which can be expressed as

$$E_1 - E_3 = \frac{K_P V_3^2}{2g} \qquad (5.13)$$

where the coefficient, K_P, has a value of about 0.35 for piers with square ends and 0.18 for rounded ends (Henderson, 1966).

Equation 5.11 is easy to solve with a calculator, but Yarnell summarised his research in chart form, which has the merit of allowing the various relationships to be easily visualised. When the flow between the piers is subcritical Fig. 5.4 should be used, but when it is supercritical Fig. 5.5 should be employed. Both charts assume that the piers have a length-to-width ratio of 4.

Table 5.2 Values of Yarnell's pier coefficients. For use with equations 5.11 and 5.12

Pier shape	*K*
Semicircular nose and tail	0.90
Lens-shaped nose and tail	0.90
Twin-cylinder piers with connecting diaphragm	0.95
Twin-cylinder piers without connecting diaphragm	1.05
90° triangular nose and tail	1.05
Square nose and tail	1.25

After Yarnell. 1934a

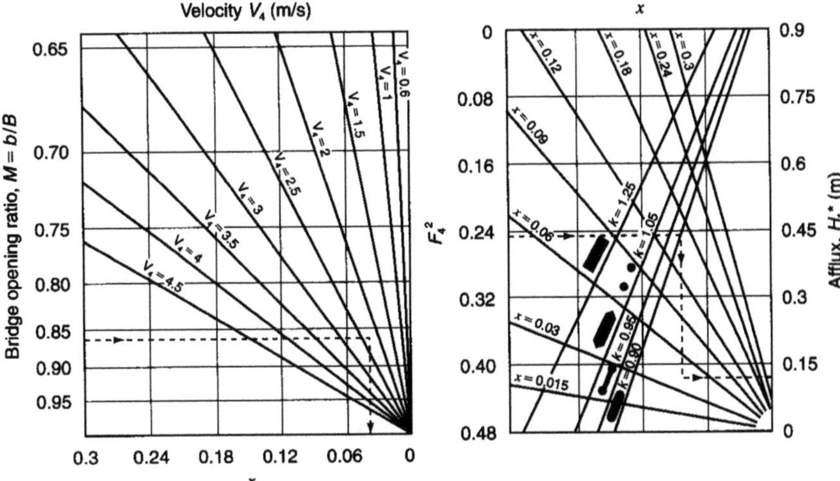

Fig. 5.4 Charts for determining afflux, H_1^* (after Yarnell, 1934a). For example, assume $M = 0.865$, $V_4 = 2.41$ m/s, $F_4^2 = 0.243$ and piers with semicircular noses and tails. Enter the left-hand chart at $M = 0.865$, move horizontally to $V_4 = 2.41$ m/s, then vertically down to read $x = 0.04$ m. Enter the right-hand chart with $F_4^2 = 0.243$, move horizontally across to the sloping $K = 0.90$ semicircular nose and tail line, then down to $x = 0.04$, and then horizontally across to obtain $H_1^* = 0.113$ m.

Figure 5.4 for subcritical flow was originally presented by Yarnell (1934a) in English units, but has been redrawn for this book in metric units. For simplicity, the metrication of the diagram assumes that 1 foot = 0.30 m. Several of the terms are dimensionless and do not need conversion; the variable x was defined by Yarnell as $x = (a + 15a^4)V_4^2/2g$ as in equation 5.12. Thus x has the unit of length and has also been converted. The diagram is used as follows.

1. Calculate M and V_4.
2. Enter the left-hand chart at the value of M, then move horizontally across to the sloping line representing the calculated value of V_4. Move vertically down the chart to obtain the value of x.
3. Calculate the value of F_4^2 (remember to square the Froude number – it is easy to forget!).
4. Enter the right-hand chart at the value of F_4^2 and move across horizontally to the sloping line representing the pier shape. Then move vertically down (or up, as the circumstances dictate) to the appropriate value of x obtained in step 2.
5. Move horizontally across to obtain the afflux H_1^* from the right-hand scale.

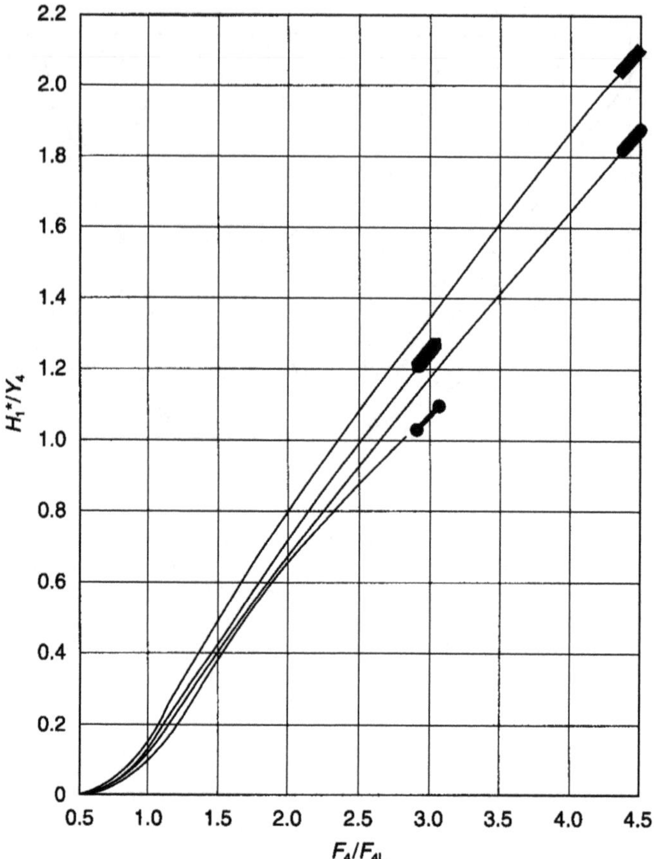

Fig. 5.5 Chart for determining the afflux caused by supercritical flow past bridge piers (after Yarnell, 1934a). As explained in the text, F_{4L} is the normal depth Froude number at the limiting contraction when the flow in the opening is at critical depth ($F_3 = 1$).

An illustration of this procedure is given in Fig. 5.4 and Example 5.2. Note that if the afflux is already known and the discharge ($Q = A_4V_4$) has to be obtained using Fig. 5.4 it will be necessary to guess a value of V_4 and then keep repeating the steps listed above until the value of H_1^* obtained from the chart equals that which was observed. When this happens V_4 is correct and Q can be calculated. Naturally the generalised charts cannot be very accurate when applied to specific situations.

Figure 5.5 was also presented by Yarnell and redrawn, but has not needed to be metricated. To use it, knowing Y_4, start by calculating the actual value of F_4 and M. Then with the actual value of M as M_L, the limiting downstream condition (F_{4L}) at which the flow between the piers will be at critical depth can be obtained from equation 3.14 or Fig 3.5. By entering the bottom scale of Fig. 5.5 at the appropriate value of F_4/F_{4L}, moving vertically up to the line representing the pier shape, and then horizontally across to the left scale, the value of $[H_1^*/Y_4]$ can be obtained. Thus the afflux $H_1^* = Y_4[H_1^*/Y_4]$.

5.4.2 *Flow through pile trestles*

In addition to the experiments on bridge piers, during 1929 and 1930 Yarnell conducted 1082 experiments on the flow of water past clean pile trestles: that is, trestles that are free of any debris. Railway bridges often utilised this form of construction (hence the reference in Table 5.3 to single- and double-track crossings), but also road bridges. Tests were conducted on models and, unusually, full-size trestles with the objective of evaluating the coefficients. A typical pile-trestle bent is shown in Fig. 5.6. The experiments included various angles of attack up to 30°. The result of this work were published in 1934.

In the calculations, the amount of channel contraction was taken as the average diameter of the piles plus the thickness of the sway bracing. When the bent was at an angle to the current, the contraction was calculated as if the bent was parallel to the flow, the effect of the angle of attack being incorporated into the coefficient, K. However, Yarnell concluded that since all bridges made of pile trestles produce practically the same relatively

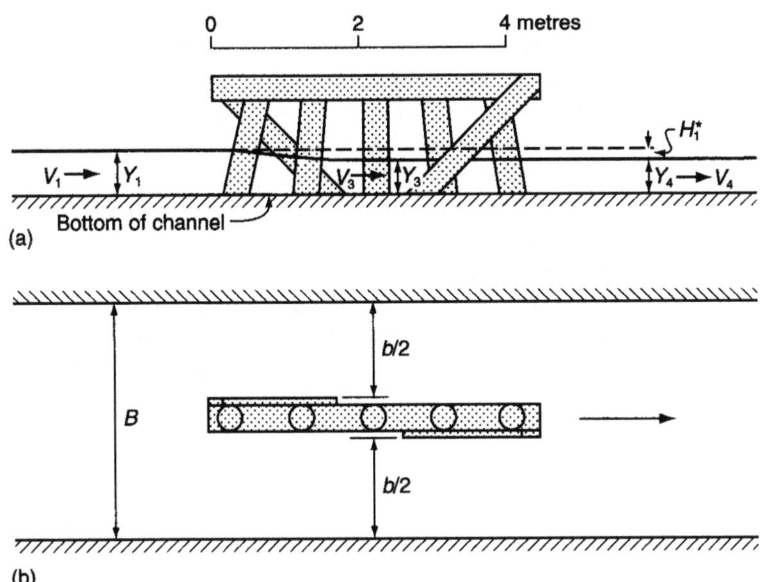

Fig. 5.6 Flow past a pile trestle bent (after Yarnell, 1934b): (a) longitudinal profile; (b) plan.

Table 5.3 Values of K_A and K_N for pile trestles. For use with equation 5.7 or 5.10

Arrangement of trestle	K_A	K_N
Bents in line with the current		
Single track 5 pile trestle bent	0.99	0.90
Double track 10 pile trestle bent	0.87	0.82
Two single track 5 pile bents offset	0.85	0.79
Bents at an angle with the current		
Single track 5 pile trestle bent at:		
10° angle	0.99	0.90
20° angle	0.96	0.89
30° angle	0.92	0.87

After Yarnell, 1934b

small amount of channel contraction, the question as to whether the pile trestle coefficient is the same for different bridge opening ratios does not arise. Consequently the coefficients presented in Table 5.3 can be used for subcritical flow (but not supercritical) without correction for the opening ratio.

Other conclusions reached by Yarnell were as follows.

- In general, the d'Aubuisson and Nagler equations could be applied 'very favourably' to most of the conditions encountered in the experiments concerning trestles.
- Until the angle of attack exceeds 10° there is little decrease in the coefficient, and hence in the discharge.
- With an angle of attack of 30° the discharge coefficient for trestle bents is about 4% less than for bents aligned with the current, while the backwater will be increased by 50–70% for a single-track crossing.
- Some beneficial effect can be obtained by setting trestle bents in echelon if a roadway crosses the river at an angle.
- A double-track trestle with the bents offset offers a somewhat greater obstruction than a double-track trestle with the piles in line.
- All of the above may be unreliable if the velocity and depth of flow met in practice are much greater than used in the model tests.

5.5 Examples

Example 5.1

A rectangular river channel 107 m wide is completely spanned by a bridge that has six equally spaced piers (but no abutments) in the watercourse. Each of the piers is 2.40 m wide and has a semicircular nose and tail.

During a flood it was observed that the afflux (H_1^*) was 0.09 m when averaged across the six piers. The average normal depth of flow downstream of the crossing ($Y_4 = Y_N$) was 2.44 m. Estimate the discharge using the d'Aubuisson and Nagler equations.

$B = 107\,\text{m}$
$b = 107 - (6 \times 2.40) = 92.6\,\text{m}$
$M = b/B = 92.6/107 = 0.865$
$H_1^* = 0.09\,\text{m}$
$Y_4 = 2.44\,\text{m}$
$Y_1 = 2.44 + 0.09 = 2.53\,\text{m}$

Note that the d'Aubuisson and Nagler equations must be solved by trial and error because the velocities V_1 and V_4 are initially unknown when the discharge is unknown. However, to shorten this example, assume that $V_1 = 2.34\,\text{m/s}$ and $V_4 = 2.41\,\text{m/s}$.

d'Aubuisson equation

$$Q = K_A\, b\, Y_4\, (2gH_1^* + V_1^2)^{1/2} \tag{5.7}$$

By interpolation, from Table 5.1 for a pier with a semicircular nose and tail $K_A = 1.04$.

$$Q = 1.04 \times 92.6 \times 2.44\,(2 \times 9.81 \times 0.09 + 2.34^2)^{1/2}$$
$$= 632.3\,\text{m}^3/\text{s}.$$

Check: $V_1 = Q/A_1 = 632.3/(2.53 \times 107) = 2.34\,\text{m/s}$ as assumed.
$V_4 = Q/A_4 = 632.3/(2.44 \times 107) = 2.42\,\text{m/s}$ – acceptable.

Nagler equation

$$Q = K_N b \sqrt{2g}\left(Y_4 - \theta\,\frac{V_4^2}{2g}\right)\left(H_1^* + \beta\frac{V_1^2}{2g}\right)^{1/2} \tag{5.10}$$

From Table 5.1 $K_N = 0.93$, assume $\theta = 0.3$ and from Fig 5.3 $\beta = 1.45$ so:

$$Q = 0.93 \times 92.6 \times \sqrt{2 \times 9.81}\left(2.44 - \frac{0.3 \times 2.41^2}{2 \times 9.81}\right)$$
$$\left(0.09 + \frac{1.45 \times 2.34^2}{2 \times 9.81}\right)^{1/2}$$
$$= 630.5\,\text{m}^3/\text{s}$$

Check: $V_1 = Q/A_1 = 630.5/(2.53 \times 107) = 2.33\,\text{m/s}$ – acceptable.
$V_4 = Q/A_4 = 630.5/(2.44 \times 107) = 2.41\,\text{m/s}$ as assumed.

Example 5.2

Using the same data as in Example 5.1 and assuming that the discharge is 630.5 m³/s but the afflux (H_1^*) is unknown, determine its value when the piers have (a) semicircular noses and tails, and (b) square noses and tails.

From Example 5.1, $V_4 = 2.41$ m/s and $Y_4 = 2.44$ m.
Thus $F_4 = V_4/(gY_4)^{1/2} = 2.41/(9.81 \times 2.44)^{1/2} = 0.493$ and so $F_4^2 = 0.243$.
$M = 0.865$ so $\alpha = (1 - M) = 0.135$ (or alternatively, $\alpha = 6 \times 2.40/107 = 0.135$).

(a) From Table 5.2 piers with round noses and tails have $K = 0.90$. Assuming subcritical flow:

$$H_1^* = K\,Y_4\,F_4^2\,(K + 5F_4^2 - 0.6)(\alpha + 15\alpha^4) \tag{5.11}$$

$$= 0.90 \times 2.44 \times 0.243\,(0.90 + 5 \times 0.243 - 0.6) \times (0.135 + 15 \times 0.135^4)$$

$$= 0.113\,\text{m}$$

The same answer can be obtained from Fig. 5.4. The caption to the diagram relates to part (a) of this example. (Note that in Example 5.1 the initial value of H_1^* was 0.090 m – different equations yield different answers.)

(b) For piers with square noses and tails Table 5.2 shows $K = 1.25$, so equation 5.11 becomes

$$H_1^* = 1.25 \times 2.44 \times 0.243\,(1.25 + 5 \times 0.243 - 0.6) \times (0.135 + 15 \times 0.135^4)$$

$$= 0.193\,\text{m}$$

Similarly using the $K = 1.25$ line in Fig. 5.4 with $F_4^2 = 0.243$ and $x = 0.04$ gives $H_1^* = 0.19$ m. (Note the substantially increased afflux compared with the semicircular nose and tail.)

References

d'Aubuisson de Voisins, J.F. (1840) *Traité d'hydraulique* (Treatise on Hydraulics), 2nd edn, Piois, Levraut et Cie, Paris.

Henderson, F.M. (1966) *Open Channel Flow*, Macmillan, New York.

Hydrologic Engineering Center (HEC) (1990) *HEC-2 Water Surface Profile User's Manual*, US Army Corps of Engineers, Davis, CA.

Nagler, F.A. (1918) Obstruction of bridge piers to the flow of water. *Transactions of the American Society of Civil Engineers*, **82**, 334–395.

Yarnell, D.L. (1934a) Bridge piers as channel obstructions. Technical Bulletin no. 442, US Department of Agriculture, Washington DC.

Yarnell, D.L. (1934b) Pile trestles as channel obstructions. Technical Bulletin no. 429, US Department of Agriculture, Washington DC.

6 How to analyse flow over embankments

6.1 Introduction

McKay (1970) described a location in Australia where an embankment had to be constructed across an 8 km (5 mile) wide floodplain, which sometimes carried enormous flows. In such cases it is often impractical to build an embankment that is higher than the flood level and uneconomic to provide as many waterway openings as might be desired purely from the hydraulic point of view. The only practicable design solution may be to allow water to pass over the embankment (Fig. 6.1). Provided the embankments are designed not to be damaged by overtopping and the road is not so important that it must be kept open at all costs, this provides a pragmatic solution to a potentially difficult problem.

Another reason for allowing water to pass over the approach embankments may be to prevent damage to the bridge. For example, consider a bridge with approach highway embankments that cross a wide floodplain (Fig. 6.2). Option A below represents a 'conventional' design approach, whereas option B is less conventional since it allows flow over the embankments.

Option A is to construct a bridge with an economically sized opening that will pass a 1 in 50 year flood, and road embankments that are above the highest recorded flood (say the 1 in 200 year flood level). This is a reasonably typical design standard. If such a flood occurred (and equation 8.19 indicates there is a 39% chance that a 1 in 200 year flood will occur in a 100 year period, which could be the life of the bridge) then the opening would have to pass a much larger flow than it was designed for. This may cause serious damage to the bridge, or possibly its destruction (see Table 1.1).

Option B is to design the opening for the 1 in 50 year flood and to keep the approach embankments just above this level over most of their length (Fig. 6.2). If the bridge superstructure is higher than the approaches to avoid it becoming submerged when the flow passes over the embankments, the greatest threat to the bridge occurs near the design flood, not the peak discharge (Richardson *et al.*, 1993; Highways Agency, 1994). Anything larger than the design flood passes relatively harmlessly over the embankments, avoiding damage to the bridge.

Fig. 6.1 Floodwater passing from right to left over the bridge approach at Chilly Bridge on the River Exe, December 1960. Preventing this flow by raising the approach embankments would increase the chance of the bridge being damaged and exacerbate flooding upstream. The object at the bottom right is reputedly a garden shed, which illustrates the hazard of debris becoming trapped in an opening. (Reproduced by permission of the Environment Agency)

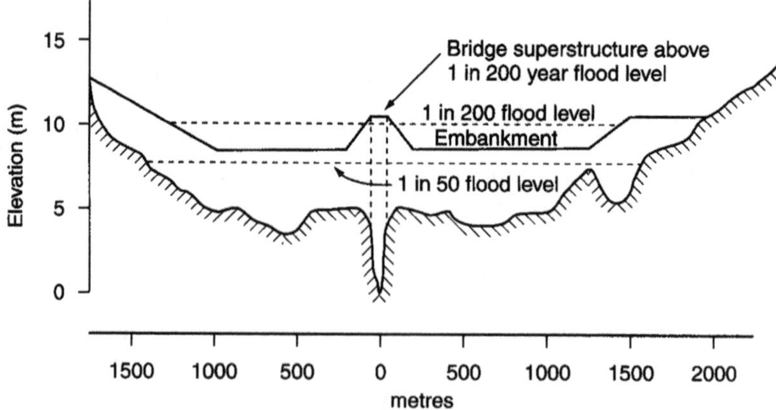

Fig. 6.2 Example of a bridge and highway approach embankments crossing a wide valley.

Option B has proved successful in many locations. Bradley (1978) quoted as an example the Nottoway River Bridge on State Route 40 in Virginia. The bridge itself was designed for a 1 in 100 year flood of around 280 m^3/s but actually coped with 740 m^3/s without incurring any damage, the excess flow passing over the embankments. After the flood, only minor repairs were required to the downstream shoulders of the embankments, a quick and cheap thing to accomplish compared with replacing a damaged or destroyed bridge. Of course, if it is known that overtopping of the embankment will happen, it can be designed accordingly (CIRIA, 1987).

If high road embankments are constructed then this will result in a greater obstacle to flow, a larger backwater and greater flooding than would occur if overtopping was allowed, as in Fig. 6.1. Another point to keep in mind is that if a road embankment is designed that retains more than a certain quantity of water on the upstream floodplain (25 000 m^3 in the UK), then the embankment is classed as a dam and becomes subject to dam safety legislation. This may impose higher construction standards initially, and then regular safety inspections throughout its life. Controlled overtopping of the embankment can avoid this problem as well.

As mentioned above, overtopping of the approach embankments can act as a 'safety valve' that prevents expensive damage to a bridge. This is illustrated by Fig. 6.3a. In the diagram, the curve ABC represents the flow through the bridge opening. This reaches a maximum at B, which is the level of the top of the embankments. From A to B all of the flow, including that on the upstream floodplains, must pass through the bridge opening. This large contraction results in a large afflux, which is needed to overcome the energy loss as the flow is funnelled through the bridge and then expands back onto the downstream floodplains. When the embankment is overtopped at B, there is no need for all of the flow to funnel through the bridge; the resistance to flow decreases and the afflux reduces

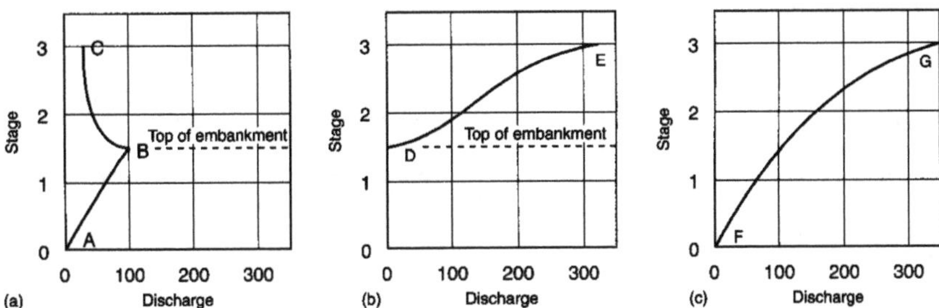

Fig. 6.3 Diagrammatic illustration of the stage–discharge curve when embankments are overtopped and form a 'safety valve'. (a) Flow through the waterway opening. Note the maximum discharge occurs at B, just before the flow spills over the embankments. (b) Flow over the embankments. (c) The combined stage–discharge curve equivalent to ABC plus DE.

as the difference in water level across the embankments diminishes. Consequently, the discharge through the bridge opening falls (BC).

Figure 6.3b illustrates the increase in discharge over the embankments with stage. Obviously, the flow is zero at D, the stage equivalent to the level of the embankments, and then rises as the head over the embankment crest increases. Figure 6.3c illustrates the combined stage–discharge relationship for the bridge opening and embankments, the curve FG being the sum of the curves ABC and DE.

The shape of the curve ABC illustrates clearly that the largest flow through the waterway opening is that just prior to the embankments being overtopped, and it is at this time that there may be the greatest risk of damage to the bridge. Subsequently the bridge passes less of the total discharge, so flow velocities and the potential for scour reduce.

6.2 Road embankments as weirs

If the embankments are either deliberately designed as a 'safety valve' or are overtopped anyway, some means of assessing their hydraulic performance is needed. Often this is built into computer software, such as HEC-2 (Hydrologic Engineering Center, 1990). Much of this is based on the work of Kindsvater (1964), who reported on the results of 936 experiments involving the discharge characteristics of 17 different model embankments, and 106 boundary layer velocity traverses on four different models.

6.2.1 *Types of flow*

Kindsvater identified two main types of flow when an embankment is overtopped: free flow and submerged flow (Fig. 6.4 and Table 6.1). In the

Table 6.1 Principal characteristics of free and submerged flow over embankments

Free flow (plunging and surface)	*Transitional flow*	**Submerged flow** (Fig. 6.4d)
Low tailwater Critical depth on the crest Discharge depends upon upstream head		High tailwater The depth on the roadway exceeds the critical depth Discharge depends on the capacity of both the tailwater channel and the upstream head Submerged flow is always on the surface and never plunges
Plunging flow (Fig. 6.4a) The jet plunges under the tailwater surface producing a submerged hydraulic jump on the downstream slope More erosive than surface flow or submerged flow		
Transitional flow		
Surface flow (Fig. 6.4b) The jet separates from the roadway surface at the downstream shoulder and 'rides' over the tailwater surface		

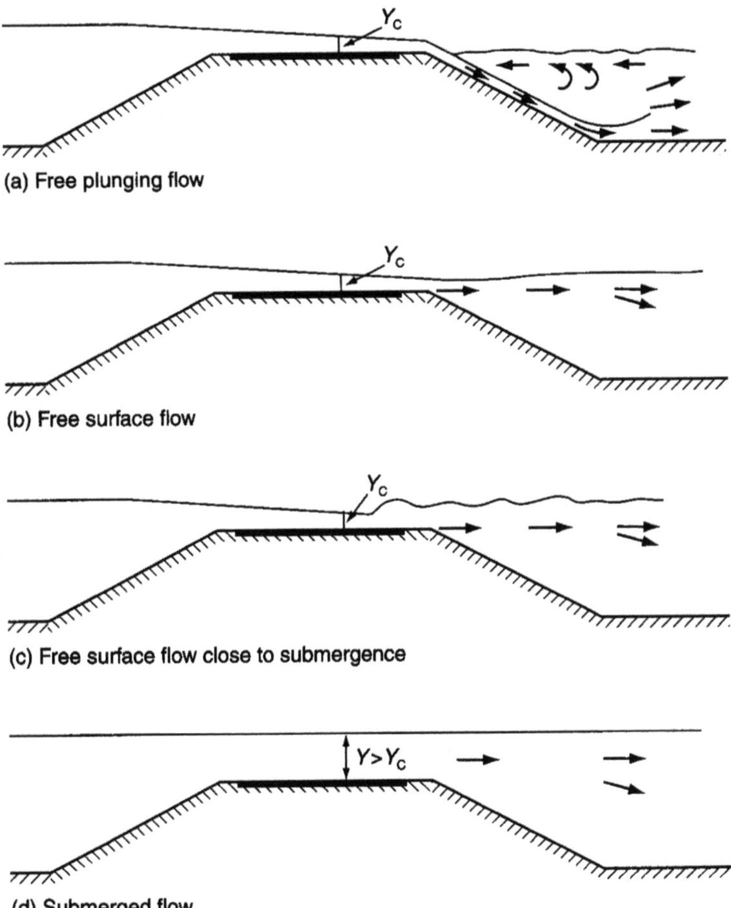

Fig. 6.4 The main types of flow over an embankment: (a) and (b) are both free flows with critical depth on the crest and a relatively low tailwater; (c) is a transitional flow; and (d) is a submerged flow with a relatively high tailwater and a depth greater than critical depth on the crest.

former case, critical depth (Y_C) always occurs on the crest, as with a broad-crested weir that is operating freely, whereas in the latter case the critical depth has been eliminated on the crest as a result of being drowned by the high tailwater level.

Free flow can be divided into free plunging flow (Fig. 6.4a) and free surface flow (Fig. 6.4b). Of the two, plunging flow is the most destructive since it consists of a high-velocity water jet flowing down the downstream face of the embankment followed by a submerged hydraulic jump. When the tailwater is sufficiently high the plunging flow goes through a transition to free surface flow. The stage at which this happens depends upon whether the water level is rising or falling. If the tailwater level rises sufficiently then a free surface flow will go through a transition (Fig. 6.4c) and become a submerged flow (Fig. 6.4d). Alternatively, if the tailwater is high to begin with then the flow may always be of the submerged type.

Which type of flow will occur at a given site depends upon several factors, including the conveyance of the downstream channel and floodplains, which governs the crucial tailwater level. It is likely that a bridge that would experience sluice gate type flow (see Section 2.5) will operate with free flow over the embankment, whereas a bridge that would experience drowned orifice flow will operate with submerged flow over the embankment. This analogy to sluice gate and drowned orifice flow extends to the factors affecting the discharge. Table 6.1 shows that with free flow the discharge depends upon the upstream head, whereas with submerged flow both the upstream and downstream heads influence the discharge.

6.2.2 *The discharge equation*

With respect to the discharge over a highway embankment, Kindsvater concluded that embankment shape and relative height have little effect, while the effect of boundary resistance is appreciable only at smaller heads. Consequently the most practical method of calculating the discharge is to consider the embankment as a broad-crested weir with critical depth (Y_C) on the longitudinal centreline, or just downstream of the centreline. This equation can then be used with experimentally determined coefficients of discharge to allow for different flow conditions and embankment roughness. If the approach embankment is in the form of a vertical curve, this can be approximated by a series of horizontal weirs with their crests at different levels. These can be analysed separately and the discharges added to get the total (Richardson *et al.*, 1990).

The cross-section of the highway is as shown in Fig. 6.5. Assuming critical depth (Y_C m) on the crest, then for continuity of flow at the critical section:

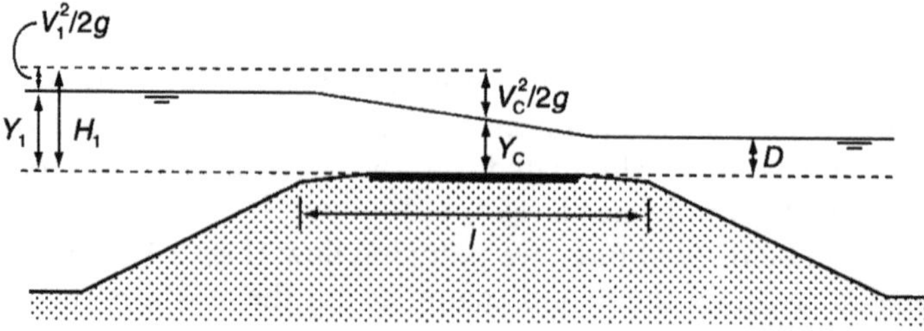

Fig. 6.5 Cross-section through an embankment defining the principal variables (not to scale).

$$Q = A_C V_C \tag{6.1}$$

$$\text{with } A_C = L_E Y_C \tag{6.2}$$

where Q is the discharge (m^3/s) over the embankment, A_C is the cross-sectional area of flow (m^2) at the critical section, L_E is the length (m) of the embankment over which the critical flow occurs, and Y_C is the critical depth (m). The critical velocity (V_C m/s) is

$$V_C = (gY_C)^{1/2} \tag{6.3}$$

$$\text{so } Q = L_E (g)^{1/2} Y_C^{3/2} \tag{6.4}$$

Using the road surface as the datum level and applying the energy equation to a section upstream (subscript 1) and to the critical section on the crest:

$$H_1 = \frac{V_C^2}{2g} + Y_C \tag{6.5}$$

With $V_C = (gY_C)^{1/2}$ from equation 6.3, then equation 6.5 becomes $H_1 = 1.5Y_C$, so:

$$Y_C = 0.667 H_1 \tag{6.6}$$

Substituting for Y_C in equation 6.4 gives

$$Q = L_E (g)^{1/2} (0.667H_1)^{3/2}$$

$$= 1.705 L_E H_1^{3/2} \tag{6.7}$$

The numerical coefficient (1.705) incorporates g and so is dimensional, the value above being for metric units (the equivalent value for English units is about 3.09). In practice, the type of flow (see below), approach conditions and embankment geometry all affect the value of the coefficient, so the theoretical value of 1.705 is often replaced with C_F, which is determined experimentally, thus:

$$Q = C_F L_E H_1^{3/2} \tag{6.8}$$

The subscript F attached to the coefficient indicates that it is for the free flow condition (Fig. 6.4a and b). Typically C_F has a value of between 1.57 and 1.71, as shown in Fig. 6.6a and b. However, the embankment can become submerged (Fig. 6.4d) in the same way that a broad-crested weir becomes submerged when a high downstream water level drowns the weir and eliminates critical flow on the crest. In this submerged condition the coefficient is C_S, which can have a value anywhere between 0.50 and 1.71, and the discharge equation (Bradley, 1978) is

$$Q = C_F L_E H_1^{3/2} \times \left[\frac{C_S}{C_F}\right] \tag{6.9}$$

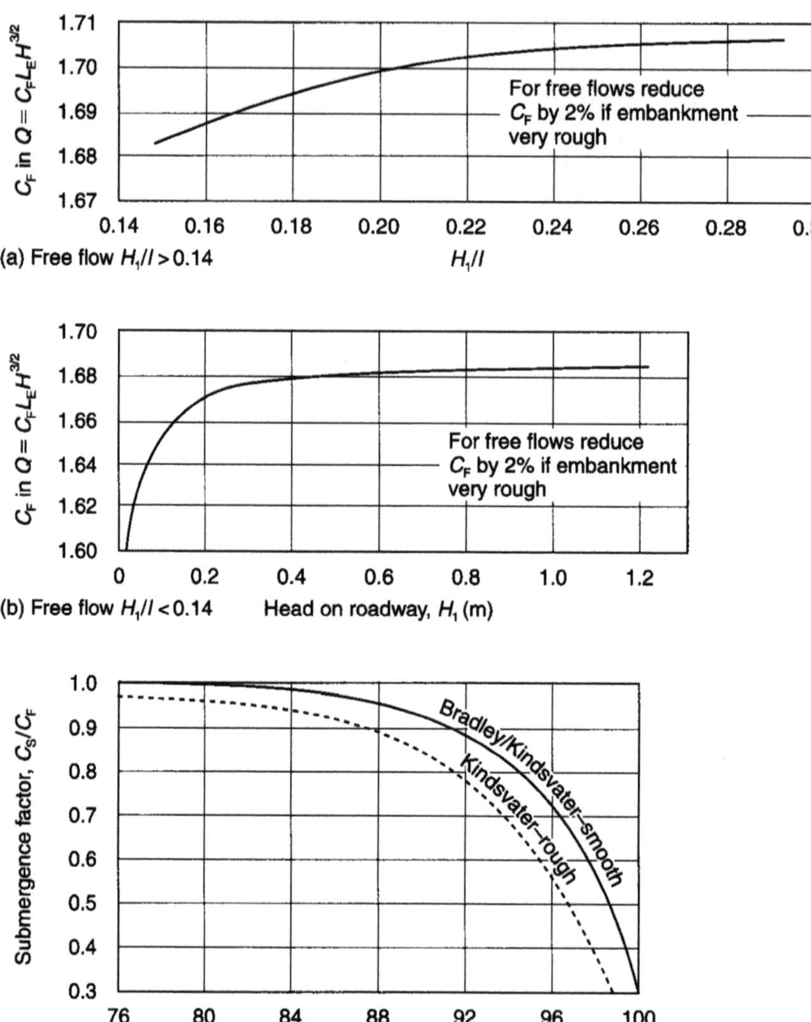

Fig. 6.6 Coefficients of discharge for embankments: (a) and (b) are used for free flows with equation 6.8; (c) is used for submerged flows with equation 6.9. (After Bradley, 1978)

The ratio in the square brackets is obtained from Fig. 6.6c and represents the amount by which the free flow coefficient (C_F) must be reduced to allow for various degrees of submergence as described by the ratio D/H_1 (Fig. 6.5). The value of $[C_S/C_F]$ is near 1.0 in the transition zone as submerged flow begins, reducing to around 0.3 when the downstream water depth (D) is about the same as the total upstream head (H_1).

When using Fig. 6.6 it should be appreciated that the graphs present a rather misleading impression of the accuracy of the coefficients. These diagrams are basically metricated versions of those presented by Bradley (1978). In the original work Kindsvater (1964) performed experiments on models with different surface roughness and obtained a number of different relationships for the variation of C_F and C_S with head. This is illustrated by the dashed line in Fig. 6.6c. With free flows, when using Figs 6.6a and b it may be prudent to reduce C_F by about 2% if the embankment is rough (e.g. 1.660 less 2% = 1.627). Rough could perhaps be considered as a gravel surface, and smooth as a paved surface. Kindsvater covered one of his models in birdshot to increase its roughness; in reality it is reasonable to expect an embankment covered in bushes to have a slightly lower coefficient than one that is smooth. What effect crash barriers will have is problematical.

***Example* 6.1**

An embankment and bridge carry a two-lane highway across the river valley shown in Fig 6.2. The width of the carriageways (l) is 35 m, and the surfaces of the embankment are relatively smooth. If the embankment is overtopped during a flood with H = 1.000 m, what is the discharge (a) assuming free flow, and (b) that the flow is submerged with D = 0.900 m?

From Fig. 6.2 the approximate combined average length of the left and right embankments at the stage involved is $L_E = 2020$ m.

(a) For the free condition, $H_1/l = 1.000/35 = 0.0286$, which is less than 0.14, so obtain the value of C_F from Fig. 6.6b: with $H_1 = 1.000$ m then $C_F = 1.684$. Substitution in equation 6.8 gives

$$Q = C_F\, L_E\, H_1^{3/2}$$

$$= 1.684 \times 2020 \times 1.000^{3/2}$$

$$= 3402\,\text{m}^3/\text{s}$$

(b) For the drowned condition, $D/H_1 = 0.900/1.000 = 0.9$ or 90%, so using the smooth curve in Fig. 6.6c, $C_S/C_F = 0.93$. Equation 6.9 gives

$$Q = C_F\, L_E\, H_1^{3/2} \times \frac{C_S}{C_F}$$

$$= 3402 \times [0.93]$$

$$= 3164\,\text{m}^3/\text{s}$$

References

Bradley, J.N. (1978) *Hydraulics of Bridge Waterways*, 2nd edn, US Department of Transportation/Federal Highways Administration, Washington DC.

CIRIA (1987) *Protection and Provision for Safe Overtopping of Dams and Flood Banks*, Construction Industry Research and Information Association, London.

Highways Agency (1994) *Design Manual for Roads and Bridges*, Vol. 1, Section 3, Part 6, BA59/94, *The Design of Highway Bridges for Hydraulic Action*, HMSO, London.

Hydrologic Engineering Center (HEC) (1990) *HEC-2 Water Surface Profile User's Manual*, US Army Corps of Engineers, Davis, CA.

Kindsvater, C.E. (1964) Discharge characteristics of embankment-shaped weirs. US Geological Survey, Water Supply Paper 1617-A, US Government Printing Office, Washington DC.

McKay, G.R. (1970) Pavement drainage. *Proceedings of the 5th Conference*, Australian Road Reasearch Board, Canberra, 5(4), pp. 305–326.

Richardson, E.V., Simons, D.B. and Julien, P.Y. (1990) Highways in the river environment. Report no. FHWA-HI-90-016, National Highways Institute/Federal Highways Administration, McLean, Virginia.

Richardson, E.V., Harrison, L.J., Richardson, J.R. and Davis, S.R. (1993) *Evaluating Scour at Bridges*, 2nd edn. Publication no. FHWA-IP-90-017, Hydraulic Engineering Circular No. 18, National Highways Institute/Federal Highways Administration, McLean, VA.

7 How to improve flow through a bridge

7.1 Introduction

When designing a new bridge there are occasions when it is desirable to optimise the hydraulic performance, perhaps because only a very small afflux can be tolerated without causing flooding upstream. This can be achieved partly by using slender, round-nosed piers (or eliminating them altogether), spillthrough rather than vertical abutments, and by avoiding skewed or eccentric openings. However, sometimes it may be necessary to adopt additional measures to improve the flow through the bridge, such as when economic or other considerations result in a less than ideal location and alignment for the crossing, or when the channel is braided or has a tendency to migrate. Under these circumstances some form of 'improvement works' may be needed to increase hydraulic efficiency at the design discharge. If possible these works should be incorporated at the design stage; it is usually less satisfactory and more expensive to add them at a later date.

Once constructed, bridges do not always behave hydraulically as their designers intended: the waterway may turn out to be too small for the floods that occur, the backwater too large, the approach channel may have changed, or scour and structural damage may be much worse than anticipated. In this case some form of 'improvement works' may have to be added.

The term 'improvement works' is a very general one, and includes such things as:

- providing either a rounded or chamfered entrance to the waterway opening;
- using wingwalls;
- using spur dykes (guide banks);
- designing a special 'minimum energy' waterway;
- using channel improvements, river training works, groynes (spurs), or channel diversions.

All are intended to increase the discharge through a bridge for a given upstream water level, or to reduce the upstream stage at a given discharge, or to reduce scour. All are described in this chapter. Entrance rounding, wingwalls and channel improvements are probably the most common means of improving the flow through a bridge, with spur dykes and minimum energy waterways being used less frequently.

Afflux (backwater) occurs because there has to be an increase in the upstream head of water to overcome the energy loss as the flow passes through the bridge opening. Inefficient waterways result in a relatively large energy loss and afflux. If the afflux is 0.5 m on floodplains that have a transverse slope of 1 in 100, then for some distance upstream this results in another 50 m of each floodplain being inundated compared with an unobstructed channel. If the afflux is larger, or the floodplains flatter, the area of innundation increases rapidly. Thus even a small reduction in afflux, perhaps through entrance rounding, can be beneficial.

7.2 Entrance rounding

For brevity, the term 'rounding' will be used to indicate that the upstream face of the waterway opening has either rounded corners or a chamfer; when necessary a distinction will be made. Entrance rounding is one of the simplest means of improving the hydraulic performance of a bridge, and has been used for centuries. Telford used a partial chamfer on some of his arches (Fig. 1.4). At Morpeth, the height of the existing streets and river banks did not allow the arches to rise well above flood level: hence the need for hydraulic efficiency. However, Ruddock (1979) wondered whether or not Telford also used a chamfer for aesthetic reasons. Some modern, rectangular openings have also been constructed with entrance rounding when hydraulic efficiency is important (Figs. 1.1 and 7.1). Entrance rounding has been used with culverts to increase the discharge by up to 25% (French, 1969).

Fig. 7.1 A modern bridge at Plympton, Devon, having a rounded soffit and a pier with chamfers.

The energy loss incurred by water flowing through a bridge opening arises as a result of the flow having to contract and then re-expand in the waterway, as described in Section 2.2. Both the contraction and expansion cause a loss of energy: the larger the contraction and subsequent expansion the larger the loss. By rounding the entrance to the opening so that it more closely matches the curvature of the streamlines, the size of the initial contraction – and hence the energy loss – can be reduced. Usually the rounding is applied to the upstream face of the bridge, there being little or no additional improvement if the downstream edges of the opening are also rounded.

In channel flow, the afflux depends partly upon the severity of the constriction as denoted by the bridge opening ratio, M. In sluice gate flow (Fig. 7.2a) where the water level rises above the top of the opening, submerging or drowning the waterway, the opening ratio becomes unimportant, so the discharge can be calculated from equation 7.1. The subscript u indicates the centreline at an upstream cross-section, either at the bridge face or one span upstream depending upon which has the largest stage.

$$Q = C_d a_W \left[2g\left(Y_u - \frac{Z}{2} + \frac{\alpha_u V_u^2}{2g}\right)\right]^{1/2} \tag{7.1/2.8}$$

where a_W is the maximum cross-sectional area of the full waterway and Z is the mean height of the opening above bed level. The coefficient of discharge (C_d) is primarily a coefficient of contraction, so if entrance rounding reduces the contraction of the jet passing through the opening there is an increase in C_d, and hence Q, for a given upstream water level (Y_u). The discharge through the bridge is determined mainly by Y_u and the geometry of the waterway (i.e. C_d, a_W, Z), so the bridge controls the flow. This is referred to as structure control. In structure control the discharge and upstream water level are independent of the conditions in the downstream channel. However, when sluice gate flow is just becoming established, the conditions in the channel may still have an influence on the upstream stage, but this influence diminishes as Y_u increases.

When both the upstream and downstream water levels are above the top of the opening (Fig. 7.2e) the flow is of the drowned orifice type, and can be described by an equation of the form

$$Q = C_d a_W (2g\Delta H)^{1/2} \tag{7.2/2.9}$$

$$\text{where } \Delta H = \left(Y_u + \frac{\alpha_u V_u^2}{2g} - Y_d\right) \tag{7.3/2.10}$$

and Y_d is the water level at the centre of the opening at the downstream face. The conditions associated with drowned orifice flow are such that the velocity head is often negligibly small, so the differential head can be taken as $\Delta H = (Y_u - Y_d)$. There is always some degree of channel control since ΔH (which includes Y_d) and the geometry of the opening (i.e. C_d, a_W) affect the discharge through the bridge. Because the waterway is flowing full there

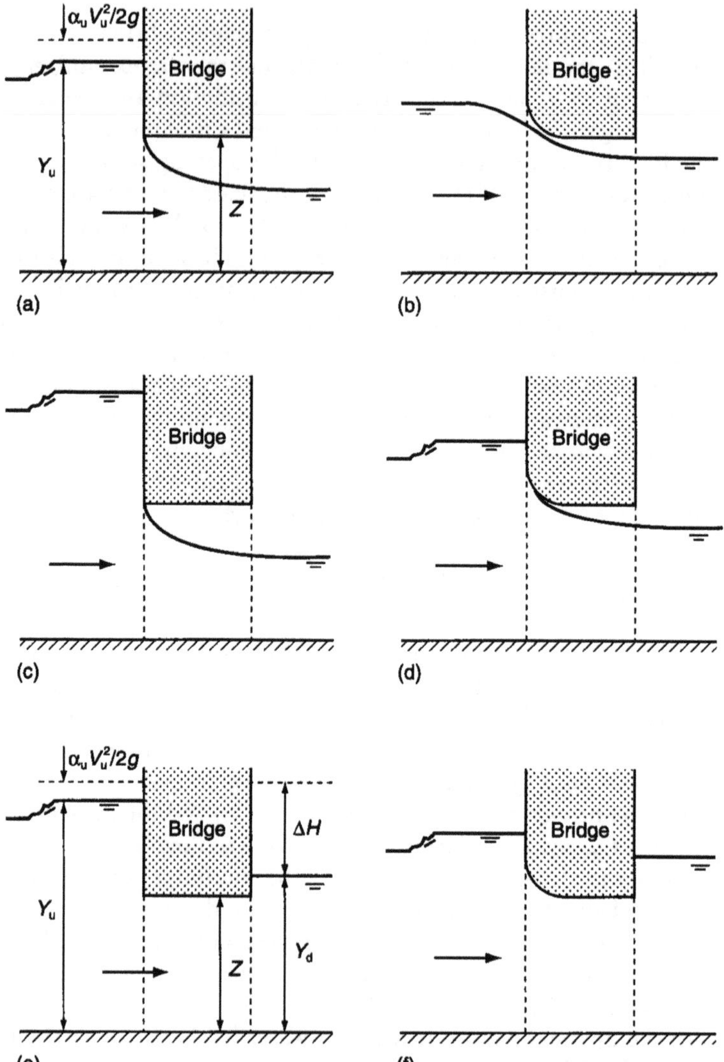

Fig. 7.2 The effect of entrance rounding (shown diagrammatically, not to scale) is generally to decrease the upstream depth (Y_u) while simultaneously increasing the downstream depth (Y_d). (a) The reduction in upstream depth is sufficient to change sluice gate flow to open channel flow through the opening (b), often resulting in a relatively large (temporary) increase in efficiency. (c) Sluice gate flow occurs initially and is maintained with the rounded entrance (d), but there is a smaller contraction from the soffit resulting in more of the waterway area being used, a reduced mean velocity in the opening, and a smaller energy loss. (e) With drowned orifice flow initially, entrance rounding (f) reduces the differential head ΔH. (After Hamill, 1997. Reproduced by permission of the Institution of Civil Engineers)

is a relatively large friction loss, while the interaction between the flow emerging from the opening and the water in the downstream channel can cause a large loss of energy and the largest afflux (Hamill and McInally, 1990). Again, the effect of entrance rounding is to reduce the contraction of the jet and to increase the value of C_d, but since the contraction is smaller than in sluice gate flow (Fig. 2.9d) there is a smaller improvement in the flow.

7.2.1 *The effect of rounding on normal depth and afflux*

The upstream depth (Y_u) consists of two components, both of which increase with increasing discharge:

$$Y_u = \text{normal depth} + \text{afflux} \qquad 7.4$$

For a particular discharge and channel, the normal depth in uniform flow is constant and can be calculated from the Manning equation, whereas in the case of a channel with an abnormal stage it is the depth that would be experienced without the bridge (Section 2.3). With low Froude numbers and/or abnormal stages the afflux may be quite small, the main cause of any flooding being the large 'normal' depth. Thus improvements to the channel (instead of to the bridge) may be the best way to alleviate flooding under these conditions.

Entrance rounding reduces the afflux but, because the normal depth is constant, the equivalent percentage reduction in Y_u is smaller (Fig. 7.3). If the normal depth is relatively large, quite a significant reduction in afflux may result in only a small decrease in Y_u. The afflux can never be reduced to zero, unless the bridge completely spans the channel and has no contact with the river ($M = 1.0$). Typically the percentage reduction in afflux as a result of entrance rounding increases relatively slowly with stage until the opening is close to submerging, after which it increases more rapidly. Eventually a stage is reached where little additional benefit accrues and the line becomes more vertical. Thus entrance rounding may be most effective when Y_u/Z is between 0.8 and 1.6 (Hamill, 1997).

When a waterway, particularly a rectangular one with a horizontal soffit, becomes submerged there is a relatively severe vertical contraction so the opening is used inefficiently (Figs. 7.2a and 2.9d). The velocity of the jet is relatively small near the top but large with a downward component at the bottom, which is why many bridges have a scour pit just downstream. As described in Section 8.3, a 60% increase in the pier scour depth may be experienced with a submerged waterway. Entrance rounding reduces the contraction from the soffit so more of the opening is used, the mean velocity of the jet is lower, and the velocity distribution is more uniform: hence there is a smaller head loss. To a lesser extent the same argument applies to the contractions from the abutments, so the overall effect of entrance rounding is either to increase the discharge through the opening for a given upstream water level or, for a given discharge, to decrease the upstream water level

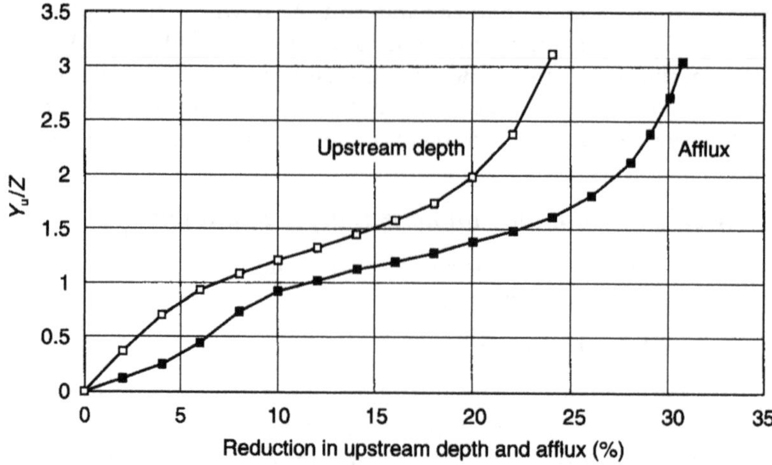

Fig. 7.3 Illustration of the reduction in afflux and upstream depth obtained by rounding the entrance to a 200 mm span model arch bridge at a slope of 1:50. The rounding was r/b = 0.125, referred to as R125 later (Fig. 7.4). Channel flow occurred up to $Y_u/Z \approx 1.1$, after which sluice gate flow prevailed. (After Hamill, 1997. Reproduced by permission of the Institution of Civil Engineers)

while simultaneously increasing the downstream water level. Depending upon the stage, the improved entrance geometry may also change the type of flow in the waterway, such as from sluice gate flow to channel flow (e.g. Fig. 7.2, a to b), or from supercritical to subcritical. Generally, entrance rounding reduces the velocity in the opening, so scour in the waterway and the channel immediately downstream may also be reduced.

7.2.2 *Full entrance rounding (Hamill, 1997)*

Hamill undertook research specifically to evaluate the effectiveness of entrance rounding to arch bridges but, to provide a comparison, also investigated rectangular openings. The aim was to present the results simply, for ease of use. To some extent, simplicity was preferred to extreme accuracy, which is probably unattainable anyway. The investigation used bridges with either a single semicircular arch or a rectangular opening (with $b = 2Z$) and a waterway length equal to the span, although the research can to some extent be cautiously applied to other geometries, multispan bridges and short culverts.

Three single-span bridges with arched openings and three with rectangular openings were tested in a 450 mm wide (= B), 15 m long tilting channel with a recirculating water supply. The bridges had spans (= b) of 200 mm, 250 mm and 300 mm (i.e. the bottom width of the arches) so the ratio M =

a/A was between 0.35 and 0.67. The normal depth flow was almost always subcritical, although supercritical flow sometimes occurred in the opening and in the channel just downstream.

For each bridge there were two sets of plywood templates of 6 mm, 12 mm and 25 mm nominal thickness, one set having a rounded edge with a radius (r) equal to the thickness of the plywood (Fig. 7.4), the second having a 45° chamfer with a width (r) equal to that of the plywood. The six different forms of entrance rounding for each bridge were identified by the ratio r/b, plus the letters R and C to denote rounded and chamfered entrances. Thus, for a 12 mm thick template and a 250 mm wide arch with a chamfer the ratio was 12/250 or C0.048. For convenience these values have been multiplied by 1000 and rounded so C0.048 becomes C50. The nine nominal ratios shown in Table 7.1 result from using three plywood thicknesses with three different bridge spans. The templates were placed against the upstream face of the bridges and the improvement in hydraulic performance measured. Initial experiments showed that with the 6 mm ply there was no significant difference between a rounded edge and a chamfer, so only the chamfered templates were used and a dual designation adopted (i.e. RC20, RC25 and RC30). Only 45° chamfers were considered; some idea of how 30° and 60° chamfers may compare can be obtained from Section 7.2.4. On a few occasions the plywood templates were placed on the downstream bridge face to see if this resulted in any improvement; as expected, it did not, so this was dismissed.

The model results were presented in terms of non-dimensional parameters to make them applicable to full-size bridges. With stage this was

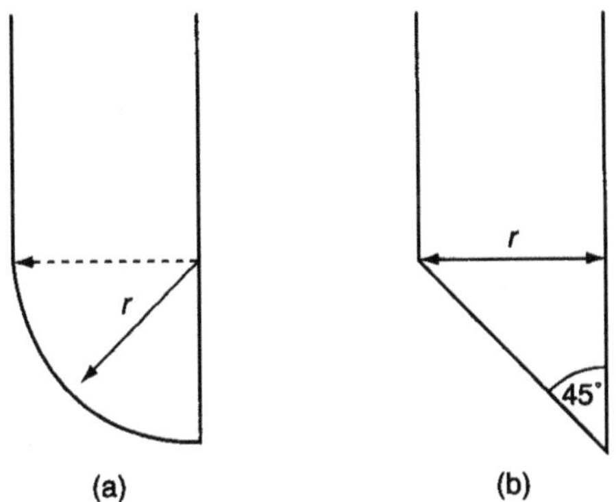

Fig. 7.4 Definition of rounded and chamfered templates. The thickness of the plywood (r) equals (a) the radius of curvature of a rounded entrance and (b) the width of the 45° chamfer. Thus for either configuration the ratio r/b describes the degree of rounding, where b is the span of the opening. For convenience these values are multiplied by 1000 in Table 7.1. (After Hamill, 1997. Reproduced by permission of the Institution of Civil Engineers)

Table 7.1 Values of r/b (× 1000) for the model spans and template thicknesses

	Span		
Template thickness (mm)	*200 mm*	*250 mm*	*300 mm*
6	30	25	20
12	60	50	40
25	125	100	85

After Hamill (1997); reproduced with permission, Institution of Civil Engineers

achieved by dividing the maximum upstream water depth by the height of the opening to obtain the proportional depth (Y_u/Z). For discharge, the actual discharge (Q) was divided by the *nominal* discharge capacity of the opening running full (Q_F) to obtain the proportional discharge (Q/Q_F). In reality there is no unique value of Q_F, because in channel flow Q_F depends upon such factors as roughness, gradient and normal depth. However, when structure control becomes fully established there are just two stage–discharge relationships, one representing sluice gate flow and one drowned orifice flow, with the latter being the higher of the two. Thus at $Y_u/Z = 2$ it was possible to identify one specific discharge (designated Q_{2F}) corresponding to sluice gate flow and one to drowned orifice flow. The nominal discharge equivalent to the opening running full (Q_F) with $Y_u/Z = 1.0$ was taken as $Q_F = Q_{2F}/2$. The resulting values are shown in Table 7.2. For real bridges the approximate value of Q_F can be calculated if it is unknown; see Examples 7.2 and 7.3.

The results from all six model bridges were superimposed using the proportional depths and discharges to produce one non-dimensional stage–

Table 7.2 Nominal capacity of arch and rectangular model waterways flowing full (Q_F)

Arch		
Span (mm)	*Q_F sluice gate flow (l/s)*	*Q_F drowned orifice flow (l/s)*
200	7.80	7.10
250	13.25	12.20
300	21.88	18.08
Rectangular		
Span (mm)	*Q_F sluice gate flow (l/s)*	*Q_F drowned orifice flow (l/s)*
200	8.90	8.50
250	16.40	14.25
300	26.30	20.70

After Hamill (1997); reproduced with permission, Institution of Civil Engineers discharge relationship for sluice gate flow (Fig. 7.5) and one for drowned orifice flow (Fig. 7.6). In the model tests, sluice gate flow almost always occurred with the normal depth Froude number $F_N > 0.25$, and drowned orifice flow with $F_N < 0.25$. The relatively wide distribution of points below $Y_u/Z = 1.1$ represents channel control. The single relationship at higher stages represents structure control; some of the scatter is due to the difficulty of measuring rapidly fluctuating water levels accurately.

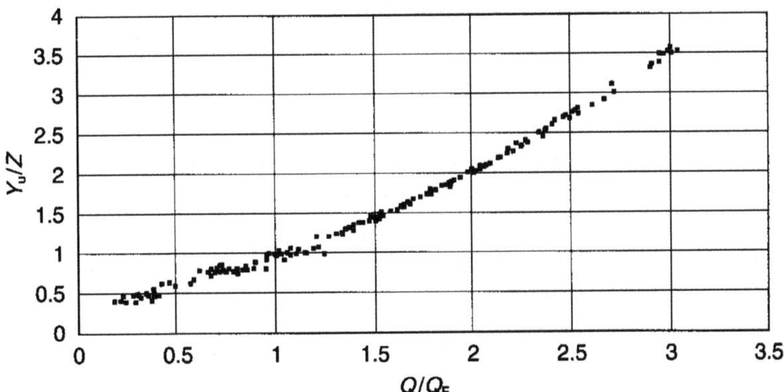

Fig. 7.5 Non-dimensional stage–discharge curve for arch and rectangular openings operating with $F_N > 0.25$ (i.e. sluice gate flow when the opening becomes submerged). The squares represent individual experimental results. (After Hamill, 1997. Reproduced by permission of the Institution of Civil Engineers)

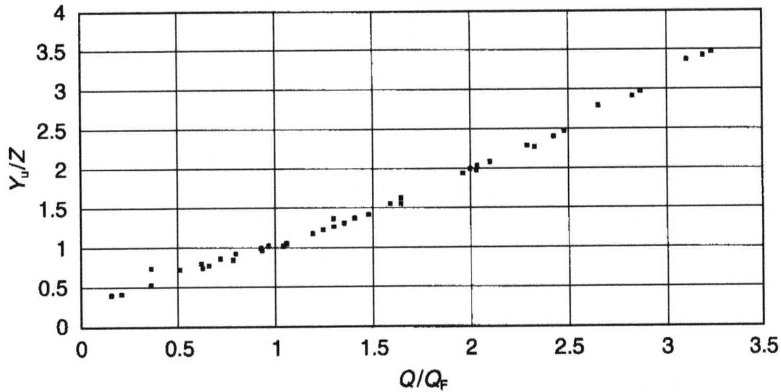

Fig. 7.6 Non-dimensional stage–discharge curve for arch and rectangular openings operating with $F_N < 0.25$ (i.e. drowned orifice flow when the opening becomes submerged). The squares represent individual experimental results. (After Hamill, 1997. Reproduced by permission of the Institution of Civil Engineers)

By placing rounded templates in front of the model bridges and then applying the procedure described above, non-dimensional stage–discharge curves were obtained for each of the new entrance geometries. By comparison with the equivalent bridge curve or by direct calculation, the reduction in upstream depth, increase in discharge and the new coefficient of discharge due to entrance rounding could be determined. The results are shown in Figs. 7.7–7.9. Note that both Fig. 7.7 and Fig. 7.8 are plotted with the initial value of the upstream depth (Y_u/Z) or initial differential head ($\Delta H/Z$) on the horizontal axis, and that $\Delta H/Z = 0$ corresponds to $Y_u/Z = 1.1$ because it is only above this stage that the opening is submerged. To determine the effect of entrance rounding, enter the graph at the appropriate Y_u/Z or $\Delta H/Z$, move vertically up to the desired r/b ratio, and then move horizontally across to obtain the associated improvement.

In most cases the increase in discharge (ΔQ) obtained by changing a square-edged opening to a rounded entrance was measured directly by keeping the upstream water level at the same value and then comparing the flow rates. This was necessary because, particularly with drowned orifice flow, when the inlet geometry was altered the downstream water level changed significantly (as well as the upstream stage, of course). Thus the calculation of the C_d values from equations 7.2 and 7.3 always had to be based on observed downstream water levels. Similarly, although it would be possible to calculate the increase in discharge resulting from entrance rounding by assuming a constant upstream stage and then using the C_d values in Fig. 7.9, this would be less accurate than obtaining ΔQ from Fig. 7.8 because it would assume that the downstream conditions remained unchanged, which would not be the case. Figure 7.9 therefore shows the *actual* value of Y_u/Z or $\Delta H/Z$ on the horizontal axis (not the initial value as in Figs 7.7 and 7.8) while the vertical scale shows the equivalent C_d (not the improvement). The improvement in C_d is apparent from the spread of the lines. For example, Fig. 7.9a indicates that with an arch, $F_N > 0.25$ and $Y_u/Z = 1.2$, the C_d of the bridge is 0.52 but with R125 rounding it is 0.59 at the same upstream stage.

Comparison of the graphs with the raw experimental data indicated good accuracy: typical differences in $\Delta Y_u/Z$ and $\Delta Q/Q_F$ were of the order of ±0.8% in channel flow, ±1.5% in drowned orifice flow and ±2.5% in sluice gate flow (see Example 7.1). Examples 7.1–7.3 illustrate that the method is reasonably accurate and easy to use, even with limited data. However, it is essential that all charts are regarded as approximate and the recommendations as guidelines, not absolute rules that are always valid under all circumstances.

Interpretation of the results

Drowned orifice flow was relatively difficult to analyse because small changes in the difference between two fluctuating water levels had to be

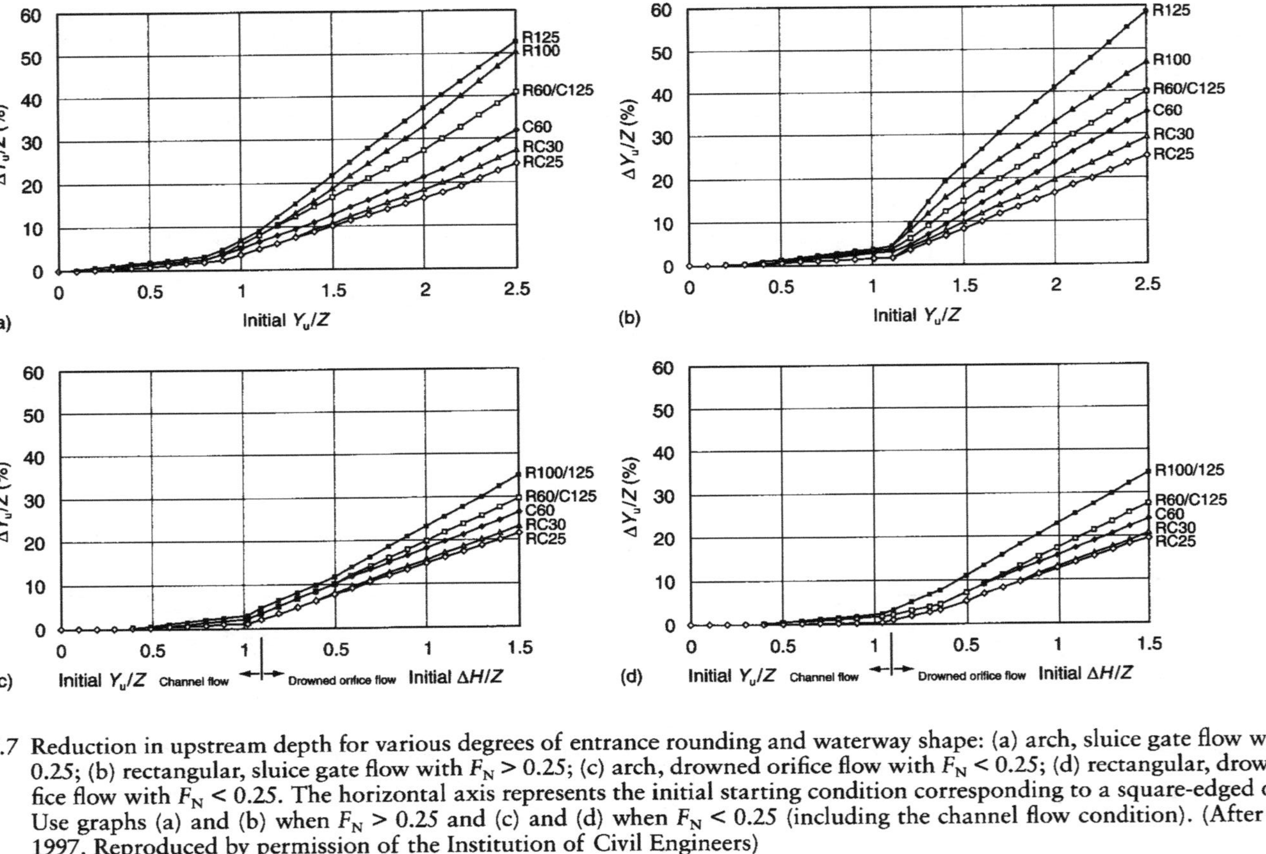

Fig. 7.7 Reduction in upstream depth for various degrees of entrance rounding and waterway shape: (a) arch, sluice gate flow with $F_N > 0.25$; (b) rectangular, sluice gate flow with $F_N > 0.25$; (c) arch, drowned orifice flow with $F_N < 0.25$; (d) rectangular, drowned orifice flow with $F_N < 0.25$. The horizontal axis represents the initial starting condition corresponding to a square-edged opening. Use graphs (a) and (b) when $F_N > 0.25$ and (c) and (d) when $F_N < 0.25$ (including the channel flow condition). (After Hamill, 1997. Reproduced by permission of the Institution of Civil Engineers)

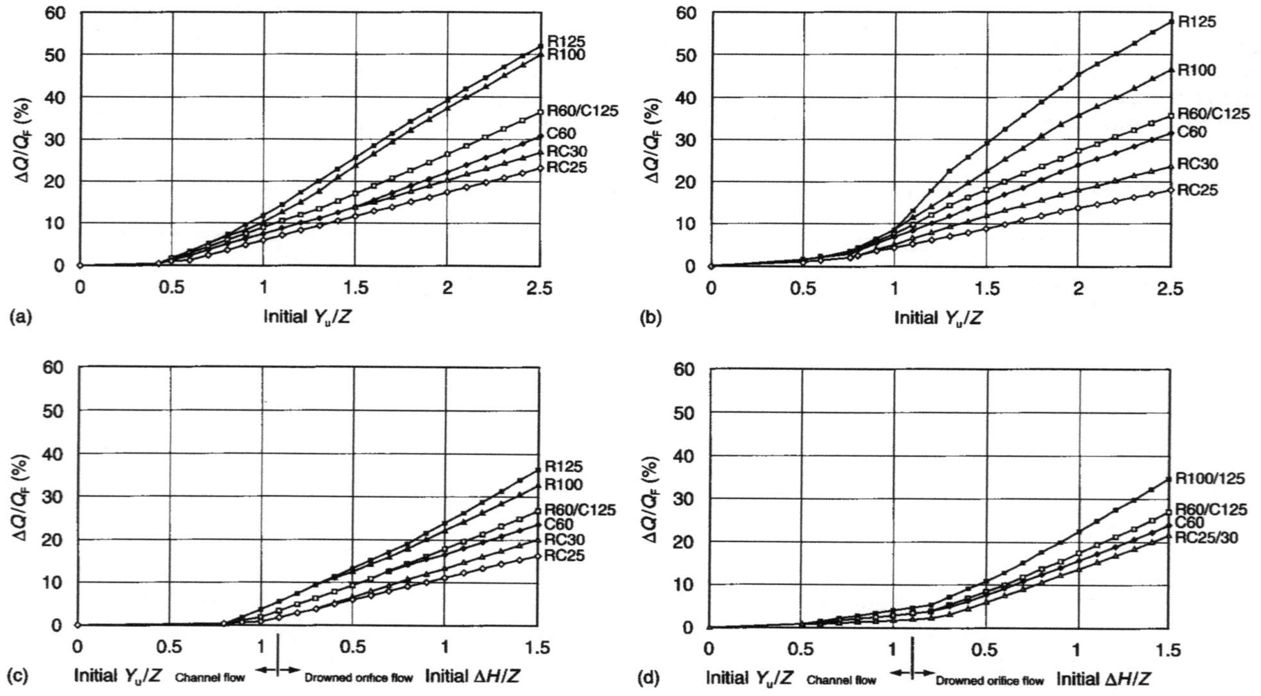

Fig. 7.8 Increase in proportional discharge for various degrees of entrance rounding and waterway shape: (a) arch, sluice gate flow with $F_N > 0.25$; (b) rectangular, sluice gate flow with $F_N > 0.25$; (c) arch, drowned orifice flow with $F_N < 0.25$; (d) rectangular, drowned orifice flow with $F_N < 0.25$. The horizontal axis represents the initial starting condition corresponding to a square-edged opening. Use graphs (a) and (b) when $F_N > 0.25$ and (c) and (d) when $F_N < 0.25$ (including the channel flow condition). (After Hamill, 1997. Reproduced by permission of the Institution of Civil Engineers)

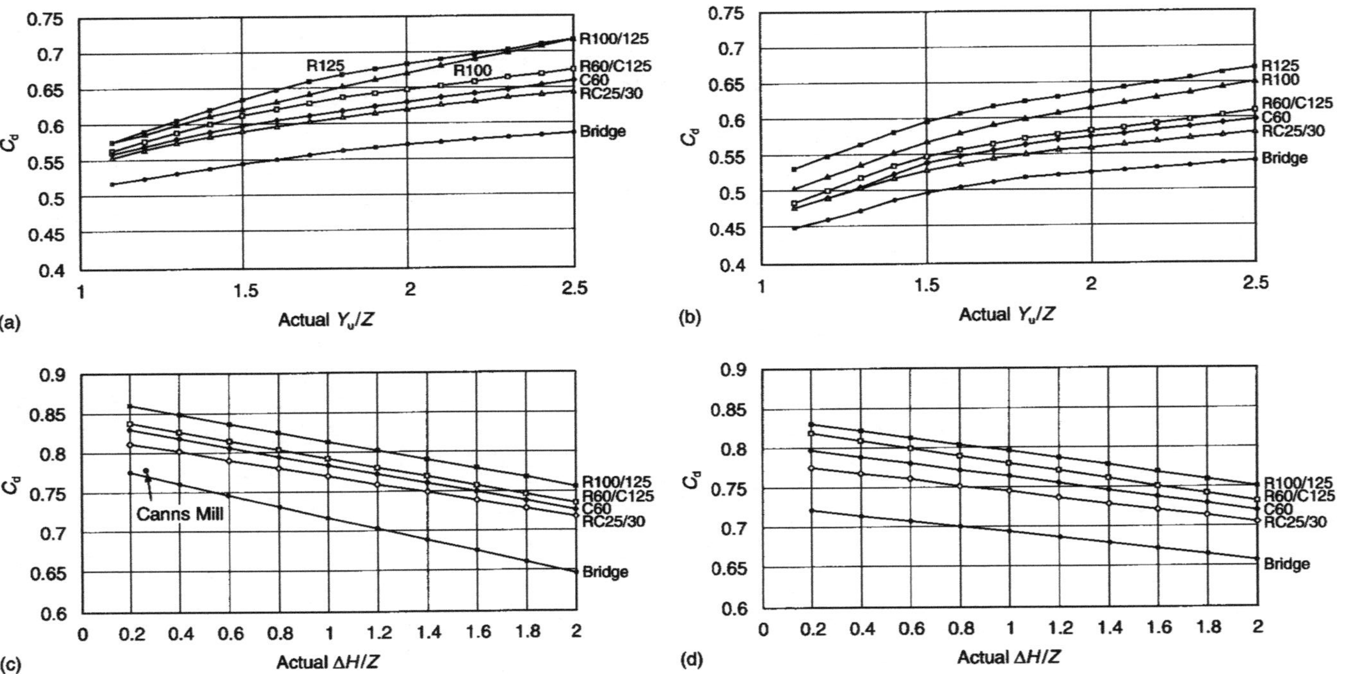

Fig. 7.9 Discharge coefficients corresponding to the actual stage observed with the rounded entrance in place: (a) arch, sluice gate flow with $F_N > 0.25$; (b) rectangular, sluice gate flow with $F_N > 0.25$; (c) arch, drowned orifice flow with $F_N < 0.25$; (d) rectangular, drowned orifice flow with $F_N < 0.25$. The line marked 'Bridge' corresponds to a normal square-edged opening. Use graphs (a) and (b) when $F_N > 0.25$ and (c) and (d) when $F_N < 0.25$. Diagram (c) also shows the data obtained in the Canns Mill Bridge field study. (After Hamill, 1997. Reproduced by permission of the Institution of Civil Engineers)

measured. Consequently it was not always possible to distinguish between RC25 and RC30, or R100 and R125. Also, with a small value of M (e.g. 0.4), there may be a period of sluice gate flow before the transition to drowned orifice flow. When applying equation 7.2 the opening must be totally submerged at the downstream face (say $Y_d \gg 1.1Z$); otherwise this adds to the scatter. This is also true of the upstream face. When Y_u/Z is between 1.1 and 1.3 the flow is always unstable and unpredictable to some extent: sometimes there was a large drawdown of the water surface as it approached the model opening, or the opening alternated between the free and submerged condition at the upstream face, and possibly the downstream face. Under these conditions model scale effects are likely to be significant, so the C_d values in this region are not reliable. Even with full-size bridges the flow in this zone is unpredicatable (see Fig 2.7).

In channel flow the increase in discharge or reduction in upstream stage that can be obtained through entrance rounding is relatively small, particularly below $Y_u/Z = 0.8$. If this improvement has to be relied upon to alleviate flooding, it may be wise to seek an alternative solution. However, when the opening is submerged and the vertical contraction is large, significant reductions in stage or increase in discharge can be achieved, particularly in sluice gate flow (e.g. Figs 7.7b and 7.8b). At times the reduction in stage that could be achieved simply by adding a template with rounded edges was quite remarkable, and almost any form of rounded entrance or chamfer with $r/b > 25$ can significantly improve the performance of a bridge. With $r/b < 30$ there was no difference between a rounded and chamfered entrance. Up to $r/b = 50$ there was little difference between the two, but over 50 a rounded entrance was always superior to a chamfer. The performance of C85 and C125 was substantially the same, so it may be assumed that there was an upper limit (85) to the width of chamfer that was effective, and that going beyond this produced a diminishing or negligible return. Generally R60 gave the same performance as C125. The rounded entrance with $r/b =$ 125 was clearly superior under all test conditions and should be adopted when the optimum performance is required. An upper limit to the effectiveness of a rounded entrance was not clear, although with the arches there was some indication that beyond R100 the increase in performance was starting to diminish. With rectangular openings, the very pronounced contraction from the horizontal soffit may mean that the limit of effectiveness is not reached so quickly.

As with all research, ideally the results should not be applied outside the range of the original observations. For instance, all the models used had a span (b) equal to $2Z$. Waterways with significantly different proportions or opening ratios may perform differently. This, coupled with different methodologies, makes comparison with other investigations difficult. For example, Matthai (1967) did not differentiate between sluice gate and drowned orifice flow, and perhaps as a consequence may have overestimated the increase in discharge due to entrance rounding under some conditions by between 2% and 10% (e.g. 12% instead of 6%). However, the two studies shared some conclusions. Mathai found that for rectangular openings and channel flow with $r/b < 50$ the percentage increase in discharge was in the same range for both rounded and 45° chamfered entrances, confirming the earlier observation that the shape is not significant below RC50. Matthai also found that with the larger ratios a rounded entrance is better than a chamfer.

For rectangular openings operating in sluice gate flow with Y_u/Z between 1.1 and 1.5, Bradley (1978) reported values of C_d between 0.37 and 0.49 (Fig. 2.11), while the corresponding model values were between 0.45 and 0.50 (Fig. 7.9b). For arched models the equivalent figures were higher, between 0.52 and 0.55 (Fig 7.9a). Arches have a higher C_d than rectangular openings, which exhibit a pronounced vertical contraction from the horizontal soffit. The hybrid, segmental arch bridge at Canns Mill had a C_d of about 0.35 and 0.46 when Y_u/Z was 1.1 and 1.35 respectively (Fig. 2.11). Thus all of the upper values are similar (allowing for the differing Y_u/Z). The largest discrepancies, predictably, are around $Y_u/Z = 1.1$ when the flow is unstable and submergence is barely established.

With drowned orifice flow the calculation of C_d was more difficult and less reliable than for sluice gate flow, so there is a good deal of scatter when C_d values are plotted. Bradley suggested that the average value was around 0.80 over a restricted range of Z/Y_u. The model results were between 0.63 and 0.76, which fits well with Bradley's observations, Canns Mill and Roughmoor Bridge when all are plotted together (Figs 2.12 and 7.9c). Roughmoor Bridge has two segmental arches, and was part of a field study (Hamill and McInally, 1990).

In conclusion, there is general agreement between the various studies undertaken, and it appears that entrance rounding can significantly increase the hydraulic performance of a bridge, particularly when it is submerged and there is a large contraction from the upstream soffit.

7.2.3 *Partial entrance rounding (Hamill)*

The research described above was concerned with rounding applied uniformly to the entire upstream face of the opening, but a limited number of experiments were conducted with the rounding applied to only part of the opening.

Chamfer on the bottom half of the arch only

The 200 mm span arch bridge was used with a 25 mm template with a 45° chamfer (C125) cut on the bottom half to achieve a geometry similar to that used by Telford for the bridge in Fig. 1.4. Up to $Y_u/Z = 0.5$ the performance was of course identical to the normal C125 template. However, as the stage increased and the square-edged upper part of the arch became submerged, the performance diminished so that at $Y_u/Z = 1.3$ it was roughly equivalent to the fully chamfered C30. At higher stages the performance was somewhat worse than the C30. It is logical to assume that a *r*/*b* ratio less than C125 would further reduce the improvement obtained.

Chamfering the bottom of the arch has the merit that this is the part that is in contact with the flow most often. The principal disadvantages are that the improvement in flow obtained at low stages is small, while at high stages it does nothing to reduce the very strong vertical contraction from the soffit. Chamfering the top part of the arch would produce a larger improvement at high stages, but this is only worthwhile if there is some certainty that floods reach this level.

Chamfer on the top half of the arch only

This template was similar to the one above but with the other (upper) half of the arch chamfered (C125) to reduce the vertical contraction. This configuration does not improve the passage of the most common floods that occupy the lower part of the waterway, only those above the level at which the chamfer starts. Consequently up to $Y_u/Z = 0.5$ there was no effect. At $Y_u/Z = 1.3$ it was less effective than the fully chamfered C30, and at $Y_u/Z = 1.7$ about the same as the C30. At higher stages ($Y_u/Z = 2.5$ to 3.5) its performance was better, being similar to the C60. Thus it was tentatively concluded that with arched openings a small chamfer over the entire opening is better than a large chamfer over part of it.

Rounding to the soffit of a rectangular opening only

The 200 mm span rectangular waterway was used with a 25 mm template that had a rounded soffit cut into it (R125) while leaving the vertical sides square-edged. This is a similar soffit geometry to that in Fig. 7.1. Openings usually drown at about $Y_u/Z = 1.1$, so up to this stage the performance was identical to the normal bridge. At normal depth Froude numbers below 0.25 (drowned orifice flow), or just above 0.25 with a low degree of submergence, the performance was between the fully rounded R30 and R60 templates. At higher F_N values the performance was between R60 and R125, usually nearer R60. Thus this configuration was quite effective, particularly at high F_N numbers, since it reduced the strong vertical contraction from the upstream soffit. Again it is logical to assume that a *r*/*b* ratio less than R125 would reduce the improvement obtained.

7.2.4 Partial entrance rounding (United States Geological Survey – USGS)

The effect of partial entrance rounding in the form of wingwalls was studied by the United States Geological Survey (USGS) as part of the work summarised in Section 4.2 (Kindsvater *et al.*, 1953; Kindsvater and Carter, 1955; Mathhai, 1967). This included rounded or chamfered (wingwall) entrances to crossings that have vertical abutments and either vertical or sloping embankments (opening types I and IV: see Figs 4.2 and 4.3c–f, and 4.11–4.13). Apparently a normal square-edged road deck was assumed, so the opening had only partial rounding. The results can be applied to rectangular openings and, to a lesser degree, arched openings (Matthai, 1967). However, this technique can be difficult to apply because it requires the evaluation of the opening ratio (M), the Froude number (F_3) at a cross-section between the abutments at the downstream face of the bridge, and the fall of the water surface (Δh) between a section upstream of the bridge and the downstream face. Despite these problems the USGS method provides a means of evaluating the effect on the discharge of either rounded abutments or wingwalls with a 30°, 45° or 60° chamfer when operating in channel flow or in the submerged condition.

The procedure is as described in Section 4.2, the only difference being that the adjustment factors k_r, k_w and k_T are included in equation 4.10, which defines the coefficient of discharge, C. For opening types I and IV the factors are evaluated as follows:

k_r	rounded entrance (Fig. 4.3)
k_w	30°, 45° or 60° chamfered entrance (Figs 4.3, 4.12 and 4.13)
k_T	submergence of the waterway (Fig 4.17)

The numerical values of C obtained for the various entrance geometries can be used in equation 4.5 or 4.9 to calculate the discharge (Q) for a given value of Δh. Since Q is directly proportional to C (and k_r, k_w) it is easy to see the effect of entrance rounding. For example, with a type I opening, $M = 0.50$, a rounded entrance and $r/b = 0.08$ then $k_r = 1.150$; with $M = 0.50$ and $w/b = 0.08$ a 30° wingwall gives $k_w = 1.066$; with a 45° wingwall $k_w = 1.106$, and with a 60° wingwall $k_w = 1.218$. Note that the coefficient increases with the angle of the wingwall, a 60° wingwall having a higher value than a rounded entrance. Similarly, with type IV wingwall embankments an angle of 60° has the largest coefficient. However, Fig. 4.3 (diagrams c–f) illustrates that there is little point in using a rounded or type I wingwall abutment with $r/b > 0.04$–0.14 (depending upon M and the type of rounding) because there will be no additional gain. As observed above, the law of diminishing returns applies.

7.3 Abutment type and extended wingwalls

As mentioned briefly in the introduction to the chapter, some types of abutment are hydraulically more efficient than others so the choice of abutment geometry will affect the performance of the bridge. The section immediately above illustrated how wingwalls can be used to increase the coefficient of discharge of an opening. Additionally, spillthrough-type abutments tend to reduce the backwater (e.g. Fig. 4.21) and the scour depth in the waterway (Table 8.8). Both of these may be important considerations at the design stage.

Many bridges are designed with wingwalls, which can have two functions. One is structural, to retain the material forming the approach embankments. The second is hydraulic, to help funnel the flow into and through the waterway opening. If this is the intention then (in plan) the wingwalls are usually curved as in Fig. 1.1, or flared outwards as in Fig. 4.11. In either case the principal effect is to form 'rounded' abutments that behave similarly to the chamfered entrances described above. However, in the examples described so far the wingwalls did not extend beyond the toe of the embankment. If they do extend beyond the toe then they start to behave as a partial transition between the opening and the full channel width, the extent of the transition depending upon how far upstream or downstream they extend. This may also slightly increase the effective length of the waterway. As described in Section 3.4, long waterways are more efficient than short ones. With a long waterway there is an initial, controlled re-expansion within the opening itself, then a secondary expansion into the downstream channel. With a short waterway there is one large expansion into the downstream channel. Thus there may sometimes be merit in trying to control the re-expansion by elongating the waterway, either by extending the wingwalls (see below and Section 7.5) or, if they are aligned to the flow, by increasing the length of the bridge piers (Fig. 1.1). However, this must be done carefully: the scour depth at a skewed pier increases with pier length, as shown in Fig. 1.15b.

There is relatively little useful information relating to the hydraulic performance of extended wingwalls. If used appropriately they may improve the flow through a bridge but, as described above, the law of diminishing returns applies to entrance rounding, and beyond a certain radius or width of chamfer there is little or no additional gain. Frequently extended wingwalls may be too short to form an effective transition so, despite their size, they are not very effective (but see spur dykes in Section 7.4). Hamill conducted some limited experiments on wingwalls, while Chow (1981) quoted the work of Formica (1955) relating to open channel transitions.

7.3.1 Extended wingwall transitions (Hamill)

This is a continuation of the work described in Section 7.2.2 and 7.2.3, a few experiments being conducted with extended wingwalls as a comparison with entrance rounding. A pair of identical, curved, extended wingwalls each 75 mm wide, 300 mm long and 400 mm high were placed against the upstream face of bridges having square-edged arched and rectangular openings. An important point is that with the 300 mm rectangular opening placed centrally in the 450 mm wide channel, in plan the wingwalls effectively formed curved abutments or transitions between the full channel width and the opening so that there was no longer a square vertical edge. However, with the 300 mm arch, as the stage increased the roof of the arch curved away from the vertical wingwalls leaving the square edge of the opening exposed. Thus the model wingwalls could be expected to be most effective with the 300 mm span rectangular bridge, less effective with the 300 mm arch, and less effective again when used with smaller-span bridges where the normal square abutments protrude beyond the wingwalls.

For the 300 mm rectanglular waterway with the perfectly fitting wingwalls the performance up to submergence at $Y_u/Z = 1.1$ was equivalent to R85, and sometimes slightly better. However, above this stage the wingwalls did nothing to reduce the strong vertical contraction from the soffit, so by $Y_u/Z = 1.2$ they decreased the upstream stage by only about 25% of the amount that the fully rounded RC20 template did. By around $Y_u/Z = 1.9$ the wingwalls had no effect. The performance of the 300 mm arch with the wingwalls was broadly similar to that described for the rectangular opening. With the 200 mm span arch the wingwalls were less effective than RC30 prior to submergence, and had no effect after submergence. Thus it would appear that extended upstream wingwalls can be used instead of entrance rounding with some success up to $Y_u/Z = 1.1$, but only if they essentially form a rounded transition from the full channel width to the opening width (see below). When used inappropriately, they may be ineffective.

7.3.2 *Open channel transitions*

Some standard transitions are shown in Fig. 7.10. According to Chow (1981), the higher losses usually occur with the contractions because they involve both an initial contraction of the flow and a subsequent re-expansion. Thus there is a conversion of energy from potential to kinetic

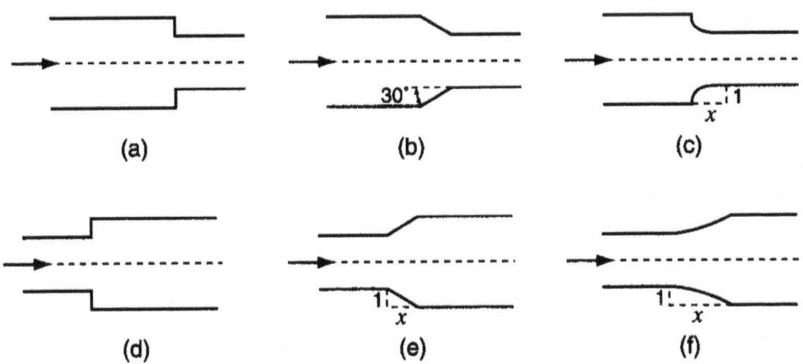

Fig. 7.10 Open channel transitions of various designs: (a) square contraction; (b) straight taper; (c) rounded taper; (d) square expansion; (e) straight expansion; (f) rounded expansion.

and back to potential, which results in less energy being recovered than when there is just the equivalent expansion. However, there is roughly twice the amount of energy lost in the expansion than in the contraction (e.g. see Section 2.3.1).

Formica (1955) investigated subcritical flow through open channel transitions of various design, the widths being 355 mm and 205 mm. Apparently the narrower section had a steeper bed slope, which may not be the case with bridges, so the results are indicative only. The difference in the energy loss between the various designs was often only fractions of one millimetre, which requires great precision and accuracy to achieve consistent results, while the relationship between the reduction in energy loss and any decrease in upstream afflux or potential increase in discharge must be inferred. Nevertheless, the work gives an indication of the effectiveness of various types of transition.

For converging flow, the square contraction in Fig. 7.10a was the worst. A rounded 1:2 taper (c) was the best, reducing the energy loss by about 70%, with a straight 30° taper (b) and a rounded 1:1 taper being marginally inferior (Chow, 1981). For expanding flow, the square expansion (d) was generally the least efficient. According to Formica a straight 1:4 expansion (e) was the best, reducing the energy loss by about 66%. A rounded 1:4 expansion (f) was also quite effective, as were hybrid designs involving an initial 1:2 or 1:3 straight expansion followed by a sudden expansion to the full width. However, there is a limit to the length of transition that is effective, beyond which there is little or no significant gain in efficiency. On the other hand, if the transition is too short or the angle of divergence too great the result may be the same or worse than the original square expansion. For this reason extended wingwalls that do not form a full and effective transition between a bridge waterway and the downstream channel are also of limited value.

Mazumder and Ahuja (1978) studied 'curved' contracting transitions in open channels. When seen in plan the geometry adopted was like the tapered neck of a white wine bottle (i.e. a 'smoother' version of Fig. 7.10f). They concluded that for the given shape an average linear contraction of 1:3 (tranverse: longitudinal) was the optimum and was practically independent of the Froude number in the throat.

It would appear that open channel transitions, or guide walls, may be effective under some circumstances if appropriately designed. However, to be successful they may have to be of a considerable size and will probably have to provide a smooth transition from the full channel width to the waterway opening and/or back to the full channel width. Ultimately this leads to the type of transitions used with minimum energy waterways (Section 7.5) that are designed to reduce or eliminate almost all energy losses except friction. Short transitions, such as small-radius rounding applied to the downstream face of a bridge opening, are not likely to have any significant effect.

7.4 Spur dykes (or guidewalls)

Spur dykes are essentially large upstream wingwalls. They are not common in Britain, simply because British rivers are relatively small. However, they are widely used throughout the world in situations where a wide floodplain exists and floods have to be funnelled through a relatively narrow bridge waterway. They are used for two reasons, the first being to improve the hydraulic efficiency of the opening: according to Brown and Recinos (1988) spur dykes can reduce bridge backwater by up to 5%, the amount increasing with the Froude number in the constriction. The second and principal reason is to reduce scour and the potential for damage to the structure.

Spur dykes are most needed where there is a tendency for water to flow along the upstream face of the highway embankment, so that close to the upstream abutments it meets at 90° the main flow passing through the bridge (Fig. 7.11a). Apart from the hydraulic inefficiency of two streams of water meeting like this, the interaction between the streams causes curvilinear flow, significantly increasing the scour potential. Bradley (1978) quoted a scour depth of 7.6 m at Big Nichols Creek, which may have been much greater since fallen piers and bridge spans were buried deep in the river bed.

The result of flow along the upstream face of the road embankment is effectively to reduce the width of the opening, since the sideflow results in a large contraction and a narrow main jet in the waterway (Fig. 7.11a): the larger the sideflows, the worse the situation. Sideflows occur most easily where there are upstream borrow pits, drainage ditches parallel to the road embankments or smooth unobstructed floodplains. The aim of the spur dyke is to guide the main flow through the opening while simultaneously reducing (or stopping) the sideflow and moving the point of interaction to the end of the spurs. The deepest scour then occurs near to the nose of the dyke, where the radius of curvature is sharpest, which is preferable to having the greatest scour in the vicinity of the abutments and any nearby piers (Highway Research Board, 1970; Neill, 1973). However, if the scour between the abutments is to be kept to a minimum, the dyke must be correctly proportioned, which means having an appropriate geometry, height and length.

Bradley (1978) recommended that the optimum shape for a dyke was a quarter of an ellipse (Fig 7.12), with a ratio of major (length) to minor (offset) axes of 2.5:1. For bridge crossings that are perpendicular to the river, the major axis should be at 90° to the highway embankment. Based on laboratory tests, this geometry worked as well as or better than any other shape. The length of the dyke required (L_S) was determined from model studies performed at Colorado State University and limited field data collected during flood. It was based largely on the spur dyke discharge ratio, Q_{fP}/Q_{30}, which is

$$\frac{Q_{fP}}{Q_{30}} = \frac{\text{flow over floodplain on one side, measured at section 1 (m}^3\text{/s)}}{\text{discharge in 30 m of opening adjacent to abutment, measured at section 1 (m}^3\text{/s)}} \qquad (7.5)$$

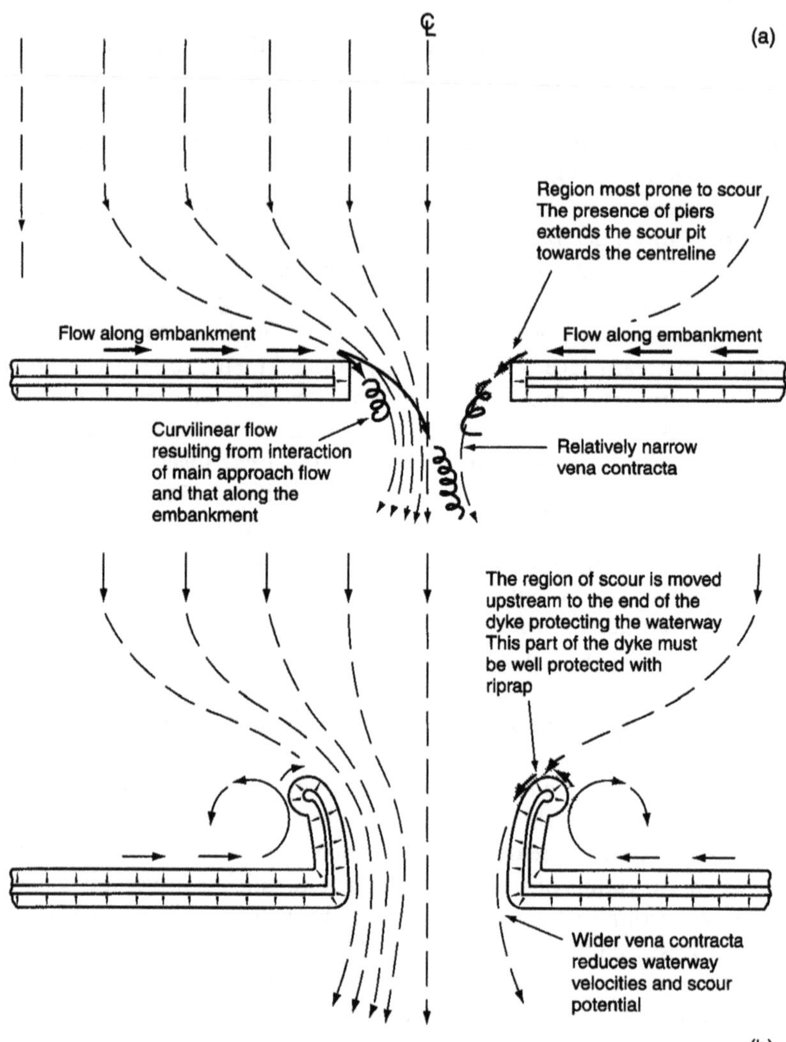

Fig. 7.11 Illustration of the effect of spur dykes when there is significant flow along the face of the highway embankments. The stippling shows the areas most affected by scour. (a) Without the spur dykes the flow along the embankment interferes with the main waterway jet, resulting in a potentially destructive curvilinear flow and high velocities. (b) By intercepting the flow along the embankment, spur dykes reduce the apparent severity of the contraction and move the most scour-prone area from the upstream corner of the embankment to the end of the spur, which is relatively unimportant.

This ratio relates the flow over one of the floodplains to the flow through the 30 m of the bridge waterway adjacent to the abutment, both measured at section 1 (Fig. 7.12). This value is combined with a representative velocity adjacent to the abutment, V_{N2} ($= Q/A_{N2}$), which is the average velocity through the opening at normal depth as in previous chapters. It can be seen from the figure that the length of dyke needed increases with both Q_{fP}/Q_{30} and V_{N2}.

For designing a spur dyke Bradley (1978) suggested the following:

- Lean towards overdesign rather than underdesign.
- If the length obtained from the chart is less than 15 m, a spur dyke is not needed.

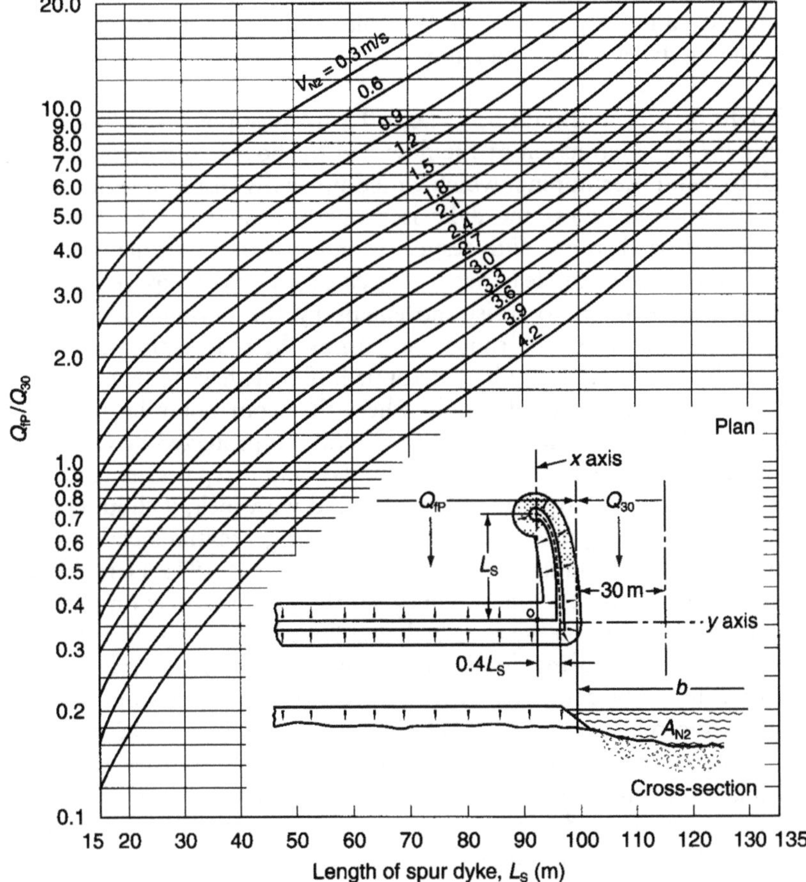

Fig. 7.12 Chart for determining the approximate length of spur dyke needed. On the inset diagram, the stippled area on the toe of the dyke illustrates the part most in need of scour protection. (After Bradley, 1978)

- If the length obtained from the chart is between 15 m and 30 m, a spur dyke at least 30 m long should be constructed so that the curvilinear flow around the end of the dyke merges with the flow in the main channel and straightens before reaching the abutment. Schneible (1966) suggested that 45 m should be the standard minimum length since this allows some damage without reducing effectiveness. Brown and Recinos (1988) reported that short spur dykes can cause an increased energy gradient and thus greater erosion or scour, so scour potential should be the controlling factor for selecting a spur length.
- The lengths obtained from Fig 7.12 may also be suitable for skewed crossings.
- There is no direct relationship between dyke length and the span of the bridge, which is why a 30 m width of the opening was considered.
- There is very limited data regarding the performance of spur dykes, so proceed cautiously and/or (if possible) undertake model tests when designing dykes for any location.

The height of a dyke should be at least 0.3 m above the design flood after allowing for the contraction of the flow and freeboard, and certainly large enough to avoid overtopping, which would normally result in serious damage or destruction unless the dyke is properly armoured or constructed from suitably sized stone. With respect to construction, Bradley suggested that the dykes may be made entirely of rock provided the exposed faces are of stone large enough to resist movement by the current. Earth dykes should be compacted to the same standard as the road embankment and should be high enough not to be overtopped. If costs have to be minimised, then rock armouring is needed only where the scour potential is greatest (Figs 7.11 and 7.12), provided that the remainder is protected by vegetation, turf and/or by the use of geotextiles or similar, and if repairs are conducted after each major flood. If earth dykes are faced with rock, it should be well graded, and have an appropriate filter blanket or geotextile membrane and riprap protection extending to below scour level (Neill, 1973; Hemphill and Bramley, 1989). With non-cohesive material the riprap protection to the toe of the dyke can be left on the river bed and 'launched' into its final position as the bed is eroded during flood, whereas with cohesive material the bank protection should be continued down to the expected worst scour level and the excavation refilled (Fig. 8.35). Other suggestions were to keep trees as close to the toe of the spur dyke as possible (to reduce transverse flow), to avoid digging borrow pits or drainage channels along the bottom of the road embankments, and to put a small pipe through the dyke to drain the area trapped between the spur and the road embankment.

Although spur dykes are often used in pairs there may be situations where only one is needed, such as where the flow in a wide channel already occurs along one well-defined river bank. It is also possible that spur geometries may have to be adjusted to take into account local conditions, such as a large bend or meander when the curvature may have to be increased on the inside of the bend and reduced on the outside (Neill, 1973). With skewed crossings the spur may be longer on the side where the flow becomes trapped, such as the right side of Fig. 3.8 (Richardson *et al.*, 1990). Richardson and Simons (1984) advised that the spur dykes should be designed to fit the streamlines of the approach flow, especially with skewed crossings, instead of automatically being set perpendicular to the embankments.

American practice generally is to use elliptical spur dykes that are convergent in plan. In India and Pakistan the spur dykes tend to be parallel to the opening and straight, except for a curved portion at the upstream and downstream ends. Typically the spurs extend $0.75b$–$1.0b$ upstream and $0.1b$–$0.25b$ downstream of a bridge of opening width b. Possibly straight spur dykes are better at producing a straight, parallel flow in the opening, which may be advantageous if there are piers. Richardson and Simons (1984) provided design guidelines for either type.

For use with the USGS procedure described in Chapter 4, Matthai (1967) presented charts for the adjustment factor, k_d, relating to straight and elliptical spur dykes used in combination with type III openings (Figs. 7.13 and 7.14). It is interesting to note that in Fig. 7.14a for normal crossings (i.e. without skew) the value of k_d is between 1.0 and 1.4 depending upon the conditions and the shape of the dyke, so spur dykes can significantly improve the coefficient of discharge and the flow through the opening. The adjustment factor is higher for the elliptical dyke, supporting Bradley's observation that this is the best shape.

The elliptical dyke in Fig. 7.13 is basically an extension of the type III abutment. The adjustment factor (k_d) in Fig. 7.14a varies with the opening ratio (M) and the length of the spur dyke L_S, which is expressed as L_S/b where b is the average width of the opening. Diagram a is for $\phi = 0°$ (i.e. no skew) whereas diagram b is for $\phi = 20°$ and gives the skew adjustment factor k_a, which is used in addition to k_d. For skew angles between 0° and 20° interpolate between the values obtained from diagrams a and b. Thus for a type III opening with an elliptical spur dyke the coefficient of discharge, $C = C' (k_\phi, k_x, k_d, k_a)$, to which may be added any other adjustment factor required as a result of deviation from the standard conditions, as in equation 4.10 and Figs 4.8–4.10. The value of x used to determine k_x should be measured on the embankment slope as if the dyke did not exist.

The straight dyke in Fig. 7.13 may be set back from the abutment by a variable distance, b_d. The adjustment factor k_d varies with L_S/b and M (Fig. 7.14c), but another factor is needed to allow for the variable offset distance b_d. This is k_b, which is obtained from Fig. 7.14d. Hence the coefficient of discharge for the type III opening with straight dykes is $C = C' (k_\phi, k_x, k_d, k_b)$ to which may be added any other adjustment factor required as a result of deviation from the standard conditions. The value of x used to determine k_x should be measured on the embankment slope as if the dyke did not exist.

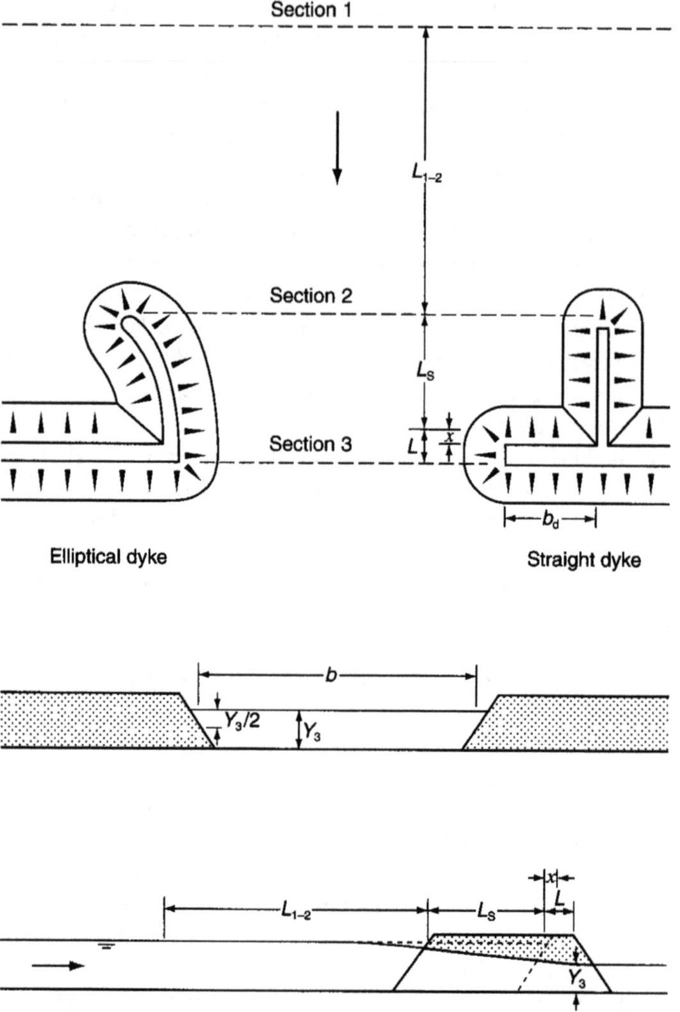

Fig. 7.13 Definition sketches of elliptical and straight spur dykes. (After Matthai, 1967. Reproduced by permission of US Geological Survey, US Dept of the Interior)

The addition of a spur dyke increases the friction loss and necessitates an alteration to equation 4.9, which is used to calculate the discharge explicitly. This involves changing the last bracket of the denominator from $\left[L + \frac{(L_{1-2}K_3)}{K_1}\right]$ to $\left[L + \frac{(L_{1-2}K_3^2)}{K_1K_S} + \frac{L_SK_3}{K_S}\right]$, where L_S is the length of the spur dyke and K_S is the conveyance at the upstream end of the spurs.

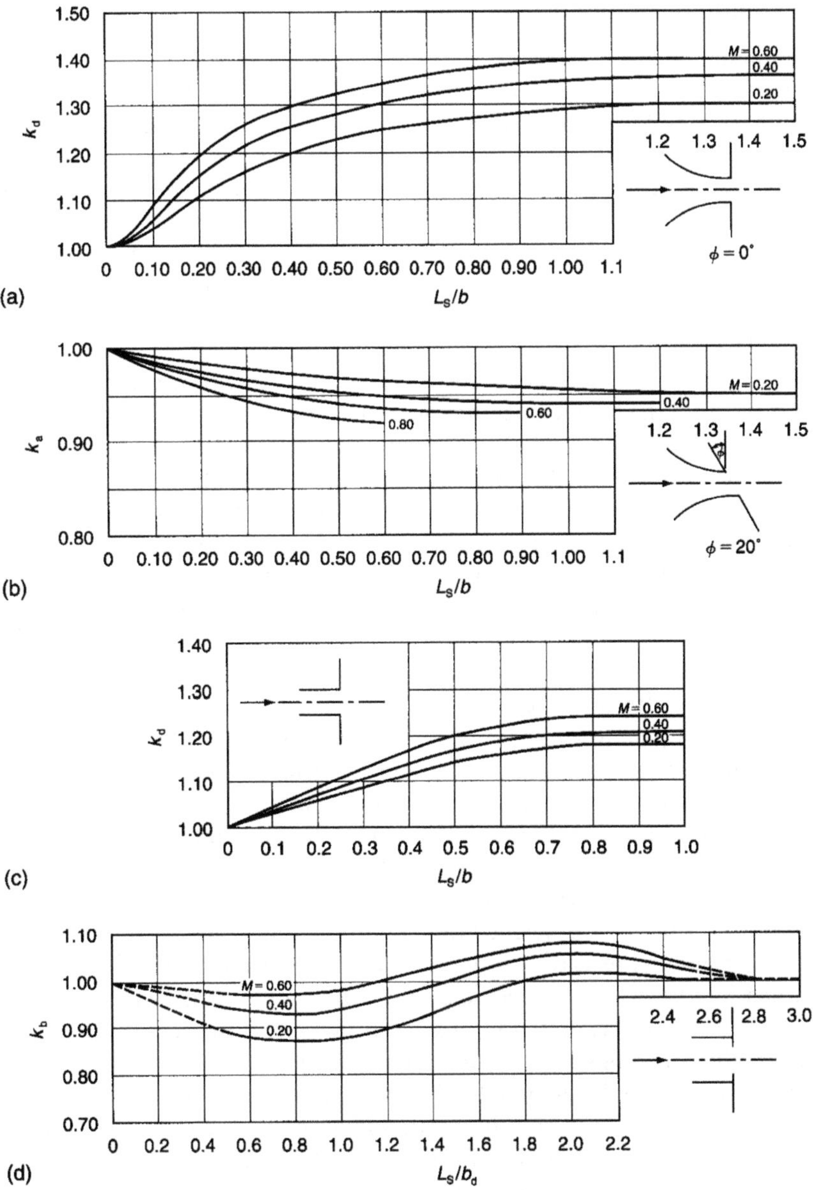

Fig. 7.14 USGS adjustment factors for spur dykes (see the text for details of their use): (a) elliptical dykes, no skew; (b) elliptical dykes, 20° skew (for other angles interpolate); (c) straight dykes, no skew; (d) adjustment factor for the offset distance, b_d, of straight dykes. (After Matthai, 1967. Reproduced by permission of US Geological Survey, US Dept of the Interior)

7.5 Minimum energy bridge waterways

A 'minimum energy' waterway is different in most respects from anything considered previously. A conventional, subcritical waterway generally consists of a sudden contraction followed by a sudden expansion, which results in a significant energy loss. To overcome this energy loss there is an increase in the upstream water level (the afflux) so that the flow accelerates through the opening. A minimum energy waterway achieves the same thing not by causing an afflux, but by using three-dimensional streamlining to accelerate the flow and to eliminate all energy losses other than friction, which is very small or negligible. With a negligible energy loss there is a negligible afflux.

Minimum energy designs consist of three main sections: the converging inlet transition; the bridge opening (or throat), which usually has a constant width and a level bed; and the diverging outlet transition. Since the plan shape is determined by drawing the flow net for two-dimensional irrotational flow, the inlet and outlet transitions are usually curved (Fig. 7.15a and b). The entire structure is streamlined to eliminate energy losses so it is often assumed that the total energy is constant throughout: hence minimum energy waterways are also referred to as 'constant energy' waterways (Fig. 7.15c).

The purpose of the narrowing inlet transition or fan is to reduce smoothly the width of the channel from its original value (B) to the width of the bridge opening (b). Ideally, to minimise the width of the opening, this should also be accompanied by a gradual fall in the elevation of the bed. The inlet serves to accelerate the flow, usually (but not necessarily) to the critical condition, so that at the design discharge, critical depth is maintained throughout the opening. Critical depth (Y_C) corresponds to the minimum specific energy that can maintain a given discharge, and so is theoretically the most efficient flow condition (Fig. 7.16). An alternative way of expressing this is that at the critical depth the maximum discharge is obtained for the available specific energy. Minimum energy waterways are designed to utilise this fact, and in appropriate circumstances can result in much smaller openings than conventional designs.

Critical depth can be achieved either by a reduction in width alone or by a combination of a reduction in width and a downward slope of the river bed as it enters the opening (Fig. 7.15b and c). By depressing or lowering the invert a narrower bridge opening is obtained than with a simple width constriction. The opening itself usually has a constant width and a flat bed (long culverts can be designed with a slope equal to the friction loss). In the outlet transition the channel widens and the bed rises back to its normal downstream width and level. Thus some of the novel design concepts usually inherent with minimum energy waterways are a constant total energy, negligible afflux, a totally streamlined design that involves a depressed bridge opening, and flow at the critical depth throughout the opening.

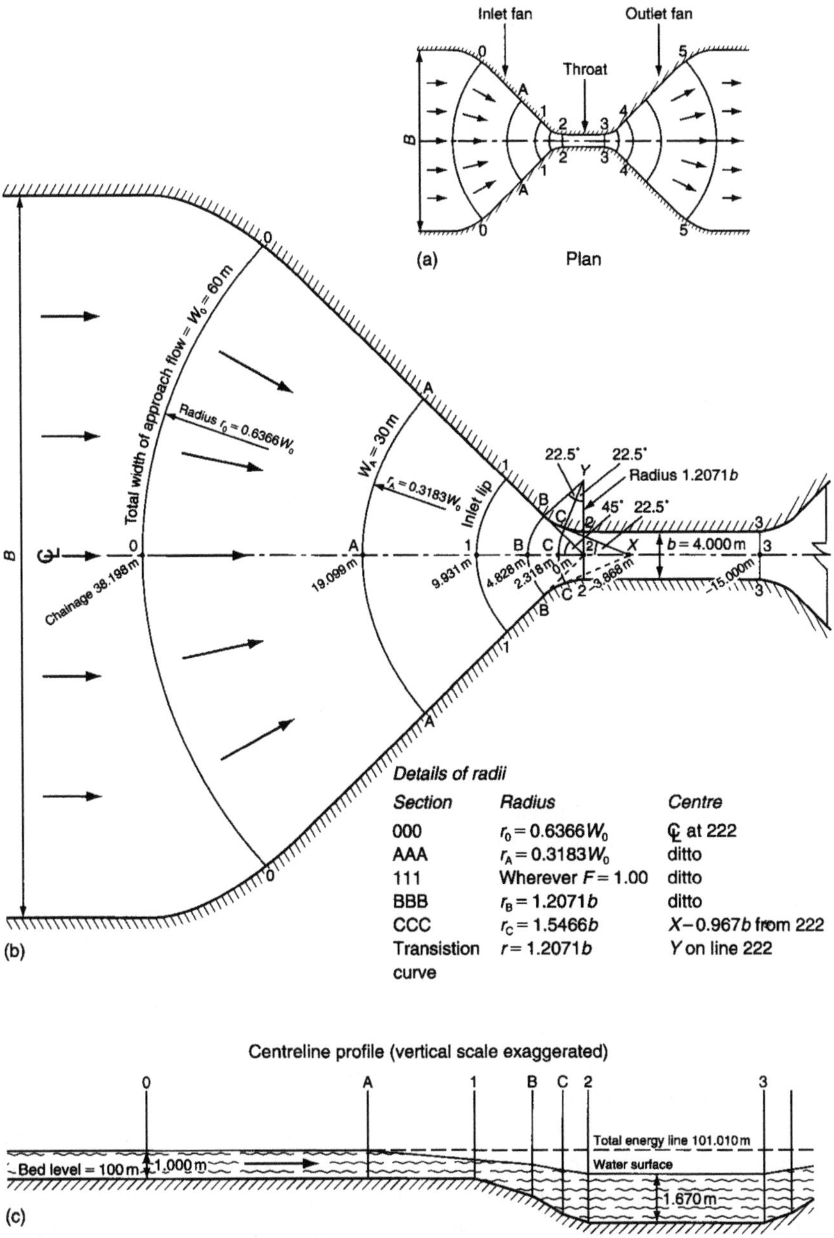

Inlet fan
Outlet fan
Throat
B
(a)
Plan

Total width of approach flow = W_0 = 60 m
Radius $r_0 = 0.6366W_0$
W_A = 30 m
$r_A = 0.3183W_0$
Inlet lip
22.5° 22.5°
Radius 1.2071b
45° 22.5°
b = 4.000 m
Chainage 38.198 m
19.098 m
9.931 m
4.828 m
2.318 m
0 m
3.868 m
15.000 m
B
(b)

Details of radii

Section	*Radius*	*Centre*
000	$r_0 = 0.6366W_0$	℄ at 222
AAA	$r_A = 0.3183W_0$	ditto
111	Wherever $F = 1.00$	ditto
BBB	$r_B = 1.2071b$	ditto
CCC	$r_C = 1.5466b$	$X - 0.967b$ from 222
Transistion curve	$r = 1.2071b$	Y on line 222

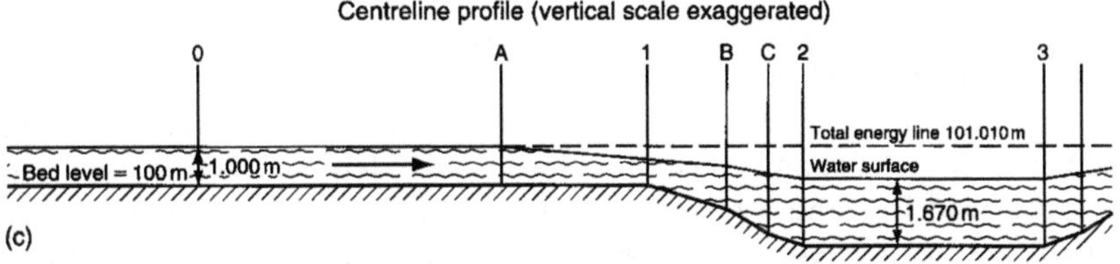

Fig. 7.15 Details of minimum energy waterway in Example 7.4: (a) plan view of inlet fan, throat and outlet fan; (b) geometry of inlet fan and throat – general details above the centreline, lengths specific to Example 7.4 below; (c) longitudinal centreline profile showing the horizontal (constant) total energy line and the depressed invert.

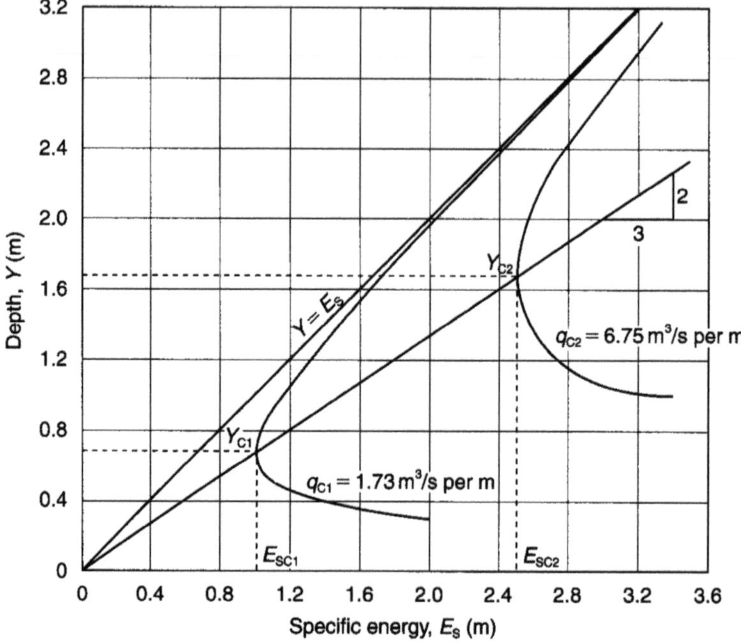

Fig. 7.16 Specific energy curves relating to Example 7.4. Note the vertical nature of the Y–E_S curve at the critical depth, which can result in flow instability.

The idea of the minimum energy waterways first appeared in Australia in the 1960s and 1970s (McKay, 1970, 1978). More or less simultaneously Cottman was puzzling over the problem of existing roads with inadequately sized waterways that crossed dry, or almost dry, river beds. These structures were prone to flood damage. The provision of adequate openings was considered too expensive, so instead attempts were made to increase the capacity of the existing openings by streamlining the approach and exit to enable floods to pass with minimum afflux, minimum damage and maximum hydraulic efficiency. Ultimately, this resulted in three-dimensional streamlining and the minimum energy waterway, which is theoretically the optimum design (Cottman, 1981; Apelt, 1981).

An analogy can be used to illustrate why minimum energy waterways are effective. In pipe flow, an orifice plate forms a sudden contraction resulting in a large head loss and a relatively low coefficient of discharge. This may be considered equivalent to a conventional bridge waterway. A streamlined Venturi meter forms a relatively small obstacle to flow, results in a small loss of head, and has a relatively high coefficient of discharge. This may be considered as equivalent to the minimum or constant energy waterway. However, this simple analogy does not include the additional advantage of maximising efficiency by inducing critical flow at the design discharge.

Although minimum energy waterways and culverts have their advocates, they have never been unanimously accepted (Griffiths, 1978). Concerns are frequently raised regarding: the wisdom of designing at the critical depth, which is relatively unstable, so that a small change in specific energy can result in a significant change in the depth of flow (Fig. 7.16); the difficulty of calculating longitudinal flow profiles accurately; the possibility of the deliberately 'undersized' opening being blocked by debris during flood; and problems relating to siltation and scour (Section 7.5.4). In defence of the concept, it is pointed out that while minimum energy structures are usually designed to operate at the critical depth because this makes the calculations easier (Section 7.5.1) and results in the smallest opening, they can be designed to operate subcritically if desired. In addition, successful minimum energy waterways have been constructed, and these have passed floods larger than the design flow and operated without trouble for decades (Cottman, 1981; Cottman and McKay, 1990). Although some minimum energy structures have been only partially successful or unsuccessful, this is usually due to a failure to understand and implement fully the design concept (e.g. by omitting the streamlined outlet fan). Thus minimum energy waterways are not something to be considered without first having consulted the full literature (not just a summary) and having decided whether the prevailing conditions are suitable.

If minimum energy waterways are to be effective they must be designed with total commitment to the principles involved, and constructed only where the conditions are suitable, success being unlikely otherwise. Cottman and McKay (1990) reported that minimum energy designs were suitable for:

- bridging or culverting relatively wide, shallow subcritical flows;
- increasing the flow capacity of bridges when the soffits are high enough and the bed can be lowered;
- passing flow under fords or floodways;
- equalising flood levels across causeways;
- controlling the level of lakes and swamps;
- enabling limited low-lying areas to be reclaimed (McKay, 1970, 1978).

Minimum energy waterways are economical, and particularly suited to streams that flow intermittently and which are dry between floods.

Minimum energy waterways are not suited to permanently flowing streams or estuaries, where a conventional piled structure would probably be cheaper. Their economical advantage is also doubtful if the abutment height exceeds 6 m or when the throat is less than 6 m long. Additionally, they are not suited to situations where:

- the flow is naturally deep, fast and well defined;
- the streams are steep, with wide supercritical flows over bedrock or bouldery beds;

- underground services make the lowering of the waterway entrance expensive;
- ponded water (that cannot be drained) in the lowered parts of the waterway is unacceptable, such as in urban areas;
- a backwater from further downstream drowns the waterway and prevents it from operating at the critical depth – if this is likely to be a problem then a good knowledge of the stage–discharge curve and flow behaviour at the site is required before considering such a structure.

The need for bridge piers may also cause problems, as described in Section 7.5.3.

The capacity of an existing bridge opening can be increased by turning it into a minimum energy waterway, provided the bed can be lowered and the existing deck is at a suitable level. This may lead to a much reduced afflux (or almost zero afflux) while also allowing a larger flood to be passed without significant scour or damage to the structure. Cottman and McKay (1990) quoted as an example the twin 7.5 m span Newington Bridge, which crossed a floodplain. Its capacity was enlarged from 23 m^3/s to a design discharge of 141.5 m^3/s (and the backwater reduced at all flows) through conversion to a minimum energy waterway. The cost involved was about one-sixth of the alternative, which was to add an extra 10 spans. However, there are situations where a minimum energy waterway may be the most economic option to begin with, namely where large, intermittent flows occur in wide channels and it is desirable to construct the smallest crossing possible. McKay (1970) described a situation in the Northern Territory of Australia where roads have to cross 8 km (5 mile) wide river valleys, which on occasions carry enormous flows at very low velocities (0.15 m/s), where any economic embankment is likely to be overtopped, and where there is a shortage of stone for protection works. The alternative to a conventional crossing was to carry the flow equivalent to the height of the embankment through minimum energy culverts without afflux, thus minimising the head difference across the embankment so that flow over the roadway could occur with relatively little damage.

7.5.1 *Equations for critical flow in a rectangular channel*

Since minimum energy waterways are normally designed to operate at critical depth at the design discharge, it is useful to review the relevant principles and equations pertaining to critical flow. Usually rectangular openings are constructed which, because the geometry of a rectangle is easy to define, simplifies the equations. When dealing with critical flow it is often most convenient to work with specific energy (instead of total energy). The specific energy (E_S) is the energy calculated above bed level:

$$E_S = Y + \frac{V^2}{2g} \qquad (7.6)$$

where Y is the depth of flow and $V^2/2g$ is the velocity head. Now in a rectangular channel the critical velocity (V_C) at which the flow becomes critical is given by

$$V_C = (gY_C)^{1/2} \tag{7.7}$$

where Y_C is the critical depth of flow corresponding to V_C. Substitution of this expression for V_C and the critical depth (Y_C) into equation 7.6 results in the equation for critical specific energy (E_{SC}):

$$E_{SC} = Y_C + \frac{gY_C}{2g} = Y_C + 0.5Y_C$$

$$\text{so } E_{SC} = 1.5Y_C \tag{7.8}$$

$$\text{or } Y_C = 0.667E_{SC} \tag{7.9}$$

Thus for a rectangular channel a characteristic of critical flow is that the critical velocity head equals $0.5Y_C$. The critical depth (Y_C) is easily calculated as two-thirds of the critical specific energy.

The design of a minimum energy waterway assumes that there is no loss of energy from inlet to outlet and that the total energy (or head) line is horizontal, so this provides a very convenient datum to work from. By calculating the specific energy at any cross-section and measuring down from the total energy line the required bed level can be obtained (Fig. 7.15c). This is illustrated in Example 7.4.

The total discharge (Q_C) in a rectangular channel (of width, B) flowing at critical depth is

$$Q_C = B\,Y_C\,V_C \tag{7.10}$$

For the specific case of a minimum energy transition, the channel width must be measured along the orthogonals to the streamlines. For example, it is along the curved line 111 in Fig. 7.15 that the flow must become critical, not along a line at right-angles to the centreline. Thus replacing the channel width (B) with the width of the curved orthogonal (W) and using equation 7.7 to replace V_C gives

$$Q_C = WY_C\,(gY_C)^{1/2}$$

$$= W(gY_C^3)^{1/2} \tag{7.11}$$

$$\text{and } Y_C = (Q_C^2/\,gW^2)^{1/3} \tag{7.12}$$

This equation illustrates that for a given value of Q_C the critical depth is inversely proportional to the channel width, so as the approach to the bridge waterway narrows the critical depth increases, then decreases again as the channel widens downstream of the opening. Equations 7.12 and 7.8 also illustrate that once the inlet fan has accelerated the flow to the critical depth with $F = 1.00$, the only way to maintain this condition is either to keep the channel width (W) constant or to simultaneously narrow the channel while dropping the invert to accommodate the increased specific energy and depth of flow. The latter is usually adopted because it allows a narrower crossing.

Equation 7.12 is often written in terms of the discharge per metre width of channel (q_C), where

$$q_C = \frac{Q_C}{W} \tag{7.13}$$

$$\text{so } Y_C = \left(\frac{q_C^2}{g}\right)^{1/3} \tag{7.14}$$

$$\text{and } q_C = (gY_C^3)^{1/2} \tag{7.15}$$

From equation 7.9, $Y_C = 0.667E_{SC}$ and $g = 9.81\ \text{m/s}^2$, so:

$$q_C = (9.81\ [0.667E_{SC}]^3)^{1/2}$$

$$\text{giving } q_C = 1.705E_{SC}^{3/2} \tag{7.16}$$

$$\text{and } E_{SC} = (q_C/1.705)^{2/3} \tag{7.17}$$

These are the critical flow equations relating the total design discharge (Q_C), or the equivalent unit discharge (q_C), to the corresponding critical depth (Y_C) and critical specific energy (E_{SC}). They are used in Example 7.4 to design a minimum energy waterway.

7.5.2 *A more detailed description of a minimum energy waterway*

One of the controversial aspects of minimum energy waterways is that they are designed to optimise the passage of a particular design flood, whereas most of the floods encountered will be of a different magnitude. There is often some concern as to how such a waterway will perform under these conditions. If $Q < Q_C$ then a subcritical flow will probably be obtained, and the actual depth of flow (Y) will be greater than Y_C. Thus when designing a minimum energy waterway it must be remembered that the soffit should be high enough to cope with subcritical depths substantially greater than Y_C. If possible, the soffit level should be above the maximum total energy level occurring upstream. If $Q > Q_C$ then initially it is likely that critical flow will be maintained through a greater length of the waterway than at the design discharge. This can often be tolerated, but it is possible that the flow might choke (Section 3.3.3). So to summarise, in a minimum energy waterway as the discharge increases from zero the depth of flow will increase to a maximum subcritical level, after which it will fall to the design critical depth and then start to rise again as the discharge continues to increase.

The shape of the whole structure is determined by drawing streamlines (Fig. 7.15), the general aim being to capture as much of the flow as required and to pass it smoothly through the waterway without causing separation of the flow, form drag or anything that will result in loss of energy (bridge piers can be a problem; see Section 7.5.3). Apelt (1981) warned that if the ratio of length/width of the inlet fan is made too small the flow will no longer be controlled by its plan shape and will enter almost as a parallel flow that chokes before it reaches the throat. A minimum length/B of about 0.5 was recommended. Cottman and McKay (1990) suggested that the overall length of the transitions needs to be from 1 to 3 times $(B - b)$, but also gave the dimensions of a generalised plan that can usually be adopted. Whether or not the fans are totally curved in plan or consist of several straight lines is not too important as long as the deflection angle between short lengths of straight guidewall is limited to 6°, the bed level is calculated accordingly, and the flow net principle is adhered to. The orthogonal lines drawn at right angles to the streamlines can be thought of as contours: the elevation of the bed should be constant along these lines. Since it is the length (W) of these lines that represents the width of the channel (not the width, B, perpendicular to the centreline), they are often assumed to be part of the perimeter of a circle for ease of calculation (Fig. 7.15b).

Initially the approaching subcritical flow must be spread uniformly across the whole width of the upstream channel at section 000, where the upstream transition begins, otherwise the flow net is invalid. If this does not occur naturally it can be achieved artificially by altering the channel, constructing some form of spreading basin, or by splitting the channel with each subchannel leading to an individually designed opening. An uneven approach flow or conditions not meeting the assumptions will lead to increased energy losses or a reduced flow capacity, or sometimes the occurrence of large-amplitude standing waves on the surface of the flow (Apelt, 1981).

The inlet and exit fans can be mirror images of each other in plan, or they can be adjusted to suit the requirements of the site so that they are very different. Similarly, although the entrance to the inlet fan and exit of the outlet fan can be designed with the same invert level, at most sites there would be a natural fall of the energy line and bed level from section 000 to 555. This can be allowed for by designing the waterway according to minimum energy principles, then rotating the design about the inlet lip. Over-rotation may lead to supercritical flow in part of the waterway. Under-rotation or the failure to allow for friction losses may lead to choking. Friction losses are normally very small or negligible, but obviously increase with the length of the waterway.

For both the inlet and outlet fans, starting with the design disharge (Q_C) the calculations can progress using either:

- the bed gradient (i.e. E_{SC}, which is the difference between the horizontal total energy line and the bed level), which enables Y_C, q_C, and W to be calculated from equations 7.9, 7.16 and 7.13; or
- the boundaries of the fans (i.e. the channel width, W, obtained from the flow net), in which case q_C , E_{SC} (and hence the bed level) and Y_C can be calculated from equations 7.13, 7.17 and 7.9.

The various parts of the waterway will now be considered in more detail.

Section 000 to 111: start of the inlet transition to the lip of the fan

This part of the inlet transition usually consists of an unlined levelled bed. The bed can be flat in the direction of flow, but if a fall is required a gradient of 1:50 is reputedly a good choice for bridges with large flows (1:30 for culverts) with a range of 1:15 to 1:100 having worked well (Cottman and McKay, 1990).

The upstream end of the fan should be wide enough to collect all of the flow from the channel that has to be funnelled through the waterway. The distribution of the flow across the channel should be approximately uniform, and the entrance velocity should be less than 1.0 m/s. In any case, at 000 the flow must be subcritical at depth Y_0. However, as the channel narrows, to maintain continuity of flow the velocity must increase, and the design necessitates that critical depth (Y_{C1}) is achieved by section 111.

Section 111 to 222: inlet lip to the inlet of the throat

This section slopes downwards to accelerate the flow, the slope being determined by the designer (in conjunction with the width, W) with the aim of keeping the flow at crititical depth with $F = 1.0$. However, the practical range of bed gradients is 1:4 to 1:15 with 1:7 being a good compromise according to Cottman and McKay (1990). They also suggest that the part of the transition near to the opening entrance (perhaps 3–10 m but rarely extending beyond the toe of the batter of the roadfill) would normally be concreted.

Although Cottman and McKay suggested that the shape of the inlet transition could be determined purely from a flow net, they also gave a detailed generalised plan with relevant dimensions and radii of transitions, as shown in simplified form in Fig. 7.15. This indicates that the outer part of the fan is a sector of a circle with an angle of 90° at the centre, changing to 45° between section 111 and 222.

Section 222 to 333: the bridge waterway

This is usually constructed with a horizontal, concrete bed and with a uniform width, b. Again the flow is critical with $F = 1.0$.

Section 333 to 444: exit of the waterway to the lip of the outlet fan

The bed of this part of the outlet transition slopes upwards while the width of the channel increases, slowing the flow. The objective is to maintain the flow at the critical depth with $F = 1.0$ up to the lip of the outlet fan. The design uses basically the same principles as section 111 to 222. The construction would also be similar to the equivalent part of the inlet.

Section 444 to 555: lip of the outlet fan to the end of the outlet transition

This is a region of decelerating flow and decreasing Froude number as the channel widens and the flow returns to the subcritical condition. The bed would usually be level, or nearly level, and constructed of natural material. This can be considered the reverse of the equivalent part of the upstream transition.

An example of the design of a minimum energy waterway is included as Example 7.4. The plan dimensions follow Cottman and McKay (1990). They presented a chart to simplify the hydraulic design, but this is not used because the equations in Section 7.5.1 provide a better illustration of the principles involved.

7.5.3 *Bridge piers*

Piers can cause a problem with minimum energy waterways (McKay, 1978; Apelt, 1981). Just outside the bridge opening the width of the channel is W (or B) while just inside the opening the width is W minus the combined width of the piers. In other words, the width of the channel is reduced instantaneously from W to b by the piers. This means that there will be an instantaneous increase in q_C, Y_C and E_{SC} (equations 7.13, 7.14 and 7.17), which requires an instantaneous drop in the level of the bed to maintain critical flow. This is impractical, because instantaneous changes in flow can rarely, if ever, be achieved. On the other hand, with a single-span waterway $W = b$ and there is no problem.

There are two possible solutions to the problem of bridge piers. The first is to split the approach channel into subchannels, each of which leads to an individually designed minimum energy waterway. Thus the piers become 'islands' between the various channels. The second solution is to introduce the piers gradually from bed level (i.e. the piers are raked in the direction of flow) while gradually lowering the bed, as shown in Fig. 7.17. According to McKay, piers of constant thickness with a half-round nose and sloping at 45° will effect a reasonably smooth transition. This does not work with pile bents, so it is logical to avoid converting an existing bridge of this type to a minimum energy waterway, while with new construction the designer can decide what type of pier to adopt.

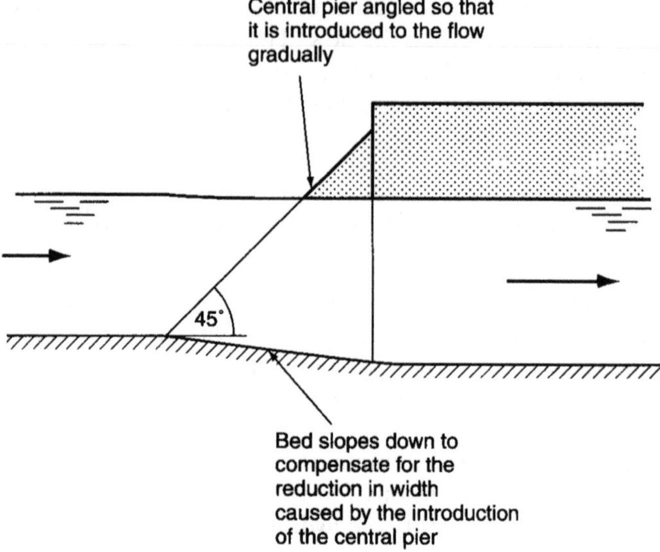

Fig. 7.17 Use of 45° raked central piers with minimum energy waterways. The pier is gradually introduced to the flow and the invert is lowered to compensate for the change in specific energy arising from the sudden reduction in channel width caused by the pier.

7.5.4 Siltation, scour and performance

The question of siltation and scour frequently arises in connection with minimum energy waterways. Siltation is perceived as a particular problem when the waterway has a depressed invert. However, this is not the case during flood when the velocity through the bridge opening is higher than in the inlet transition, so anything that is carried in should be carried out, and possibly away (McKay, 1970). It is possible that some siltation may occur on a falling stage, but unless the sediment is cohesive, becomes baked hard or covered in vegetation, this will normally be removed again when the flow increases. Scour is usually assumed to be a problem because velocities can be relatively high, perhaps 5 m/s or so in the throat of a minimum energy waterway. McKay (1978) argued that if form losses are eliminated entirely through streamlining then the system becomes incredibly smooth, so there is no large-scale turbulence and hence no scour despite the high velocities.

Some measure of independent assessment of minimum energy structures was provided by Loveless (1984), who used relatively small-scale model tests to compare the performance of a standard and minimum energy culvert design, including the effects of sedimentation and scour. The standard culvert had wingwalls set at 30° to the centreline and a level bed throughout.

The minimum energy design had a depressed waterway, but unfortunately the outlet transition had to be truncated slightly, so the design was not ideal. This should be borne in mind when interpreting the results.

At the design discharge (Q_C) it was found that the afflux of the minimum energy culvert was 28% of that for the conventional culvert design. Theoretically the afflux should have been zero. Loveless attributed the difference to the truncated outlet fan, and the fact that the theory *assumes* that critical depth will occur in the control sections rather than showing that it must. At $Y_u/Z = 0.4$ the minimum energy structure passed 100% more flow than the conventional culvert, and at $Y_u/Z = 0.9$ it passed 80% more flow (Q_C corresponded to about $Y_u/Z = 0.73$). By $Y_u/Z = 1.2$ the discharge was about 31% greater than Q_C and the advantage of the minimum energy design had diminished significantly, but it still discharged much more than the standard culvert.

With respect to siltation, Loveless assumed that the worst condition for the minimum energy culvert would be when the lowered invert had been completely filled with sediment, effectively turning it back into a conventional structure. Two types of sediment were used: sand and gravel. The sand began to move in the throat at a flow equal to 5% of Q_C; it was substantially cleared at 20% of Q_C and was completely cleared at 30% of Q_C. The gravel began to move at 30% of Q_C; it was substantially cleared at 44% of Q_C and completely cleared at 65% of Q_C. It was concluded that provided that the sediment is not cohesive, even large boulders would be removed from the depressed culvert throat well below the design discharge. It was also pointed out that when the discharge is about 66–90% of the design value the bed profile of a minimum energy culvert is identical to that which would be obtained if the bed was erodible.

With respect to scour, even with the gravel the conventional culvert produced a deep scour hole at flows greater than 45% of the design discharge, Q_C. In contrast, the minimum energy design resulted in no scouring of the gravel below a discharge of $1.18Q_C$, and relatively slight scouring of the sand bed at the design discharge.

The conclusions drawn by Loveless are worth repeating:

- A well-designed minimum energy culvert can discharge twice the flow of a standard design of the same size.
- Although siltation may occur in the lowered invert of a minimum energy culvert at discharges less than 30% of Q_C, all non-cohesive sediments will be removed by 65% of Q_C.
- Wide three-dimensional inlet and outlet transitions must be provided for minimum energy structures to perform well.
- Minimum energy culverts are less vulnerable to scour problems at the outlet.
- Theoretically, minimum energy culverts can be designed that will result in zero afflux.

From the limited data available it appears that minimum energy waterways are worthy of further investigation, and can be used under appropriate conditions to solve quite severe flooding problems.

7.6 Channel improvements

Instead of improving the hydraulic efficiency of the bridge waterway, in some situations it may be more productive to improve the channel, or to improve both the bridge and channel. It should also be remembered that erosion arising from lateral movement of the channel is perhaps the greatest cause of structural damage and bridge collapse (Brice, 1984; Trent and Brown, 1984). Therefore channel stabilisation, as well as flow improvement, may be the objective of channel improvement works (Sections 8.2.2 and 8.9.3). A few typical scenarios are described below.

Reduction of abnormal stage

At sites that experience low Froude numbers, large normal depths or abnormal stages, perhaps as a result of a backwater emanating from further downstream, channel improvements may result in lower stages and an increased discharge (Section 7.2.1). Typical improvements to a channel are the removal of bends, to shorten it and increase its gradient, and the removal of obstructions to produce a uniform channel with a reduced Manning's n value. If taken to the ultimate the improvement works may result in a straight, artificial channel of uniform section, which is effectively a new river channel with an increased conveyance. This may be desirable from the hydraulic perspective, but environmental considerations must also be weighed carefully. Care must also be taken to optimise the alignment of a new channel and to stabilise it: rivers have sometimes displayed a tendency to seek out their old course during flood, while the removal of bends can cause degradation (Fig. 8.8).

Improvement to the bridge approach

Ideally a bridge should be constructed perpendicular to a stable river channel. However, river morphology or economics may not allow this, so something has to be done to optimise a less than ideal approach. If the channel meanders, shifts its course regularly or approaches the bridge at an angle, river training works can be employed to establish a permanent channel. This may be achieved by stabilising the banks using riprap, gabions, or any number of proprietary protection systems. Under some circumstances it may be considered appropriate to cut through meanders to create a new channel with the ideal approach. This may affect other reaches (Fig. 8.8), so caution is necessary, and environmental factors must also be considered.

Dykes can be used to reduce the effects of skew (Fig 3.8) while groynes (or spurs) can be employed to reposition the flow in the channel (Figs 7.18 and 7.19). For example, if a bridge is located on a bend it may be necessary to prevent the natural migration of the channel by pushing the flow away

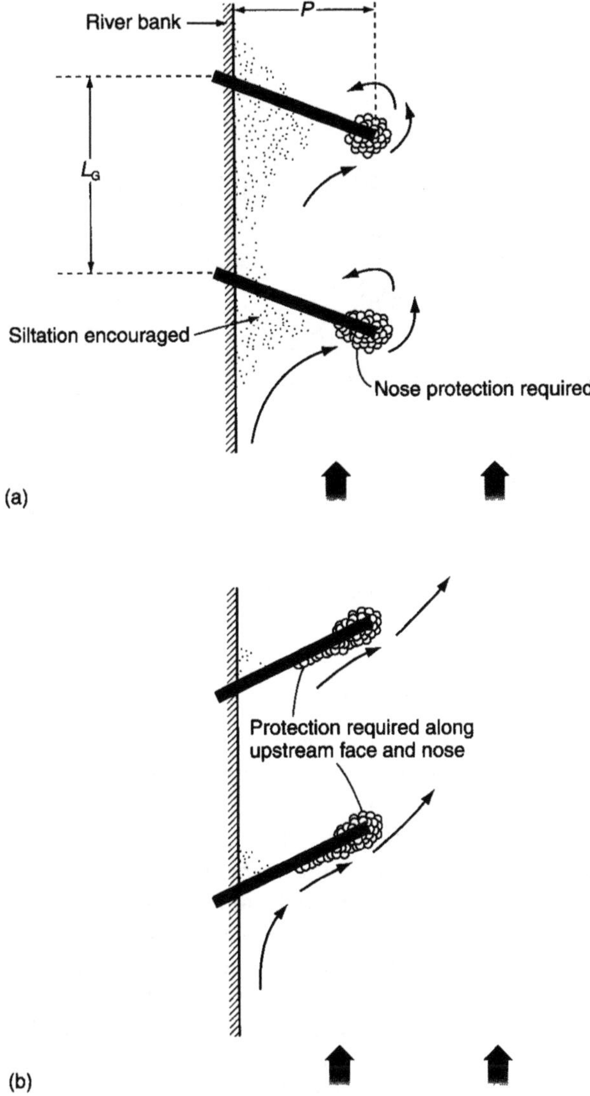

Fig. 7.18 Use of groynes to control the flow: (a) retardance groynes angled upstream, resulting in areas of relatively stationary water in the upstream corners with some siltation; (b) diverter groynes angled downstream so that there is flow along the length of the groyne, deflecting the current towards the centre of the channel.

Fig. 7.19 Groynes on the River Loire, France, with a concrete revetment on the bank. The flow is away from the camera. Note the size (the people on the groyne are barely visible) and the siltation, especially on the opposite bank.

from the outside of the curve. Groynes can be used in conjunction with, or instead of, spur dykes.

Improvement of the bridge opening ratio, M

The hydraulic performance of a bridge depends to a large extent upon the value of the opening ratio, $M = b/B$. If the flow in the river channel is very wide (B) and shallow while the opening is relatively narrow (b), the large contraction will result in poor hydraulic efficiency. As an alternative to increasing b it is possible to use river training works to reduce B instead. In the extreme, a narrower channel with new artificial banks could be created, but the consequences of this would have to be carefully assessed. A less extreme solution may be to use groynes to push the flow into the desired part of the channel (Figs 7.18 and 7.19). Novak *et al.* (1990) suggested that longitudinal dykes (training walls) constructed within the river channel roughly parallel to, but some distance from, the existing bank can be more economical than groynes, and more effective if properly positioned.

Control of erosion

If there is a danger of bank erosion in the vicinity of the bridge, groynes can be used to give some degree of protection, either by repositioning the flow in the channel or by encouraging siltation in protected areas (Neill, 1973; Petersen, 1986). There are also many other protective measures that can be adopted, as listed in Section 8.9.3. However, since groynes have been mentioned above several times, their use is described in more detail below.

7.6.1 Groynes (or spurs)

Groynes may be used to stabilise eroding streambanks in the vicinity of bridges or to reposition the flow in the channel. Basically groynes (also called spurs) are long, narrow embankments or walls that protrude from the bank into the river channel (Fig. 7.18). They may be used singly, but are normally used in groups because a single groyne can severely disturb the flow, resulting in deep scour at its outer end. The type of groyne adopted depends upon whether the primary function is to protect the existing bankline (perhaps by reducing velocities), to encourage sedimentation, to re-establish a previous stream alignment, to create a new stream alignment, or to narrow the channel by controlling the flow (Brown, 1984, 1985). These functions can be classified as either retardance or flow diversion (or a combination of both).

Retardance groynes should normally be perpendicular to the primary flow direction or point upstream so that they create a region of still water that is trapped in the corner of the upstream face encouraging siltation (Fig. 7.19). Scour tends to be concentrated around the nose of the groyne, so this must be adequately protected. In terms of construction, bank protection groynes may be permeable or impermeable, the permeable variety being useful in encouraging sedimentation when the river carries a lot of silt. Permeable groynes can be cheaply constructed from a double row of timber piles filled with trees and brushwood or stone, or wood and wire fences that collect debris and silt. Permeabilities can be between 35% (significant velocity reduction and flow control) and 80% (mild velocity reduction and flow control): the greater the permeability the smaller the area and depth of scour downstream. Retardance jacks (metal-wire frames or tetrahedrons) may be used in some parts of the world.

If the groynes are angled in the downstream direction the result will be to deflect or divert flow along the groyne towards the centre of the channel, so protection may be required along the full length of the upstream face. Groynes used to divert the flow have to be impermeable, a stone embankment or solid wooden structure being a typical construction.

Brown (1985) analysed over one hundred field sites and gave some general guidelines regarding the applicability of groynes.

- Groynes are not well suited to channels less than 46 m wide.
- They are not well suited to bends with a radius less than 107 m.
- Most groynes are best suited to protecting channels with banks less than 6 m high, and are particularly suited to protecting the lower part

of the bank from toe scour and in reducing undermining and bank failure.

- Groynes can be cost effective and economical where irregular channel banks have to be protected, especially if other protective measures require significant site preparation.
- They can be used in rivers with a wide variety of sediment types, but inclusion of the sediment type in the design process is critical to achieving effectiveness.
- Groynes can be a hazard to recreational river use.
- A common mistake is to provide streambank protection too far upstream and not far enough downstream.

The length of river bank protected by an individual groyne increases on the inside of a bend and decreases on the outside, but a typical protected length for an individual groyne that projects a distance, P, into the channel is roughly P to $2P$ both upstream and downstream. With groups of four or more groynes the protected length increases. Thus an economic and hydraulic decision has to be taken as to whether to use short, closely spaced groynes or longer, more widely spaced groynes. Long groynes will project further into the main channel, where the flow will be faster, so greater nose protection will be needed. Of course, the groynes should never be so long as to cause hydraulic problems on the opposite bank, while the passage of boats, ice and logs may also have to be considered. Brown suggested that the projected length of impermeable groynes should be less than 15% of the bank full channel width, with the equivalent length for permeable groynes being between 15% and 25% for permeabilities of $< 35\%$ and 80% respectively.

Farraday and Charlton (1983) gave the following equation as an indication of the spacing required for a group of groynes:

$$L_G < \frac{CY^{1.33}}{2gn^2} \tag{7.18}$$

where L_G is the spacing between groynes (m), C is a constant with a value of about 0.6, Y is the mean depth of flow (m), g is the acceleration due to gravity (9.81 m/s^2), and n is the Manning coefficient (s/m$^{1/3}$). However, this probably gives a false impression of precision and certainty; unless the problem is a simple one the effectiveness of groynes is difficult to assess, and hydraulic model tests are advised. Apart from orientation to the flow, length and spacing, other factors that have to be considered are the crest height with respect to flood flows and bank level, the variation of the crest height within a group (i.e. all the same, or increasing or decreasing in the direction of flow), and whether the crest height of a particular groyne is constant or slopes down to the nose. With complex stream geometries the orientation may have to be decided on a groyne-by-groyne basis with the aim of guiding the flow through the channel while protecting the banks. One suggestion is that the tip of the most upstream diverter groyne should be at around 150° to the main flow, each subsequent groyne having a smaller angle until the last one is perpendicular to the flow. If the design stage is lower than the river bank, the minimum height for the groynes should be 1.0 m below the design stage. If the design stage is above the river-bank then the groynes should be built to bank level. If the groyne is intended to be above flood level then the freeboard is typically between 0.5 m and 1.0 m (Brown, 1985). Embankment groynes typically have side slopes of 1 in 3 and a crest width of about 2.0–2.5 m.

7.7 Examples

The first three examples use laboratory data so that the calculated and actual result [in brackets] can be compared. Example 7.1 shows how easy the charts in Figs 7.5–7.9 are to apply if Q_F is known, and illustrates the typical range of error. Examples 7.2 and 7.3 show how the charts can be used even when little is known about the actual stage–discharge relationship.

Example 7.1

A model 300 mm span rectangular opening has Z = 150 mm, F_N > 0.25, Q_F = 26.3 l/s and Y_u/Z = 0.99 (channel flow). What reduction in upstream depth (for the same discharge) and increase in discharge (for the same upstream depth) does R85 give?

From Fig. 7.7b, with $Y_u/Z = 0.99$, interpolating for R85, $\Delta Y_u/Z = 0.035$ (i.e. 3.5%).
Thus $\Delta Y_u = 0.035 \times 150 = 5.3$ mm [5.0 mm].
Typical error in $\Delta Y_u/Z$ is ±0.8%, which gives an error in $\Delta Y_u = \pm 0.008 \times 150 = \pm 1.2$ mm.
From Fig. 7.8b, with $Y_u/Z = 0.99$, interpolating for R85, $\Delta Q/Q_F = 0.075$ (i.e. 7.5%).
Thus $\Delta Q = 0.075 \times 26.3 = 1.97$ l/s [1.76 l/s].
Typical error in $\Delta Q/Q_F$ is ±0.8%, which gives an error in ΔQ of $\pm 0.008 \times 26.3 = \pm 0.21$ l/s.

Example 7.2

A model has a 200 mm span arch, Z = 100 mm, F_N > 0.25 but Q_F is unknown. (a) When Y_u/Z = 1.30, what increase in Q is possible for the same stage? (b) When Q = 14.10 l/s, what increase in Q is possible for the same stage?

Q_F can be calculated from equation 2.8. Assume that $Y_u = 2Z$, the velocity head is negligible and that $C_d = 0.57$ (from Fig. 7.9 with $Y_u/Z = 2.0$).

Thus: $Q_{2F} = 0.57\, a_W\, [2g(2Z - Z/2)]^{1/2}$ giving $Q_{2F} = 15.36$ l/s.

$Q_F = Q_{2F}/2 = 7.68$ l/s [7.80 l/s].

(a) If $Y_u/Z = 1.30$, with R125 Fig. 7.8a gives $\Delta Q/Q_F = 0.20$ so $\Delta Q = 0.20 \times 7.68 = 1.54$ l/s [1.48 l/s].

(b) If $Q = 14.10$ l/s then $Q/Q_F = 1.83$. From Fig 7.5, $Y_u/Z = 1.80$.

From Fig. 7.8a with R125 and $Y_u/Z = 1.80$, $\Delta Q/Q_F = 0.34$ so $\Delta Q = 0.34 \times 7.68 = 2.61$ l/s [2.58 l/s].

Example 7.3

A model 250 mm span rectangular opening has $Z = 125$ mm and $F_N < 0.25$ (drowned orifice flow). As a result of gauging it is known that $Q = 19.87$ l/s when $Y_u = 172$ mm, but nothing else is known. When $Q = 27.62$ l/s, what increase in discharge can be achieved using R25 and R100?

$Y_u/Z = 172/125 = 1.376$. From Fig. 7.6 this is equivalent to $Q/Q_F = 1.4$.

Thus $Q_F = 19.87/1.4 = 14.19$ l/s [14.25 l/s].

When $Q = 27.62$ l/s, $Q/Q_F = 27.62/14.19 = 1.946$. From Fig. 7.6 this gives $Y_u/Z = 1.95$.

Thus $Y_u = 1.95 \times 125 = 244$ mm [242 mm].

To calculate ΔH, use equation 2.9: $Q = C_d a_W [2g\Delta H]^{1/2}$, where $Q = 27.62$ l/s, $Z/Y_u = 125/244 = 0.51$ and from Fig. 2.12 the provisional value of C_d is 0.71. Assume the velocity head is negligible. Solving gives $\Delta H = 79$ mm [80 mm].

Thus $\Delta H/Z = 79/125 = 0.63$.

From Fig. 7.9d the value of C_d is 0.71, so the provisional value is accurate.

From Fig. 7.8d with $\Delta H/Z = 0.63$, R25 gives $\Delta Q/Q_F = 0.080$.

Thus $\Delta Q = 0.080 \times 14.19 = 1.14$ l/s [1.15 l/s].

Similarly, R100 gives $\Delta Q/Q_F = 0.14$. Thus $\Delta Q = 0.14 \times 14.19 = 1.99$ l/s [1.90 l/s].

Example 7.4: minimum energy waterway

Produce an outline design of a minimum energy waterway capable of discharging 27 m^3/s given that at section 000 the length of the curved section (W_0) is 60 m, the bed level is 100 m above datum, and the depth of flow is 1.00 m. The proposed opening width = 4.00 m and its length is 15.00 m. In the absence of site details it is not possible to draw a flow net so the geometry of the transitions is taken from Cottman and McKay (1990), and it is assumed that the minimum energy waterway is symmetrical about both axes. Hence only the upstream transition is designed below, the downstream transition being the mirror image.

Section 000: start of the inlet fan, $W_0 = 60\,m$

Assume a 90° sector of a circle with its centre located at the entrance of the opening (222) and on the centreline of the bridge waterway (Fig. 7.15).

If the length of the sector's perimeter is $W_0 = 60\,\text{m}$ then $2\pi r_0/4 = 60\,\text{m}$ and hence $r_0 = 38.198\,\text{m}$ (i.e. $r_0 = 0.6366W_0$).
Thus the chainage of section 000 is 38.198 m from section 222 measured along the centreline.
At section 000 the total energy $E_0 = Z_0 + Y_0 + V_0^2/2g$, where $Z_0 = 100\,\text{m}$, $Y_0 = 1.000$ m and $V_0 = 27/(60 \times 1.000) = 0.450\,\text{m/s}$ ($< 1.00\,\text{m/s}$ as required).
Thus $V_0^2/2g = 0.010\,\text{m}$ and $E_0 = 100.000 + 1.000 + 0.010 = 101.010\,\text{m}$ above datum.
$F_0 = V_0/(gY_0)^{1/2} = 0.450/(9.81 \times 1.00)^{1/2} = 0.143$

Intermediate section AAA, $W_A = 30\,m$

This is half the perimeter length, and $2\pi r_A/4 = 30\,\text{m}$ gives $r_A = 19.099\,\text{m}$ (i.e. $r_A = 0.3183W_0$).
Thus the chainage of section AAA is 19.099 m from section 222 along the centreline.
Assume there is no fall in bed level between 111 and AAA, that Y_A is about 1.000 m, with $W_A = 30\,\text{m}$ then $V_A = 27/(30 \times 1.000) = 0.900\,\text{m/s}$.
$F_A = V_A/(gY_A)^{1/2} = 0.900/(9.81 \times 1.00)^{1/2} = 0.287$

Section 111: inlet lip, $F_1 = 1.00$

The chainage of the start of the inlet lip measured along the centreline from section 222 is a variable that must be determined by calculation. Basically the position of 111 (and the corresponding value of W_1) depends upon the design discharge and depth of flow and must be located so that $F_1 = 1.00$ (see the hydraulic calculations below). However, if the length of the inlet lip is W_1 then the chainage $r_1 = 0.6366W_1$.

Section 111 is where the flow must become critical, and where the assumption of constant total energy ($Z = 101.010\,\text{m}$ in this example) becomes important. If the bed is still horizontal at 100.00 m above datum then the specific energy $E_{SC1} = 101.010 - 100.000 = 1.010\,\text{m}$.

From equation 7.9, $Y_{C1} = 0.667 \times 1.010 = 0.673\,\text{m}$.
From equation 7.7, $V_{C1} = (9.81 \times 0.673)^{1/2} = 2.569\,\text{m/s}$.
From equation 7.16, $q_{C1} = 1.705E_{SC}^{3/2} = 1.705 \times 1.010^{3/2} = 1.731\,\text{m}^3/\text{s}$ per metre.
Using equation 7.13, $W_1 = Q/q_{C1} = 27/1.731 = 15.600\,\text{m}$.
Check: $F_1 = V_{C1}/(gY_{C1})^{1/2} = 2.569/(9.81 \times 0.673)^{1/2} = 1.00$ as required.

Chainage of section 111 from 222, $r_1 = 0.6366W_1 = 0.6366 \times 15.600 = 9.931$ m.

Intermediate section BBB: start of the transition curve to the bridge waterway

This section marks the end of the 90° sector and is defined as $r_B = 1.2071b$, where b is the width of the bridge opening. Thus $r_B = 1.2071b = 1.2071 \times 4.00 = 4.828$ m, so this is the chainage of the section measured along the centreline from 222.

The length of the curved perimeter of the sector $W_B = 2\pi r_B/4 = (2\pi \times 4.828/4) = 7.584$ m (i.e. $0.948b$).
If $W_B = 7.584$ m then from equation 7.13, $q_{CB} = 27/7.584 = 3.560\,m^3/s$ per metre.
From equation 7.17, $E_{SCB} = (3.560/1.705)^{2/3} = 1.634$ m.
Assuming the total energy line is constant at 101.010 m then the elevation of the bed $Z_1 = (101.010 - 1.634) = 99.376$ m above datum (i.e. 0.624 m below initial bed level).
From equation 7.9, $Y_{CB} = 0.667 \times 1.634 = 1.089$ m.
From equation 7.7, $V_{CB} = (9.81 \times 1.089)^{1/2} = 3.268$ m/s.
Check: $F_B = V_{CB}/(gY_{CB})^{1/2} = 3.268/(9.81 \times 1.089)^{1/2} = 1.00$ as required.
Check: $Q = W_B\, Y_{CB}\, V_{CB} = 7.584 \times 1.089 \times 3.268 = 26.990\,m^3/s$ OK

Intermediate section CCC

This is the midpoint of the transition curve and is located on the 45° sector. It is defined by $r_C = 1.5466b$. The centre of the 45° sector is located on the longitudinal centreline $0.967b$ downstream of section 222, so this distance must be subtracted from r_C to obtain the centreline chainage. Thus in this case $r_C = 1.5466 \times 4.00 = 6.186$ m so the chainage is $(6.186 - 0.967 \times 4.00) = 2.318$ m.

The length of the curved perimeter of the sector, $W_C = (2\pi r_C/8) = (2\pi \times 6.186/8) = 4.858$ m (i.e. $1.215b$).
The hydraulic calculations are as above, and are summarised in Table 7.3.

Section 222: opening entrance

This is the end of the transition coinciding with the entrance to the waterway and chainage 0. The waterway is rectangular in section with parallel sides, and the width of the waterway is b.

Here $b = 4.000$ m, so $q_C = 27/4.00 = 6.750\,m^3/s$ per metre.
From equation 7.17, $E_{SC2} = (6.750/1.705)^{2/3} = 2.503$ m.
Assuming the total energy line is constant at 101.010 m then the elevation of the bed $Z_2 = (101.010 - 2.503) = 98.507$ m above datum (i.e. 1.493 m below initial bed level).

Table 7.3 Minimum energy waterway design calculations for Example 7.4

Section	*Chainage (m)*	*Total energy, H (m)*	*Specific energy, E_S (m)*	*Bed level, Z (m)*	*Width, W (m)*	*Water depth, Y (m)*	*Velocity, V (m/s)*	*Discharge per m width, q (m^3/s per m)*	*Froude number, F*
000	38.198	101.010	1.010	100.000	60.000	1.000	0.450	0.450	0.143
AAA	19.099	101.010	(1.010)	100.000	30.000	(1.000)	0.900	0.900	0.287
111	9.931	101.010	1.010	100.000	15.600	0.673	2.569	1.731	1.000
BBB	4.828	101.010	1.634	99.376	7.584	1.089	3.268	3.560	1.000
CCC	2.318	101.010	2.198	98.812	4.858	1.465	3.791	5.558	1.000
222	0	101.010	2.503	98.507	$b = 4.000$	1.670	4.047	6.750	1.000
333	−15.000	101.010	2.503	98.507	$b = 4.000$	1.670	4.047	6.750	1.000

Values in brackets are assumed

From equation 7.9, $Y_{C2} = 0.667 \times 2.503 = 1.670$ m.
From equation 7.7, $V_{C2} = (9.81 \times 1.670)^{1/2} = 4.047$ m/s.
Check: $F_2 = V_{C2}/(gY_{C2})^{1/2} = 4.047/(9.81 \times 1.670)^{1/2} = 1.00$ as required.
Check: slope from 111 to 222 = (100.000 − 98.507)/9.931 = 1:6.7, as recommended in Section 7.5.2.

Section 333: opening exit

The waterway length is 15 m, so this is chainage − 15.000 m.
The hydraulic data are as at section 222.

Check length of transition

Length/$B \approx 38/60 = 0.63$. A minimum value of 0.5 was suggested in the text as one 'rule of thumb'.

References

Apelt, C.J. (1981) Hydraulics of minimum energy culverts and bridge waterways, in *Proceedings of a Conference on Hydraulics in Civil Engineering,* Sydney, 1981, pp. 39–43.

Bradley, J.N. (1978) *Hydraulics of Bridge Waterways,* 2nd edn, US Department of Transportation/Federal Highways Administration, Washington DC.

Brice, J.C. (1984) Assessment of channel stability at bridge sites, in *Transportation Research Record 950,* Second Bridge Engineering Conference, Vol. 2, Transportation Research Board/National Research Council, Washington DC, pp. 163–171.

Brown, S.A. (1984) Design guidelines for spur-type flow-control structures, in *Transportation Research Record 950,* Second Bridge Engineering Conference, Vol. 2, Transportation Research Board/National Research Council, Washington DC, pp. 193–201.

Brown, S.A. (1985) Streambank stabilisation measures for highway stream crossings – executive summary. Report no. FHWA/ED-84/099, Federal Highways Administration, McLean, VA.

Brown, S.A. and Recinos, S. (1988) Effect of spur dikes on bridge backwater: laboratory report. Report no. FHWA-RD-88-270, Federal Highways Administration, McLean, VA.

Chow, V.T. (1981) *Open-Channel Hydraulics,* International Student Edition, McGraw-Hill, Tokyo.

Cottman, N.H. (1981) Experiences in the use of minimum and constant energy bridges and culverts, in *Proceedings of a Conference on Hydraulics in Civil Engineering,* Sydney, 1981, pp. 44–48.

Cottman, N.H. and McKay, G.R. (1990) Bridges and culverts reduced in size and cost by use of critical flow transitions. *Proceedings of the Institution of Civil Engineers,* Part 1, **88**, June, 421–437.

Farraday, R.V. and Charlton, F.G. (1983) *Hydraulic Factors in Bridge Design,* Hydraulics Research Station, Wallingford, England.

Formica, G. (1955) Preliminary tests on head losses in channels due to cross-sectional changes. *L'Energia ellectrica*, **32**(7), 554–568.

French, J.L. (1969) Nonenlarged box culvert entrances. *Proceedings of the American Society of Civil Engineers, Journal of the Hydraulics Division*, **95**(HY6), 2115–2137.

French, R.H. (1986) *Open-Channel Hydraulics*, International Student Edition, McGraw-Hill, Singapore.

Griffiths, W.T. (1978) *Problems Associated with Minimum Energy Design.* Workshop on minimum energy design of culverts and bridge waterways, Melbourne, December 1978, Australian Road Research Board Publication (18pp).

Hamill, L. (1993) A guide to the hydraulic analysis of single-span arch bridges. *Proceedings of the Institution of Civil Engineers, Municipal Engineer*, **98**, March, 1–11.

Hamill, L. (1997) Improved flow through bridge waterways by entrance rounding. *Proceedings of the Institution of Civil Engineers, Municipal Engineer*, **121**, March, 7–21.

Hamill, L. and McInally, G.A. (1990) The hydraulic performance of two arch bridges during flood. *Municipal Engineer*, 7, October, 241–256.

Hemphill, R.W. and Bramley, M.E. (1989) *Protection of River and Canal Banks*, CIRIA/Butterworths, London.

Highway Research Board, (1970) Scour at bridge waterways. National Cooperative Highways Research Program, Synthesis of Highway Practice 5, Highway Research Board, Washington DC.

Kindsvater, C.E. and Carter, R.W. (1955) Tranquil flow through open-channel constrictions. *Transactions of the American Society of Civil Engineers,* **120**, 955–992.

Kindsvater, C.E., Carter, R.W. and Tracy, H.J. (1953) *Computation of Peak Discharge at Contractions.* Circular 284, United States Geological Survey, Washington DC.

Li, R.M. MacArthur, R., and Cotton, G. (1989) Sizing riprap for the protection of approach embankments and spur dikes and limiting the depth of scour at bridge piers and abutments. Report no. FHWA-AZ89-260 II, Arizona Department of Transport, Phoenix.

Loveless, J.H. (1984) A comparison of the performance of standard and novel culvert designs including the effects of sedimentation, in *Channel and Channel Control Structures* (ed. K.V.H. Smith), Proceedings of 1st International Conference on Hydraulic Design in Water Resources Engineering: Channel and Channel Control Structures, Computational Mechanics Centre, Southampton, pp. 183–193.

Matthai, H.F. (1967) Measurement of peak discharge at width contractions by indirect methods, in *Techniques of Water Resource Investigations of the United States Geological Survey,* Ch. A4, Book 3, *Applications of Hydraulics*, US Government Printing Office, Washington DC.

Mazumder, S.K. and Ahuja, K.C. (1978) Optimum length of contracting transitions in open channel subcritical flow. *Journal of the Institution of Engineers (India), Civil Engineering Division*, **58**, March, 218–223.

McKay, G.R. (1970) Pavement drainage. *Proceedings of the 5th Conference*, Australian Road Research Board, Canberra, 5(4), pp. 305–326.

McKay, G.R. (1978) Design principles of minimum energy waterways. Workshop on minimum energy design of culverts and bridge waterways, Melbourne, December 1978, Australian Road Research Board Publication.

Neill, C.R. (ed.) (1973) *Guide to Bridge Hydraulics,* Roads and Transportation Association of Canada, University of Toronto Press, Toronto.

Novak, P., Moffat, A.I.B., Nalluri, C. and Narayanan, R. (1990) *Hydraulic Structures*, Unwin Hyman, London.

Petersen, M.S. (1986) *River Engineering*, Prentice-Hall, Englewood Cliffs, NJ.

Richardson, E.V. and Simons, D.B. (1984) Use of spurs and guidebanks for highway crossings, in *Transportation Research Record 950*, Second Bridge Engineering Conference, Vol. 2, Transportation Research Board/National Research Council, Washington DC, pp. 184–193.

Richardson, E.V, Simons, D.B. and Julien, P.Y. (1990) Highways in the river environment. Report no FHWA-HI-90-016, National Highways Institute/Federal Highways Administration, McLean, VA.

Ruddock, E. (1979) *Arch Bridges and Their Builders, 1735–1835*, Cambridge University Press.

Schneible, D.E. (1966) Field observations on the performance of spur dykes at bridges. ACSE Transportation Engineering Conference, Philadelphia, PA, 17–21 October 1966.

Tracy, H.J. and Carter, R.W. (1954) Backwater effects of open channel constrictions. *Transactions of the American Society of Civil Engineers*, separate 415.

Tracy, H.J. and Carter, R.W. (1955) Backwater effects of open channel constrictions. *Transactions of the American Society of Civil Engineers*, 120, 993–1018.

Trent, R.E. and Brown, S.A. (1984) An overview of factors affecting river stability, in *Transportation Research Record 950*, Second Bridge Engineering Conference, Vol. 2, Transportation Research Board/National Research Council, Washington DC, pp. 156–163.

8 How to evaluate and combat scour

8.1 Introduction

Scour can be defined simply as the excavation and removal of material from the bed and banks of streams as a result of the erosive action of flowing water. In the context of this book, it is assumed that this erosive action may potentially expose the foundations of a bridge. Scour is usually considered to be a local phenomenon, but includes degradation that can cause erosion over a considerable length of a river. Some of the observable effects of scour are shown in Figs 8.1 and 8.2.

It is sometimes assumed that scour will be a problem only when the bed material consists of fine cohesionless material. This is not true: ultimately the scour depth in cohesive or cemented soils can be just as large, it merely takes longer for the scour hole to develop. For example, under constant flow conditions, scour will reach maximum depth in sand and gravel in a matter of hours (perhaps during one flood); in cohesive materials it will take days; in glacial tills, sandstones and shales it will take months; in limestone years; and in dense granite centuries (Richardson *et al.*, 1993). However, the biggest and most frequently encountered scour-related problems usually concern loose sediments that are easily eroded.

Scour is a very serious problem. Floods that result in scour are the principal cause of bridge failure (Table 1.1). In 1973 in the USA a national study of 383 bridge failures caused by catastrophic floods showed that around 25% involved pier damage and 72% abutment damage (Chang, 1973). In 1985, some 73 bridges were destroyed by floods in Pennsylvania, Virginia and West Virginia, while during the spring floods of 1987 17 bridges in New York and New England were damaged or destroyed by scour. With about 485 000 bridges spanning rivers in the US National Bridge Inventory, it is likely that hundreds of these structures will encounter a 1 in 100 year flood during any twelve month period, and that some will be damaged or destroyed. On a worldwide scale the problem is even larger. Many countries have programmes that are designed to identify the bridges that are at risk from scour (i.e. scour critical) with the dual aim of ensuring the safety of users and preserving the affected structures.

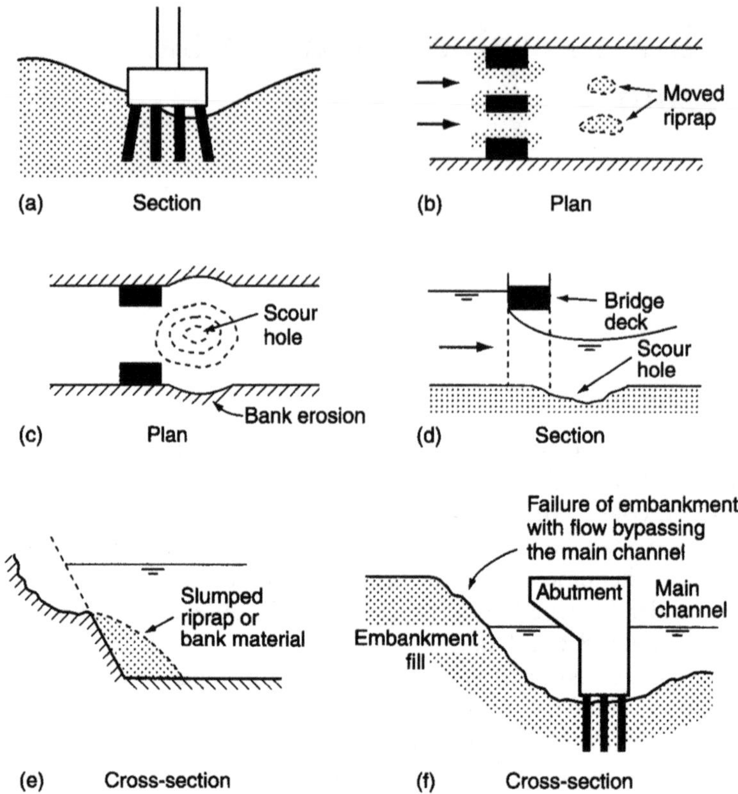

Fig. 8.1 Some observable effects of scour: (a) pier piles and pile cap exposed; (b) pier and abutment riprap moved downstream; (c) downstrean scour hole and bank erosion; (d) downstream scour hole arising from submergence of the opening (pressure flow); (e) slumped material at the toe arising from failure of the riprap or bank; (f) erosion (mass wasting) and failure of the highway embankment with flow on both sides of the abutment.

It is not just old structures, such as nineteenth century rail bridges, that are at risk; new structures can also be susceptible if not properly designed with scour in mind. Of course, it is easy to say that the foundations of all new structures should be made so deep as to eliminate any potential problems relating to scour, just as it can be said that the bridge opening should be made large enough to pass any flood that occurs, but in reality things are not this simple, and economic factors must also be considered. If unnecessary expense is incurred by making all of the foundations significantly deeper than the probable scour depth, the cumulative cost will be very substantial because a large number of bridges are involved. In the State of Washington, for example, environmental regulations designed to preserve the ecology of the stream require that any construction activity that disturbs riverbed material must take place within a watertight enclosure such

Fig. 8.2 Bank erosion at Stoneyford Bridge, Honiton, July 1968. (Reproduced by permission of the Environment Agency)

as a cofferdam. The cost of this enclosure varies with the plan area and the depth of excavation, so deliberately designing very deep foundations will complicate construction and add significantly to the cost (Copp and Johnson, 1987). On the other hand, it should be remembered that the total cost of a failure may be of the order of two to ten times that of the original structure, allowing for disruption to transport and commerce. Thus it is necessary to strike a balance, setting the foundations deep enough to resist the scour that can reasonably be expected at the site without going so deep as to incur additional unnecessary expense. Unfortunately, when deciding just how deep is deep enough the equations that are available to predict the depth of scour are very numerous and contradictory. In 1987, Copp and Johnson reported that 35 different formulae for scour estimation at piers had been proposed since 1949, almost one per year! Most of these equations were of the form

$$\frac{d_{SP}}{b_P} = K\left(\frac{Y}{b_P}\right)^n \tag{8.1}$$

where d_{SP} (m) is the predicted pier scour depth, b_P (m) is the width of the pier, K is a dimensionless factor that allows for pier geometry and orientation to the flow, Y (m) is the depth of the approach flow, and n is a factor reflecting the erosive characteristic of the streambed.

Many of the equations for scour were derived from laboratory studies, for which the range of validity is unknown; some were verified using very limited field data, which itself may be of doubtful accuracy. In the field, the scour hole that develops on the rising stage of a flood, or at the peak, may be filled in again on the falling stage so that the maximum depth cannot be assessed easily after the event. Measurement or observation during flood using divers is not safe or practical, but it is sometimes possible to detect the maximum scour depth afterwards. For instance, if a cohesive material is scoured and then subsequently the pit is filled with an incohesive material, by probing it should be possible to detect the change in the strata. Similarly, with cohesionless material it may be possible to detect changes between the fill and the underlying bed material. New instruments are being developed to detect and monitor scour as it occurs (Fig. 8.3 and Table 8.1), so better information may be available in future (Apt *et al.*, 1992).

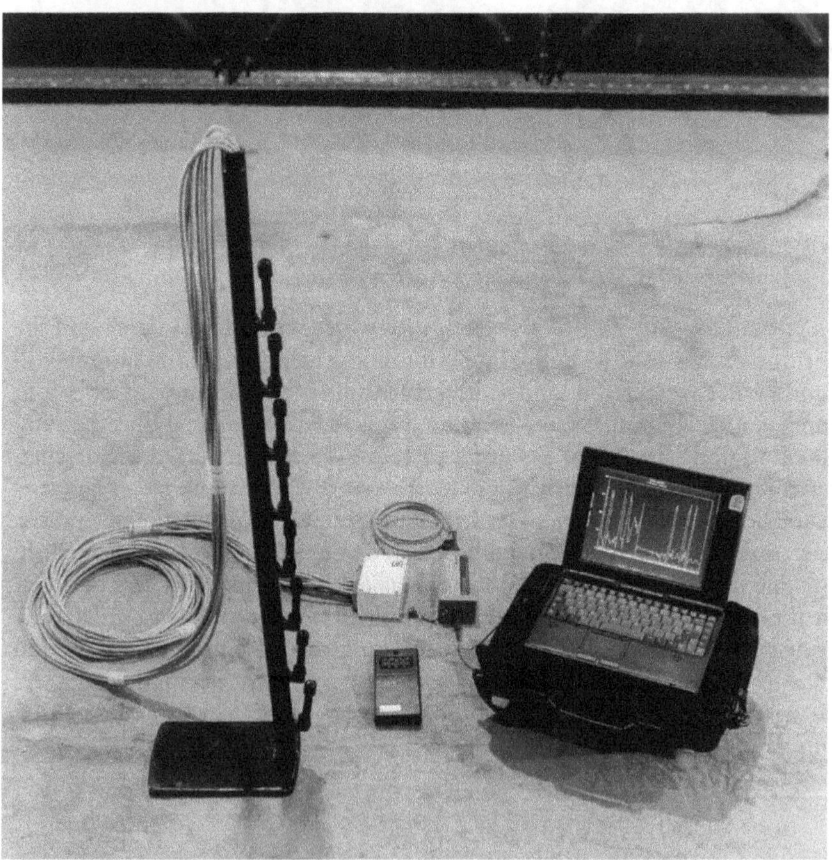

Fig. 8.3 The HR Wallingford 'Tell-Tail' scour monitoring system. One or more motion sensors are buried at different levels within the river bed, adjacent to the structure. Under normal conditions a buried sensor does not move, but when exposed by scour it begins to oscillate in the flow, triggering an alarm at the surface. The system also allows a real-time assessment of scour depth, and indicates whether scour hole refill has occurred. (Reproduced by permission of HR Wallingford Ltd)

Table 8.1 Scour detection and monitoring equipment

Device	*Brief description and limitation*	*RT or I*	*M or F*
Cable and lead weight	Essentially for determining the profile of the bed surface in deep water post-flood. Limited, and does not indicate infilling.	I	M
Penetration testing	Essentially consists of probing with a rod to find the interface between the unscoured bed material and the infill material, which is assumed to be of different density or resistance to penetration. Expensive and time consuming if a large area has to be assessed.	I	M
Ground penetrating radar (GPR)	Electromagnetic pulses (80–800 MHz) are reflected from the scour interface. Not suitable for or limited effectiveness with highly conductive materials such as clay, salt water, sediments saturated with salt water, and water depths over 7.5 m. Can cover a large area quickly with good resolution under favourable circumstances.	RT/I	M
Tuned transducer	Uses a reflected low-frequency acoustic wave (20 kHz) to locate the scour interface, which gives better penetration but lower resolution than high-frequency equivalents. Can cover a large area quickly with adequate resolution under favourable circumstances.	RT/I	M
Colour fathometer	A variable-frequency acoustic wave is reflected from the scour interface, measured in decibels, digitised, and displayed on a colour monitor. Similar to the tuned transducer.	RT/I	M
Black and white fathometer	Uses a 200 kHz acoustic wave to obtain a plot of the channel bottom. Very limited penetration in most conditions but is relatively simple, quick, and can be effective, particularly if used with other methods.	RT/I	M
Fixed sonic fathometers	Essentially as above, but cheaper versions fixed to bridge piers and connected to datalogging or telemetry systems. Effectiveness can be limited by debris, ice or water with entrained air.	RT	F
Sounding rods	A rod resting on the bed drops vertically as the scour hole develops, allowing depth to be determined manually or electronically. May be limited by length of rod and difficulty of avoiding the rod sinking into the bed under its own weight.	RT	F
Buried instrumentation	Instruments may be buried where scour is expected and work by sensing the channel bed–water interface by means of electrical conductance, sliding collars, piezoelectric strips, tip switch or vibrations. Must be in the correct place and robust enough to withstand prolonged submergence and subsequent exposure.	RT	F

Note that scour monitoring is the term for the real-time (RT) measurement of scour depth as it occurs. Scour inspection (I) is a post-flood activity. Equipment may be mobile (M) or fixed (F). Clearly if a mobile device is used for RT monitoring a difficulty may be getting it to the right place at the right time. The technology of these devices is advancing rapidly so details above will change.

Based on Richardson *et al.* (1993)

Even if the conditions relating to the model tests are known and there are accurate field data for comparison, there is still the difficulty of relating small-scale model results to prototype conditions (Johnson, 1995). The only way to determine which of the many alternative equations is 'the best' is to compare the scour predictions with comprehensive field measurements. Unfortunately, it is only recently that such data have started to become available. In the USA over 20 states have been undertaking scour investigations (e.g. Copp and Johnson, 1987; Tyagi, 1989; Miller *et al.* 1992; Bryan *et al.* 1995). This has resulted in a useful and authoritative guide to evaluating scour at bridges, namely Hydraulic Engineering Circular 18

(HEC 18) produced by the US Federal Highways Administration/ Department of Transportation (Richardson *et al.* 1993). The first edition was published in 1991; the second in 1993 incorporated the results of further research. The UK's guidelines appear to be largely based on this work (Highways Agency, 1994). Additionally, Jones (1984) and Melville (1988) provided a good introduction to the problem of scour and the background to many of the equations.

There are so many equations that it is not possible to present them all with a detailed description of where they should be used, and to caution when and where the calculated depth is likely to be exceeded. Consequently the equations in this chapter are those recommended by Richardson *et al.* (1993). Some are effectively the enveloping curve to many (not all) of the other equations, so in most situations they are conservative and tend to overestimate the scour depth, particularly in the UK but also in the USA. Laursen (1984) remarked that if some of them predicted the true state of affairs, there would be very few bridges still standing in Arizona! However, with new bridges and where public safety is involved it can be argued that it is preferable to design for something approaching the worst-case scenario and experience relatively little, rather than the other way around. When assessing the potential for scour at existing bridges the conservative nature of the equations can make it difficult to assess the true risk. A few alternative equations are included in Appendix B, but these should be applied cautiously with reference to the full literature because the indicated scour depth may be exceeded. Johnson and Ayyub (1996) described how fuzzy regression can be used to quantify the bias that results when laboratory-based scour models are used for field applications. For very busy bridges engineers can select a high degree of belief as well as a high degree of reliability and thus obtain a suitably conservative estimate of scour depth, whereas for a rural bridge a lower reliability requirement would result in a smaller depth.

Scour can also cause problems with the hydraulic analysis of a bridge. As described in Section 3.11, scour may considerably deepen the channel through a bridge and effectively reduce or eliminate the backwater. Sometimes a negative backwater can be obtained, which may be regarded as desirable, but only if the bridge and its foundations are designed to withstand the scour depths involved. This reduction in backwater should not be relied on because of the unpredictable nature of the processes involved.

When considering scour it is normal to distinguish between non-cohesive or cohesionless (alluvial) sediments and cohesive material. The former are usually of most interest and will be considered in Sections 8.2–8.6, the last describing the regime theory of alluvial channels. Cohesive materials will be considered briefly in Section 8.7. In all cases it will be assumed that the bridge is located on a non-tidal stream. The special case of scour in tidal areas is considered in Section 8.8. Scour prevention measures, notably the use of riprap, are covered in Section 8.9. Before applying any of the equations it is essential to have an appreciation of the types of scour and the factors that influence scour depth. These are described below.

8.2 Types of scour and its classification

So far, for simplicity, scour has been considered as a single entity. However, from the point of view of bridge crossings it is usually classified under the headings shown in Table 8.2.

8.2.1 Clear-water scour and live-bed scour

The first major division is between clear-water scour and live-bed scour. The critical issue here is whether or not the mean velocity (V m/s) of the flow upstream of the bridge is less than or larger than the scour-critical velocity (V_S m/s) needed to move the bed material. If $V < V_S$ then the bed material upstream of the bridge is at rest: this is referred to as the clear-water condition because the approach flow is clear and does not contain sediment. This means that any bed material that is removed from a local scour hole is not replaced by sediment being transported by the approach flow. The maximum local scour depth is achieved when the size of the scour hole results in a local reduction in velocity and the flow can no longer remove bed material from the scoured area.

Live-bed scour occurs where $V > V_S$ and the bed material upstream of the crossing is moving. This means that the approach flow continuously transports sediment into a local scour hole. By itself, a livebed in a uniform channel will not cause a scour hole; for this to be created some additional increase in velocity is needed, such as that caused by a contraction (natural or artificial, such as a bridge) or a local obstruction (e.g. a bridge pier). The equilibrium scour depth is achieved when material is transported into the scour hole at the same rate at which it is transported out.

The clear-water and live-bed conditions are significant because to some degree the growth of the scour hole will depend upon whether or not the bed material is already in motion (Fig. 8.4). With clear-water scour the growth is slightly slower but more uniform than in live-bed scour, which shows fluctuations (dotted line) about the equilibrium depth (solid line). The fluctuations are due to the migration downstream of bed features such

Table 8.2 Classification of scour

Type of scour (see eqns 8.2 and 8.3 for V_S)	*Clear-water scour ($V < V_S$)*	*Live-bed scour ($V > V_S$)*
Degradation or aggradation	Assessed over the life of the bridge – see Section 8.2.2	
Contraction scour (e.g. bridge opening)	Eqns 8.10–8.12	Eqns 8.13 and 8.11
Local scour – piers	Eqn 8.14 and Fig. 1.15	Eqn 8.14 and Fig. 1.15
– abutments	Eqn 8.17 or 8.18	Eqn 8.17 or 8.18

as ripples and dunes (see Fig. 8.9). Note that typically the maximum local clear-water scour is about 10% larger than the equilibrium local live-bed scour, clear water having a greater sediment-carrying capacity. For this reason the formulae used to estimate scour depth are classed as either clear-water or live-bed equations, and generally each group of equations should be applied only to their respective condition. Conditions that favour clear-water scour are: channels with flat bed slopes during low flows; a coarse bed material that is too large to be transported (riprap is an artificial example); channels with natural vegetation or artificial reinforcement where velocities are only high enough to cause scour near piers and abutments; and flow over floodplains (assuming they are grassed). To determine whether the flow condition is clear-water or live-bed, Neill (1968) suggested that equation 8.2 be applied to the unobstructed flow. If the average velocity ($V = Q/A$ m/s) in the unobstructed approach section is greater than V_S (m/s) the scour will be live bed.

$$V_S = 1.58\,[(s_S - 1)\,g\,D_{50}]^{1/2}\left(\frac{Y}{D_{50}}\right)^{1/6} \tag{8.2}$$

where s_S is the (dimensionless) specific gravity or relative density of the sediment or bed material, g is $9.81\,\text{m/s}^2$, D_{50} is the median diameter (m) at which 50% of the bed material by weight is smaller than the size denoted, and Y is the average depth (m) in the upstream channel. Usually s_S is about 2.65, so that equation 8.2 becomes

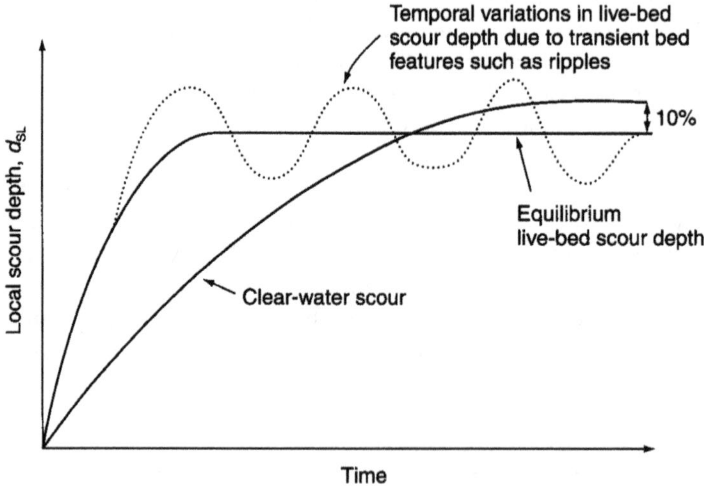

Fig. 8.4 Diagrammatic illustration of the increase in local scour depth (d_{SL}) with time for clear-water and live-bed conditions. The oscillations for the live-bed condition (dotted) are due to transient bed features such as ripples. The final clear-water scour depth exceeds the equivalent equilibrium live-bed depth by about 10%.

$$V_S = 6.36\ Y^{1/6}\ D_{50}^{\ 1/3} \tag{8.3}$$

At any particular location both clear-water and live-bed scour may be experienced. During a single flood the mean velocity will increase and decrease as the discharge rises and falls, so it is possible to have clear-water conditions initially, then a live bed, then finally clear water again (Fig. 8.5). The maximum scour depth may occur under clear-water conditions, not at the flood peak when live-bed scour is experienced. Similarly, relatively high velocities can be experienced when the flow is just contained within the banks, rather than spread over the floodplains at the peak discharge. It is also possible to have the clear-water and live-bed conditions occurring at the same time. For example, if the floodplains are grassed or composed of material that is larger in diameter than that in the main channel, then clear-water conditions may occur on the floodplain with live-bed conditions in the main channel. Thus the problem may not always be as simple or as well defined as would be convenient.

8.2.2 *Degradation*

The next classification in Table 8.2 is degradation, which can only occur with a live bed ($V > V_S$). Degradation is not the result of bridge or embankment construction. It is more widely defined as an adjustment of the bed elevation over a large area due to changes in hydrology, hydraulics or sediment transport. This can be illustrated by the following equation (Lane, 1955; Bryan *et al.* 1995):

$$QS_O \propto Q_S D_{50} \tag{8.4}$$

where Q is the water discharge, S_O is the channel bed slope, Q_S is the bed material discharge, and D_{50} is the median grain size of the bed material. A

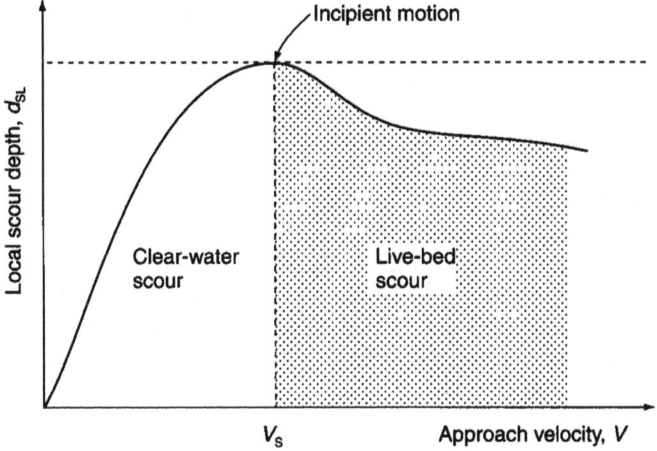

Fig. 8.5 Variation of local scour depth (d_{SL}) with approach velocity.

change to one variable on either side of the equation will affect the stability of the channel, and can lead to degradation or aggradation (deposition of sediment). While aggradation sounds harmless, it should be remembered that this could reduce the size of the bridge opening and make it more likely that debris will become trapped (perhaps against the upstream soffit), so increasing scour potential.

Urbanisation has the effect of increasing flood magnitudes and causing hydrographs to peak earlier, resulting in higher stream velocities and degradation. Channel improvements or the extraction of gravel (above or below the site in question) can alter water levels, flow velocities, bed slopes and sediment transport characteristics and consequently affect scour. For instance, if an alluvial channel is straightened, widened or altered in any other way that results in an increased flow-energy condition, the channel will tend back towards a lower energy state by degrading upstream, widening and aggrading downstream. The construction of dams stops sediment transport (Q_S) and results in an artificial discharge (Q) of clear water downstream, which has a greater erosion potential. The Hoover Dam resulted in a 7.1 m degradation of the sand and gravel bed of the Colorado River, its effect extending for 111 km (Melville, 1988).

The significance of degradation scour to bridge design is that the engineer has to decide whether the existing channel elevation is likely to be constant over the 100 year life of the bridge, or whether it will change. If change is probable then it must be allowed for when designing the waterway and foundations.

The lateral stability of a river channel may also affect scour depths, because movement of the channel may result in the bridge being incorrectly positioned or aligned with respect to the approach flow (Trent and Brown, 1984). This problem can be significant under any circumstances but is potentially very serious in arid or semi-arid regions and with ephemeral (intermittent) streams (Richardson *et al.*, 1990). As an illustration, the Cimarron River in Kansas, USA, was 15 m wide around the year 1900 but as a result of a series of major floods in the 1930s its width increased to around 365 m, essentially occupying the entire valley bottom. The Kosi River in India can move laterally by up to 760 m per year, while parts of the Colorado and Mississippi Rivers in the USA have rates of about 3–46 m and 48–192 m respectively. Lateral migration rates are largely unpredictable: sometimes a channel that has been stable for many years may suddenly start to move, but significant influences are floods, bank material, vegetation of the banks and floodplains, and land use (see Fig. 1.6). In a survey of 224 bridge sites in the USA it was found that hydraulic problems attributable to lateral erosion by the stream occurred at 106 of them (Brice, 1984). The next largest problem was contraction scour, which was found at 55 sites, followed by local scour, which occurred at 47. Channel degradation was found at 34, and the accumulation of debris caused problems at 26.

8.2.3 Contraction scour

Contraction scour occurs over a whole cross-section as a result of the increased velocities arising from a narrowing of the channel by a constriction such as a bridge. In general, the smaller the opening ratio ($M = q/Q$ or b/B) the larger the waterway velocity and the greater the potential for scour. If the flow contracts from a wide floodplain, considerable scour and bank failure can occur. According to Blodgett (1984), during the first two or three years after the construction of a bridge that forms a constriction bed levels may be reduced by around 0.5 m or more ('several feet'). Relatively severe constrictions may require regular maintenance for decades to combat erosion. Needless to say, one way to reduce contraction scour is to make the opening wider.

8.2.4 Local scour

Local scour arises from the increased velocities and associated vortices as water accelerates around the corners of abutments, piers and spur dykes.

The flow pattern around a cylindrical pier is shown in Fig. 8.6. The approaching flow decelerates as it nears the cylinder, coming to rest at the centre of the pier. The resulting stagnation pressure is highest near the water surface where the approach velocity is greatest, and smaller lower down. The downward pressure gradient at the pier face directs the flow downwards. Local pier scour begins when the downflow velocity near the stagnation point is strong enough to overcome the resistance to motion of

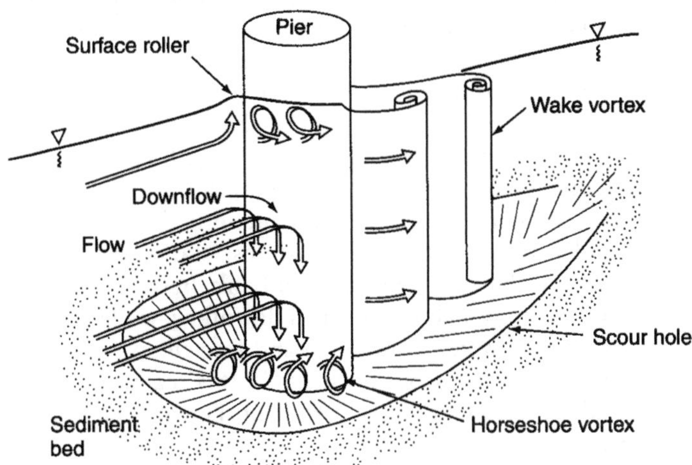

Fig. 8.6 The flow pattern and scour hole at a cylindrical pier. The downflow, horseshoe vortex and wake vortex are the principal cause of local bed erosion. (After Melville, 1988. Reproduced by permission of Technomic Publishing Company Inc., Lancaster, PA, USA)

the bed particles. Without a scour hole the maximum downward velocity is about 40% of the mean approach velocity (V), the maximum strength of the downflow being recorded just below bed level. When scour occurs the maximum downflow velocity is about 80% of V (Copp and Johnson, 1987; Melville, 1988). The impact of the downflow on the bed is the principal factor leading to the creation of a scour hole. As the hole grows the flow dives down and around the pier producing a horseshoe vortex, which carries the scoured bed material downstream. The combination of the downflow with the horseshoe vortex is the dominant scour mechanism. As the scour hole becomes progressively deeper the downflow near the bottom of the scour hole decreases until at some point in time equilibrium is reached and the depth remains constant.

At the sides of the pier flow separation occurs, resulting in a wake vortex whose whirlpool action sucks up sediment from the bed. As the vortices diminish and velocities reduce, the scoured material is deposited some distance downstream of the pier.

For piers that are essentially rectangular in plan and aligned to the flow the basic scour mechanism is similar to that just described, albeit rather more severe because of the square corners. However, as the angle of attack to a rectangular pier increases so does its effective width, so the scour depth increases and the point of maximum scour moves downstream of the nose to a point on the exposed side. With a large degree of skew the maximum scour may occur at the downstream end of the pier. If the flow direction is likely to change there is merit in using cylindrical piers to avoid these complications.

The scour mechanism at a bridge abutment is similar to that at a pier, although the boundary layer at the abutment or channel wall may result in an additional deceleration of the flow compared with a central pier. The approach flow can be considered as separating into an upper layer, which forms an upflow surface roller on hitting the abutment, and a lower layer, which becomes the bottom or principal vortex (Fig. 8.7). Viewed in plan, the upper layer divides or separates, with part of the flow accelerating around the upstream corner of the abutment into the bridge waterway while the remainder slowly rotates in an almost stationary pool trapped against the face of the abutment and the river bank. In the bottom layer, the flow near the bank forms an almost vertical downflow, while that nearer to the end of the abutment accelerates down and into the waterway, forming the principal vortex. Usually scouring starts in this region of accelerating flow and grows along the faces of the abutment. Downstream of the abutment wake vortices form.

The basic scouring process is the same for most types of abutment, although with wingwall and vertical wall types (as opposed to the spillthrough type in Fig. 8.7) the stagnation region is larger, and scour is most severe near the end of the abutment where the principal vortex is concentrated. Scour at spur dykes was considered briefly in Section 7.4.

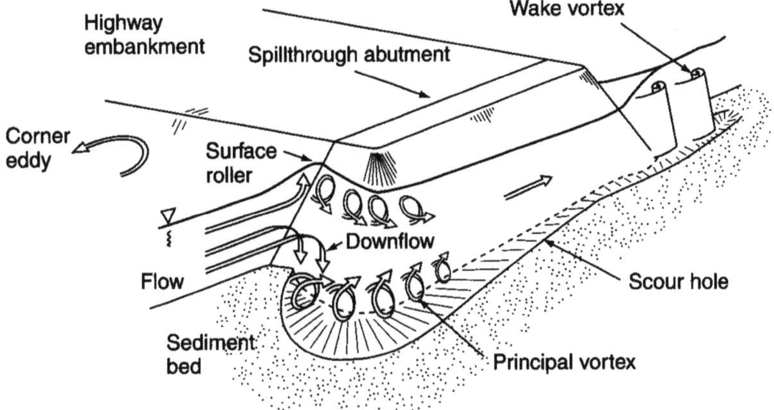

Fig. 8.7 The flow pattern at a spillthrough abutment. The downflow and principal vortex are the main causes of local bed erosion. (After Melville, 1988. Reproduced by permission of Technomic Publishing Company Inc., Lancaster, PA, USA)

8.2.5 *Total scour depth, d_S*

Degradation, contraction and local scour are additive, but only where the scour holes overlap (Table 8.2). For instance, contraction scour may have to be added to pier or abutment scour to get the total scour depth. However, pier scour and abutment scour would not be added unless the two scour holes overlap. This usually has to be determined by drawing a cross-section through the waterway and superimposing the scour depths. If the holes do overlap (see equation 8.16) the resultant scour depth is often larger than the two components, but difficult to predict. Nevertheless, as a general reminder:

$$\text{total scour } (d_S) = \text{degradation } (\Delta d) + \text{contraction scour } (d_{SC}) + \text{local scour } (d_{SL}) \quad (8.5)$$

Some of the causes of degradation and scour are depicted in Fig. 8.8. Before considering how to calculate scour depths it is beneficial to consider the factors that influence scour.

8.3 Factors affecting scour in cohesionless material and associated difficulties

Table 8.3 shows the scour depths recorded in 1986 during high flows in streams in Oklahoma (Tyagi, 1989). The table is included to give a broad indication of the range of scour depth that may be encountered, and to indicate that the depth is not just a function of the bed material. The large number of variables (listed below) that ideally should be included in the

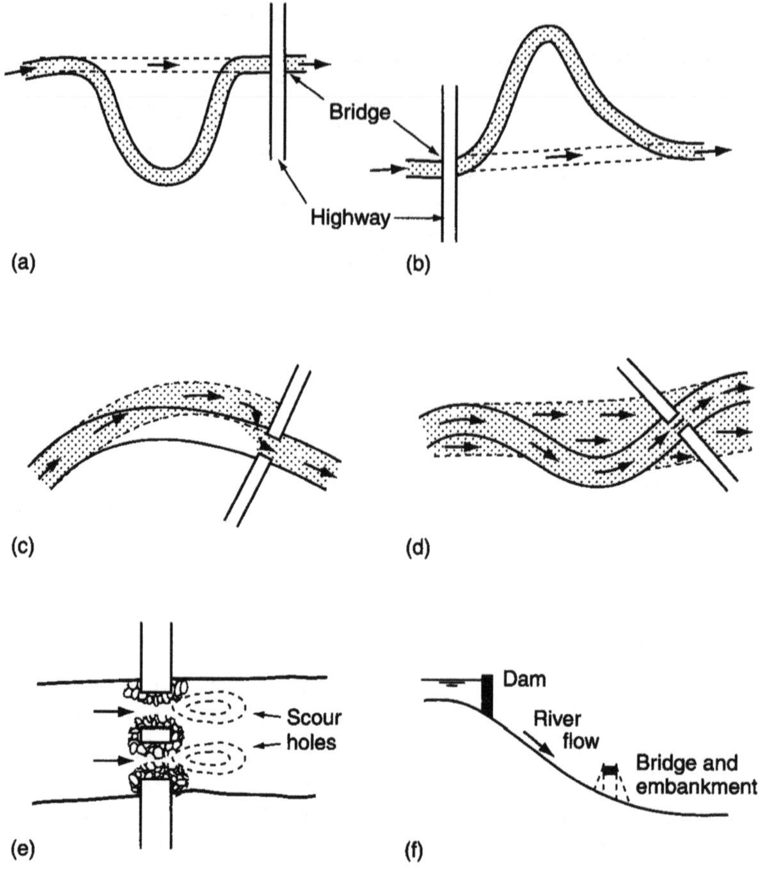

Fig. 8.8 Some typical causes of degradation and scour. (a, b) The removal of the bend (naturally or artificially, as a channel improvement measure) leads to an increased hydraulic gradient, higher velocities and degradation at the bridge. (c) Bank erosion and migration of the bend results in the bridge not being properly aligned to the flow, possibly with the result shown in Fig. 8.1f. (d) The bridge is perpendicular to the main river channel but not to the overbank flow, resulting in a reduced hydraulic performance during floods. (e) A narrow bridge waterway or excessive use of large stones to protect piers and abutments leads to high velocities and downstream scour. (f) The construction of a dam retains sediment and causes increased clear-water scour and degradation downstream at the bridge.

scour depth equations illustrates the difficulty of deriving simple but accurate formulae. However, an understanding of the significance of these inter-related factors will help in producing a sound estimate of scour depth.

1. Hydraulic variables including the stream discharge (Q m^3/s), approach depth and mean velocity (Ym and Vm/s), the water's density and viscosity (ρ kg/m^3 and ν m^2/s), and the Froude number (F). These factors (in conjunc-

Table 8.3 Scour depths at some bridge sites in Oklahoma during 1986 and 1988 floods and low flows

Bridge site	*River*	*Flow depth (m)*	*Scour depth (m)*	*Channel characteristics*
Cleo Springs	Cimarron	0.61	0.18	Boulders and sand
Ringwood	Cimarron	0.23	0.23	
Lacey	Cimarron	0.52	2.71	Coarse sand with wide floodplain
Cimarron City	Cimarron	0.61	0.91	
Guthrie	Cimarron	0.31	2.44	
I-35	Cimarron	0.31	3.96	
Cayle	Cimarron	0.31	4.88	
Perkins	Cimarron	0.61	5.49	
Ripley	Cimarron	4.27	3.35	
Cushing	Cimarron	5.18	2.44	
Oilton	Cimarron	4.57	5.18	
Oilton	Cimarron	7.62	0.91	
Sand Springs	Arkansas	4.57	2.74	Mostly fine sand to silt
Tulsa	Arkansas	4.27	0.61	
Ponca City	Arkansas	1.22	0.91	
Bartlesville	Caney	1.13	0.85	Clay sediment
Collinsville	Caney	Piers out of water	0.85	
Asher	South Canadian	0.76	2.23	Medium to fine sand
Calvin	South Canadian	1.16	2.71	
White Field	South Canadian	1.31	2.96	
Pauls Valley	Washita	0.94	2.47	Silt
Davis	Washita	1.01	2.71	
Gene Autry	Washita	1.12	2.80	

After Tyagi (1989)

tion with the characteristics of the bed material) determine whether clear-water or live-bed scour will occur. With shallow flows an increase in flow depth can increase the local depth of scour in the vicinity of piers by a factor of 2.0 or more, and for abutments by a factor of 1.1–2.15 depending upon shape (Richardson *et al.*, 1993). However, as the water depth increases the scour depth becomes almost independent of flow depth. Figure 1.14 illustrates the effect of depth on the velocity required for bed movement. Scour increases with velocity, and may be influenced by whether the flow is subcritical or supercritical. There is relatively little research concerning scour in supercritical flow, so this condition is often best avoided (for several reasons).

2. The bed configuration of channels formed from sand influences scour depths. Sediments with a diameter $D > 0.7\,\text{mm}$ do not form bed ripples, while those with $D < 0.7\,\text{mm}$ can have various bed configurations: ripples, dunes, plane bed or antidunes (see Fig. 8.9). During a flood the configuration may change within minutes with changing flow, or as a result of other factors. Typically ripples may develop at mean flow velocities of 0.3–0.6 m/s but are replaced by dunes at 0.6–0.9 m/s (Fig. 8.10). As the velocity increases

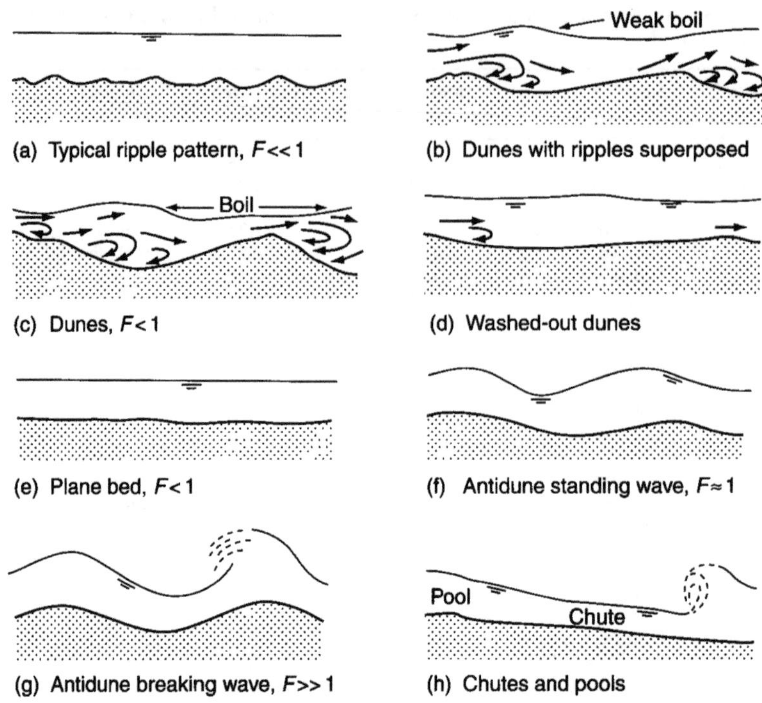

Fig. 8.9 Illustration of different bed configurations in sand channels: (a) typical ripple pattern, $F \ll 1$; (b) dunes with ripples superposed; (c) dunes, $F < 1$; (d) washed-out dunes; (e) plane bed, $F < 1$; (f) antidune standing wave, $F \approx 1$; (g) antidune breaking wave, $F \gg 1$; (h) chutes and pools. (After Richardson *et al.*, 1990)

further the dunes change character and offer a decreased resistance to flow (often resulting in a stepped or discontinuous stage–discharge curve) before finally tending towards a plane bed at velocities of around 0.9–1.5 m/s. Richardson *et al.* (1990) quoted the following comparative values of Manning's *n* in fine ripple-forming sediments: plane bed 0.010–0.013, antidunes with standing waves 0.010–0.015, antidunes with breaking waves 0.012–0.020, chutes and pools 0.018–0.035, ripples 0.018–0.028 and dunes 0.020–0.040.

3. **Bed material characteristics**, such as grain size distribution, grain diameter (D m), sediment density (ρ_S kg/m) and cohesive properties. Equations often use the median diameter at which 50% of the material by weight is smaller than the size denoted (i.e. D_{50}) or sometimes the effective mean diameter ($D_M = 1.25D_{50}$). It should be appreciated that even determining something as simple as the D_{50} value may not be entirely straightforward. For example, how is the sample to be obtained? If the material is stratified, samples of every layer may be needed so the equations can be applied to each in turn. Bagging with a shovel is often as good as anything,

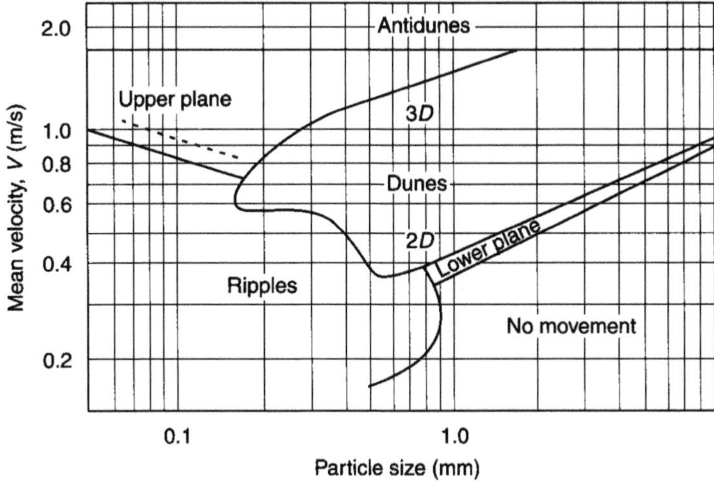

Fig. 8.10 Illustration of the effect of mean flow velocity and bed material particle size on the bed configuration. (After Raudkivi, 1997; Ripples on stream bed, *Journal of Hydraulic Engineering*, ASCE. Reproduced by permision of ASCE)

but some form of mechanical grab may be needed if a sample has to be obtained from the maximum likely scour depth. If overbank flow occurs it may be necessary to sample both the main channel and the floodplains. If samples are taken from within the flowing stream, then some of the fine material may be lost. A few large boulders can be ignored, but a large number mixed with mainly fine material may make it difficult to obtain a truly representative sample.

The diameter of the bed material also affects the bed configuration and scour depth (Figs 8.9 and 8.10). Figure 8.11 shows the variation of the relative scour depth (d_{SP}/b_P) for a cylindrical pier with uniform coarse non-ripple-forming and ripple-forming sediments. Under clear-water conditions ($V/V_S < 1$) with coarse sediments the local scour depth ratio increases rapidly to around 2.3 at the scour-critical velocity, then decreases under live-bed conditions ($V/V_S > 1$) when there is a flow of sediment into the scour hole from upstream, and then increases again. The dashed lines show the fluctuation in relative scour depth due to the passage of transient bed features. The live-bed peak may be the most significant for design purposes since the clear-water condition may not last long enough for the full scour depth to be attained. Of course, for non-cylindrical piers and piers not aligned with the flow relative scour depths may be very different from those illustrated.

With $D_{50} > 0.7$ mm there is a fairly clear relationship between the relative maximum clear-water local scour depth (d_{SP}/b_P) and the size of the bed material relative to pier width (b_P/D_{50}), as shown by the line in Fig. 8.12a, but when $D_{50} < 0.7$ mm the data are scattered. Figure 8.12b shows the data plotted to a log scale and illustrates the following relationships:

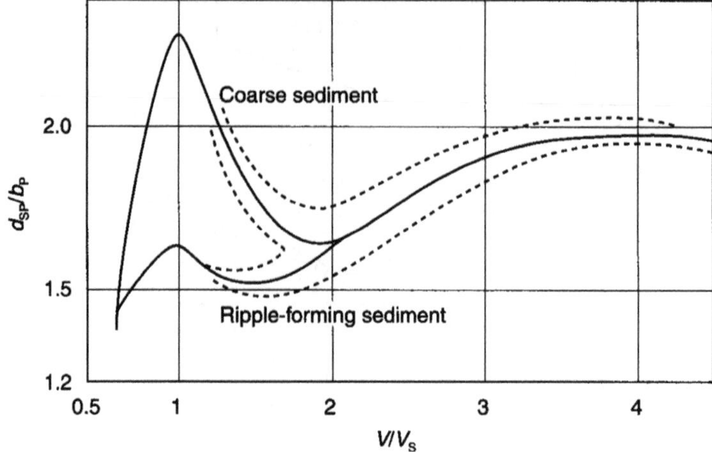

Fig. 8.11 Diagrammatic illustration of the variation of scour depth (d_{SP}) at a cylindrical pier (of width b_P) in a uniform sediment with velocity. Note that V/V_S = 1.0 is the threshold for bed movement. (After Melville, 1988. Reproduced by permission of Technomic Publishing Company Inc., Lancaster, PA, USA)

$D_{50} < 0.7$ mm: ripple forming, no relationship

$D_{50} > 0.7$ mm: $(d_{SP}/b_P) = 0.5(b_P/D_{50})^{0.53}$ when $(b_P/D_{50}) < 18$

$$(d_{SP}/b_P) = 2.3 \text{ when } (b_P/D_{50}) > 18$$

One of the limitations of most equations for scour depth is that they make no allowance for the self-armouring characteristics of graded material. For instance, if the bed consists of a mixture of sand which can be eroded and cobbles which cannot, then it is possible that once the top layer of sand has been scoured away the bed will be covered with a layer of cobble anchor stones and larger-diameter particles, which will 'seal' the bed and reduce or prevent further erosion (Chin *et al.*, 1994).

Figure 8.13 shows results obtained in a recirculating flume with a continual supply of material from upstream, as may occur in a river. It appears that pier scour is significantly reduced when non-uniform or graded materials are involved (the dashed line is for a zero sediment supply from upstream). Thus many of the equations that assume a uniform grain diameter overestimate the depth of scour when applied to a graded material. Based on studies of clear-water scour depths at cylindrical piers, Ettema (1976, 1980) suggested that under some circumstances scour depths in river gravels with $\sigma = (D_{84}/D_{16})^{0.5} > 4.0$ may be only about 20% of the depths found in uniform sediments (Fig. 8.14). The diagram shows the correction factor (K_σ) that relates scour depth in a graded material to that in a uniform sediment. This factor has not been adequately verified

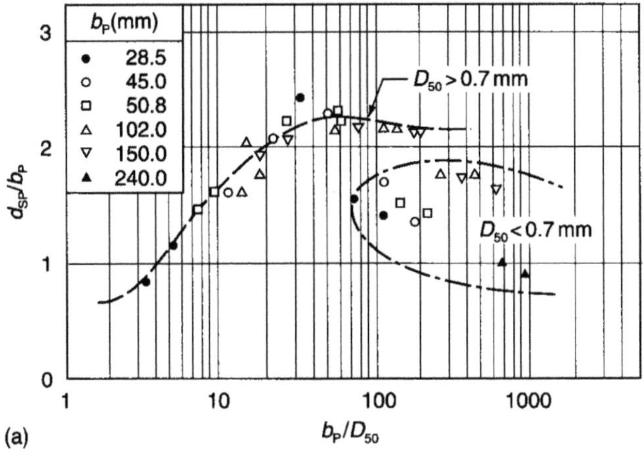

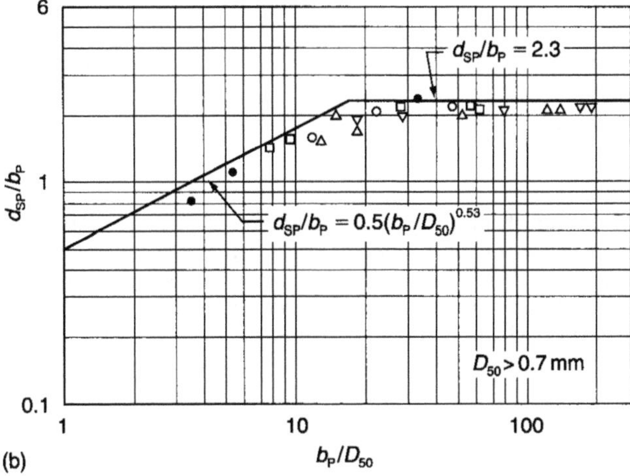

Fig. 8.12 (a) Variation of the local clear-water pier scour depth (d_{SP}/b_P) with the relative median bed material diameter (b_P/D_{50}); and (b) a logarithmic plot of the data in (a). (After Copp and Johnson, 1987. Reproduced by permission of Washington State University and Washington State Department of Transportation)

by field measurements and so is not included in the equations quoted below; if it is cautiously adopted, it is prudent to adopt a factor of safety (Copp and Johnson, 1987; Richardson *et al.*, 1990). Under clear-water conditions sediment grading affects the local scour depth at wingwall and spillthrough abutments in the same way as at circular piers.

4. Channel evolution and secondary flow features. Many alluvial channels in broad valleys follow a cycle of degradation (down-cutting, widening, caving of the banks), aggradation (widening, becoming shallower, deposition), and restabilisation (Bryan *et al.*, 1995). This may result in both

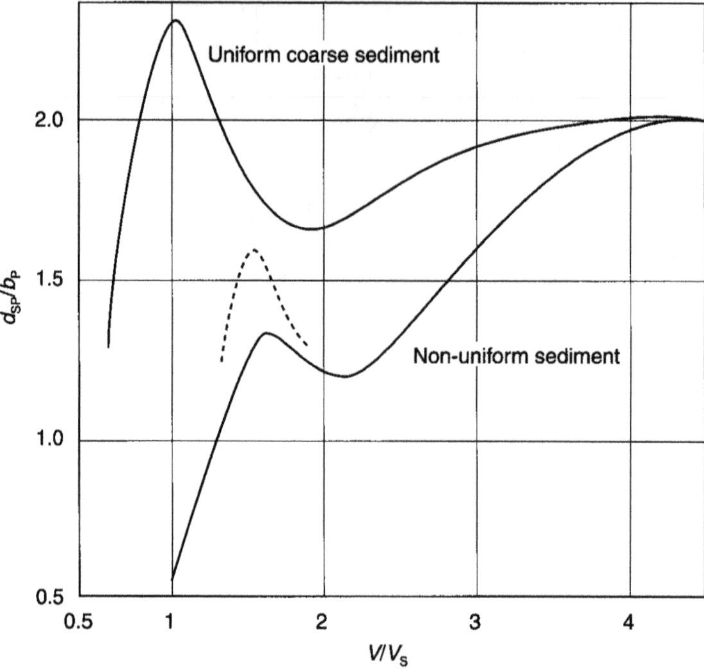

Fig. 8.13 Variation of the local pier scour depth (d_{SP}/b_P) in uniform and non-uniform sediments with velocity expressed as a proportion of the scour-critical value (V_S). The full lines are for recirculating sediment transport; the dashed line shows the armour peak with diminishing sediment transport. (After Melville, 1988. Reproduced by permission of Technomic Publishing Company Inc., Lancaster, PA, USA)

the thalweg and the channel itself shifting course. In plan, the bends of meandering 'S-shaped' channels move laterally and downstream, while some bends may become cut-off. Such changes have to be extrapolated over the life of the bridge and are often best determined by comparing time sequential maps and aerial photographs (see Fig. 1.13).

In a braided stream that contains numerous channels the deepest scour occurs where two or more channels converge or where the flow comes together downstream of an island or bar. Under such circumstances the scour depth has been observed to be 1.0–2.0 times the average flow depth (Richardson *et al.*, 1993), or possibly even 4.0 times under some circumstances (Melville, 1988). Similarly, at bends the velocity on the outside of the curve can be 1.5–2.0 times the mean velocity, which increases the scour potential, as does the secondary (transverse) flow that arises from the super-elevation of the water surface and the non-uniform velocity distribution (e.g. see Chadwick and Morfett, 1993). Brice and Blodgett (1978) analysed 224 bridge sites in the USA where scour had occurred and attributed the hydraulic problems at 106 of them mainly to lateral bank erosion. They

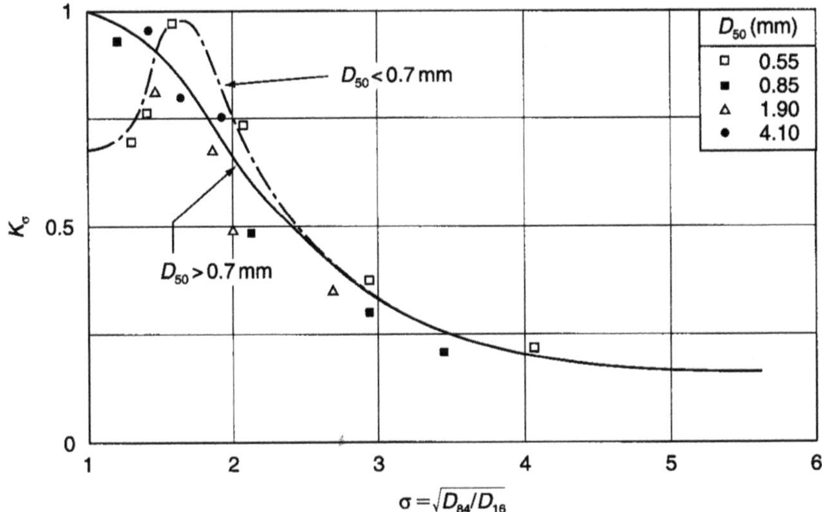

Fig. 8.14 The effect of bed armouring, showing the variation of the coefficient K_σ with grading, expressed as $\sigma = (D_{84}/D_{16})^{1/2}$. K_σ relates the scour depth in a graded sediment to that in uniform material. (After Melville, 1988. Reproduced by permission of Technomic Publishing Company Inc., Lancaster, PA, USA)

concluded that this was the most common cause of scour. Brice (1984) found a tendency for erosion rates to increase with the width of alluvial channels, the problem being greatest with braided channels and least with sinuous canal forms. Johnson and Simon (1997) provided an interesting case history, which combined modelling the stage of channel evolution with the probability of failure of piers set at different levels in a situation where a bridge had previously failed.

5. **Changes in hydrology, hydraulics or sediment transport.** Changing catchment use may alter runoff and flood magnitudes, while channel improvements may alter flow depths and velocities, all of which will change the depth of scour experienced at a site. This was considered under degradation in Section 8.2.2.

6. **The severity of the constriction** and the length of the abutments in the direction of flow. This can be crudely summarised as the smaller the span of the bridge (b) relative to the channel width (B) the greater the contraction of the flow and the greater the scour depth, which is why many equations incorporate the opening ratio ($M = b/B$ or q/Q). A discrepancy arises between model results obtained with relatively uniform flow across a laboratory channel and results obtained in the field where the flow is not uniformly distributed but often concentrated in a main channel. Thus laboratory-based equations may overestimate scour. If the abutment is set back from the river bank by less than 3–5 times the depth of flow through

the bridge there is a possibility that the combination of contraction and abutment scour may destroy the bank. Consequently the use of guidewalls or riprap protection to the bank and bed in the vicinity of the bridge should be considered (Richardson *et al.*, 1993).

Many laboratory-based equations were derived using a long constriction in which normal depth was established in the contracted section (Fig. 8.15), whereas with bridges this is usually not the case. This creates additional uncertainty when applying these equations. However, generally it appears that the assumption of a long constriction produces a conservative over-estimation of the scour depth, which is usually regarded as desirable.

7. **Abutment geometry.** Abutments with vertical walls parallel to the flow will produce scour depths approximately double that of spillthrough (sloping) abutments. These types of abutment are illustrated in Fig. 4.21.

8. **Pier geometry.** Scour depth increases with increasing pier width (b_P). Pier shape affects scour, square corners increasing scour depths by 10–30% (see Table 8.6 and Fig. 1.15a) while streamlining reduces it. Pier length may

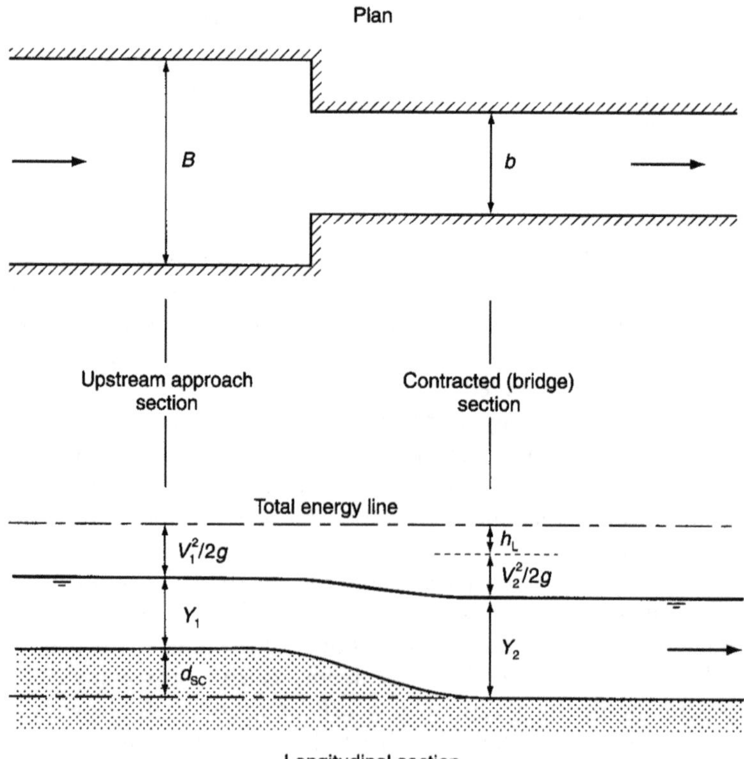

Fig. 8.15 Definition of scour depth in a long constriction that has uniform flow in the contracted section. The contraction scour depth is assumed to be $d_{SC} = Y_2 - Y_1$.

significantly affect scour depths, particularly when the pier is at an angle to the flow. Some piers that are inclined outwards towards the top by more than 20° to the vertical may increase the scour depth by a factor of two under some conditions (Neill, 1973; Farraday and Charlton, 1983).

9. The angle of attack of the approach flow. The scour depth at piers and abutments increases as the angle of skew increases, as shown in Figs 1.15b and 8.20. Abutment or embankment scour is usually regarded as being reduced when embankments are angled downstream and increased when angled upstream (Fig. 8.21). A 45° angle downstream may reduce the maximum scour depth by 20% whereas a 45° angle upstream may increase the maximum depth by 10% (Ahmad, 1953). However, this effect may diminish if a flood lasts long enough (Melville, 1988). Although it is usual to design piers and abutments so that they point directly into the approach flow it should be remembered that the principal flow direction during flood may differ from that during low stages, while the thalweg may shift or meandering channels may migrate (Figs 8.8 and 8.16). Consequently, scour depths

Fig. 8.16 Halfpenny Bridge on the River Torridge at Weare Gifford, December 1979. The presence of levees poses some questions with respect to the hydraulic analysis of the bridge. Is there a significant flow between the levees and the outside of the floodplain, or is the water trapped and static? What is the principal direction of the flow approaching the bridge? In this case it looks as though it is parallel to the levees, but this could change if they were overtopped at higher stages. There is some flow over the road embankments, so how much flow bypasses the bridge? Note that if the road embankments were made higher to prevent bypass flow or to keep the road open, this could make scour in the bridge waterway worse and increase the upstream afflux. (Reproduced by permission of the Environment Agency)

can be larger than anticipated because the angle of attack is greater than planned.

10. The type of flow encountered at the bridge, namely channel flow, sluice gate flow or drowned orifice flow (see Figs 2.6 and 7.2). The last two types occur when the waterway opening is submerged so that a pressurised flow is obtained through the bridge. Often the vertical contraction from the soffit deflects the jet down towards the bed, increasing scour potential. Little research has been conducted regarding scour depths under these conditions, but it has been suggested that the calculated pier scour depth may have to be multiplied by a coefficient ranging from 1.0 for low approach Froude numbers ($F = 0.1$) to 1.6 for high approach Froude numbers ($F = 0.6$). If the bridge is overtopped the depth used to calculate F should be that to the top of the bridge deck or guard rail, depending upon what causes an obstruction to the flow (Richardson *et al.*, 1993).

11. Ice jams may have the effect of changing abutment and pier geometry, reducing the effective area of the waterway, and deflecting the flow downwards. This may be most dangerous when the ice accumulation is near to the bed. Limited field measurements of scour at ice jams have shown that scour depths of 3–6 m are possible. Neill (1973) suggested that a round-nosed, vertical or slightly raked pier is best for discouraging accumulations of drifting ice. This is largely unpredictable and unresearched, but if ice is likely to be a problem then the pier should be assumed to be wider than it really is at the design stage.

12. Debris trapped on the abutments and piers (Figs 1.10 and 1.11). The effect is very similar to ice jams. To a limited extent this can be allowed for in calculations by adjusting the shape and effective width of piers and the effective length of abutments, and by estimating the likely decrease in waterway area (Melville and Dongol, 1992). However, this is difficult to do with confidence.

Table 8.4 illustrates the use of some of the factors listed above, and their relative importance, in assessing the scour potential of bridges in Tennessee (for the full methodology and explanation see Bryan *et al.*, 1995). Given the large number of factors that can contribute to scour potential, it is extremely difficult to include all of them adequately in an equation to predict the maximum scour depth. Consequently the investigation of scour has often proceeded using dimensional analysis and simple laboratory tests that involve uniformly graded material and known conditions. Dimensional analysis leads to a very general functional relationship that summarises some of the factors influencing scour depth, d_S:

$$d_S = f\,(\rho, \nu, g, D, \rho_S, Y, V, b_P) \tag{8.6}$$

where the notation is as before; g is the acceleration due to gravity (m/s^2) and ρ_S is the density of the sediment forming the bed (kg/m^3).

With all studies of scour it is difficult to define or to observe just when incipient motion occurs (i.e. the bed material first moves) and to relate this

Table 8.4 Mean values for the contribution of various factors to scour potential in Tennessee rivers

Variable	*Maximum permissable contribution*	*Mean value for contribution*	*Mean percentage contribution*
Bed material	4	1.94	18.3
Stage of evolution of channel	4	1.50	14.1
Presence of bank erosion	4	1.97	18.6
Meander impact (on pier or abutment, channel shifting)	3	0.97	9.1
High flow angle (high angle of attack)	3	0.43	4.0
Bed protection (reinforcement to bed/banks, or lack of)	3	1.62	15.3
Severity of channel constriction	4	0.33	3.1
Channel piers (number of piers in main channel)	2	0.86	8.1
Presence of skewed piers	9	0.46	4.3
Debris blockage	4	0.25	2.4
Mass wasting with pier (erosion of banks near a pier)	6	0.29	2.7
Totals	46	10.62	100.0

To assess scour potential at a bridge the site is marked against the maximum permissible contribution (e.g. 1.94 out of 4 for bed material). The mean percentage contribution is the mean individual value divided by the total score (e.g. 1.94/10.62 = 18.3%). The figures above are the average of the four regions used in the original study. Obviously the percentage contribution will change from river to river and from region to region depending upon the nature of the bed material and channel. The table is intended to show only that there are many important factors that must be considered when assessing scour.

After Bryan *et al.* (1995)

to the mean velocity (V), the significant bed velocity (U) adjacent to the material that is available to be transported or eroded, or the shear velocity (U^*, see below). The bed velocity is less than the mean velocity ($V = Q/A$) and the surface velocity (e.g. see Fig. 2.9a). Often the bed velocity is unknown, but it is this value that should be compared with the values in Table 1.3a, for example. Recognising this difficulty, Hjulström (1935) felt compelled to use V (which was assumed to be about 40% greater than the bottom velocity when the depth of flow exceeded 1 m) when he investigated the erosion, transportation and sedimentation characteristics of uniformly graded loose particles on a smooth bed (Fig. 8.17). The diagram shows a partial line separating sedimentation from transportation, and a broad zone in which transportation changes to erosion. With transportation, material is moved downstream in traction, saltation and suspension, but is generally replaced by material carried from upstream so that the bed elevation remains constant, whereas with erosion it is assumed that there is a net loss of material and a reduction in bed level.

Hjulström found that 0.25 mm diameter particles are the most easily eroded, starting at a mean velocity of 0.2 m/s. For smaller particles the velocity required is higher because of the need to overcome electrochemical cohesion, while for larger particles a higher velocity is needed to roll or lift the heavier material. Figure 8.17 also incorporates the work of Shields (1936) and Sundborg (1956). Although this (and Fig. 1.14) are useful in giving some idea of the mean velocities at which erosion may be expected, they are not very accurate. For example, there is no fixed relationship between bed velocity and mean velocity; bed material often consists of a range of sizes, which behave differently; the existence of sand ripples on the bed will reduce the erosion velocity, while fine cohesive material may break away in lumps rather than being moved grain by grain (Rice, 1977).

Frequently the significant stream velocity (U or V) is replaced with the shear velocity (U^* m/s). While U^* has the dimensions of velocity, it is not a real velocity although it is related to the real fluid velocity and to the shear stress at the bed. The shear velocity is defined as

$$U^* = (gYS_O)^{1/2} \tag{8.7}$$

where S_O is the bed slope. If U_S^* is the scour-critical shear velocity at which motion of the bed begins, then by applying a few simplifying assumptions to equation 8.6 it is possible to obtain (Copp and Johnson, 1987)

$$\frac{d_S}{b_p} = f\left(\frac{U^*}{U_S^*}, \frac{U^*}{(gY)^{1/2}}, \frac{Y}{b_p}, \frac{D}{b_p}\right) \tag{8.8}$$

All of the terms in this equation are dimensionless, with $U^*/(gY)^{1/2}$ being a kind of Froude number. With non-uniform bed materials another dimensionless factor should be introduced to represent the variation of particle size. This may be $\sigma = (D_{84}/D_{16})^{1/2}$. Even with this extra term, equation 8.8 does not include all of the factors that affect scour depth, so any formula based on it can at best be an approximation.

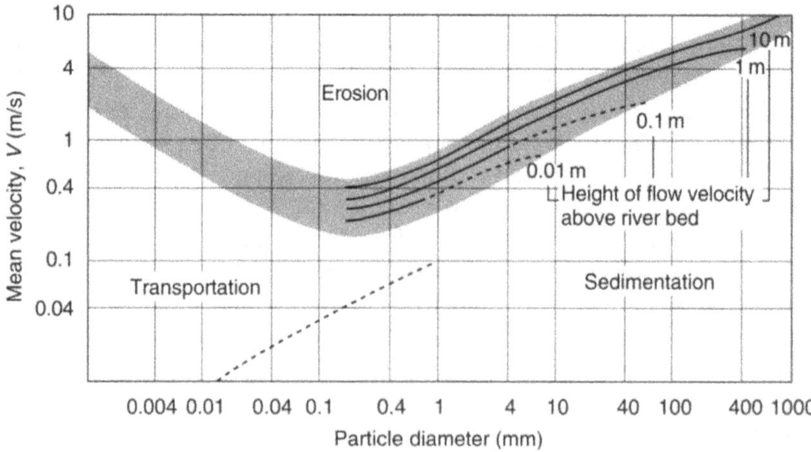

Fig. 8.17 Stream velocities required to set channel material in motion. (After Rice, 1977. Reproduced by permission of Addison Wesley Longman Ltd)

Obviously predicting scour depths is difficult, and there are two basic approaches. The first is to use laboratory-based equations for scour in a long constriction, perhaps in conjunction with a factor of safety, although this does not guarantee that either an accurate or a safe estimate will be obtained. This approach is explored in more detail in Sections 8.4 and 8.5 below. The second option is to estimate how the natural width and depth of the channel will be affected by the construction of a bridge. This is explored in Section 8.6, where regime theory is considered. Under appropriate conditions this can also be used to estimate degradation and aggradation.

8.4 Estimation of scour depth in cohesionless material

The degradation scour depth (Δd) has to be evaluated as part of the design process, but the two types of scour that are the direct result of bridge construction are contraction and local scour, which can occur with either clear-water or live-bed conditions, as described earlier and summarised in Table 8.2. Each condition may have its own set of equations that should be used to estimate the scour depth.

8.4.1 Contraction scour depth (d_{SC})

Contraction scour occurs as a result of the narrowing of a channel by a bridge and its highway embankments, but this effect can be exacerbated by natural stream constrictions, islands or sand banks, vegetation growth in the channel, debris accumulation, ice jams and bends. Bends produce a non-uniform flow, so the contraction scour may be concentrated on the outside of the bend, or possibly nearer the centre during flood when the thalweg may shift.

The effect of any constriction is to reduce the flow area and to increase the velocity. The higher velocity results in an increased erosive force so that more bed material is removed from the contracted reach than is transported into it from upstream, where there may be no transport at all (clear-water conditions) or a lower transportation rate (live-bed conditions). However, as the bed elevation is lowered and a scour pit develops, the cross-sectional area increases and the velocity falls again so that some form of equilibrium is reached. This can be achieved either when the velocity in the contracted section falls below the critical value required to initiate motion (clear-water scour) or when the rate of transport of material into and out of the scour hole are equal (live-bed equilibrium scour).

A quick but conservative method of estimating the scour depth at a contraction of fixed width such as a bridge is to estimate the mean flow velocity and depth in the opening ($V = Q/a$ and Y), and the median diameter of the bed material by weight (D_{50}). From Fig. 1.14 the competent mean velocity (V_{SC}) of cohesionless material can be obtained; this is the velocity at which the flow is just competent to move the exposed bed material at the scoured depth. By increasing a (via Y) the depth at which the bed material can resist being moved can be determined, and thus the average contraction scour depth, d_{SC}. The answer obtained is obviously approximate, and can be seriously affected by local conditions such as eddies and bends. For example, at a bend the scour depth may be much greater on the outside of the curve, and this should be allowed for by adjusting either the average velocity or the shape of the scoured cross-section to obtain the maximum scour depth (see point 4 of Section 8.3). This simple technique is likely to overestimate when the approach flow carries a large bed load (live-bed scour). For cohesive materials compare V with the values in Table 1.3b.

At any particular location a combination of clear-water and live-bed scour may occur. A typical example would be a bridge with several openings, some in the main channel (which are likely to experience live-bed conditions) and one or more relief openings on the floodplain (which is likely to be grassed and to experience clear-water conditions). Each opening must be analysed separately using the equations, bed material and hydraulic parameters appropriate to that particular opening. For either the clear-water or live-bed condition the equations below, which are based on a simplified transport function, are the starting point recommended by Richardson *et al.* (1993).

Clear-water contraction scour

Laursen (1963) presented the following equation for the scour-critical velocity, V_S (m/s) at which the bed material will start to move with clear-water scour:

$$V_S = 6.0\ Y^{1/6}\ D_{50}^{1/3} \qquad (8.9)$$

where Y is the depth (m) in the upstream channel. This is basically the same as equation 8.3 except for the numerical coefficient. If it is written in terms of the discharge by employing the continuity equation then the expression in equation 8.10 is obtained. This can be applied where there is a constriction involving only the main channel ($B > b$, $Q_1 = Q_2$), or to an overbank constriction where only the floodplains are narrowed by the highway embankments ($B = b$, $Q_1 < Q_2$), as illustrated in Fig. 8.19. Sometimes a constriction may be a combination of the two types. The equations can also be used to analyse relief openings on floodplains or bridges on secondary channels, provided the flow is clear water and the variables are estimated accordingly.

$$Y_2 = \left(\frac{Q_2^2}{36b^2(D_M)^{2/3}}\right)^{3/7} \qquad (8.10)$$

where Y_2 is the average depth of flow (m) in the bridge opening (or on the overbank at the bridge), Q_2 is the corresponding discharge (m^3/s) through the bridge opening (or on the overbank), b is the bottom width (m) of the bridge opening less pier widths (or overbank width), and D_M (= $1.25D_{50}$ m) is the effective mean diameter of the bed material in the bridge opening (or on the floodplain). Note that for a single-span bridge Q_2 will equal the total upstream discharge (Q_1) unless there is flow around or over the structure being analysed. With multispan bridges Q_2 is that part of the total flow passing through the opening under consideration. The original equations presented by Laursen used the median diameter of the bed material (D_{50}) not D_M. This modification was introduced by Richardson *et al.* (1993) as a result of further research with the effect that the calculated clear-water scour depths are reduced. With stratified material the equation would have to be applied to the layer with the smallest D_{50} to obtain the worst result, or alternatively applied to each layer in turn.

If the average depth of flow in the main approach channel prior to scour is Y_1 (m), if the head loss (h_L) is small and if the velocity heads are approximately equal so the water surface through the bridge waterway is almost horizontal (Fig. 8.15), then the average depth of contraction scour (d_{SC} m) is

$$d_{SC} = Y_2 - Y_1 \tag{8.11}$$

Of course, if the depth of flow remains constant and follows the scoured bed profile through the contraction (Section 3.11) then this equation is not valid.

By assuming continuity between the upstream main channel (subscript 1) and the contracted section (subscript 2) equation 8.10 can be written in terms of the approach velocity (V_1 m/s):

$$Y_2 = Y_1 \left(\frac{B}{b}\right)^{6/7} \left(\frac{V_1^2}{36(Y_1)^{1/3}(D_M)^{2/3}}\right)^{3/7} \tag{8.12}$$

where B is the bottom width of the upstream main channel (m) and b is the botttom width (m) of the channel in the contracted section. Either bottom or top channel widths can be used as long as there is consistency.

Example 8.1

When unobstructed, the discharge in a 7.5 m wide channel is 6.9 m^3/s when the depth is 1.0 m. The bed material has a specific gravity of 2.65 and a D_{50} = 0.01 m. If all of the flow has to pass through a rectangular bridge opening with b = 4.0 m, calculate the scour depth.

Approach average flow velocity, $V = 6.9/(7.5 \times 1.0) = 0.92$ m/s.
From equation 8.3, the scour-critical average velocity, $V_S = 6.36\ (1.0)^{1/6}\ (0.01)^{1/3} = 1.37$ m/s.
Since $V < V_S$ the condition would be clear water so equations 8.10 and 8.11 are applicable.

The bed material in the waterway has $D_M = 1.25D_{50} = 1.25 \times 0.01 = 0.0125$ m.

$$Y_2 = \left(\frac{Q_2^2}{36b^2(D_M)^{2/3}}\right)^{3/7} = \left(\frac{6.9^2}{36 \times 4.0^2 \times 0.0125^{2/3}}\right)^{3/7} = 1.20\,\text{m}$$

$d_{SC} = Y_2 - Y_1 = 1.20 - 1.00 = 0.20$ m.

Live-bed contraction scour

The following equation can be used to estimate live-bed contraction scour (Laursen, 1962):

$$Y_2 = Y_1\left(\frac{Q_2}{Q_1}\right)^{6/7}\left(\frac{B}{b}\right)^{k_1}\left(\frac{n_2}{n_1}\right)^{k_2} \tag{8.13}$$

where Y_2, Y_1, B and b are as above, Q_1 is the flow (m^3/s) transporting sediment in the upstream main channel (i.e. floodplain flows are not included), Q_2 is the discharge (m^3/s) through the contracted channel (of width b), k_1 and k_2 are exponents determined from Table 8.5 and Fig. 8.18 according to the mode of bed material transport, and n_1 and n_2 are Manning's roughness coefficients for the upstream main channel and the contracted section respectively (see below). Note that in simple cases typically involving a single opening with no bypass flow $Q_1 = Q_2$ as in Example 8.2, but $Q_1 > Q_2$ if some of the live-bed upstream flow passes around or over the bridge instead of through the waterway. However, $Q_1 < Q_2$ when the flow is spread out on an upstream floodplain which experiences clear-water scour and only the live-bed discharge in the main live-bed channel (ie Q_1) is considered, as in

Table 8.5 Exponents for determining live-bed contraction scour for use with equation 8.13. Calculate U_1^*/w using the footnotes, then determine the appropriate values of k_1 and k_2 to be used in equation 8.13

Value of U_1^*/w	k_1	k_2	*Mode of bed material transport*
< 0.50	0.59	0.07	Mostly contact bed material
0.50–2.0	0.64	0.21	Some suspended bed material discharge
> 2.00	0.69	0.37	Mostly suspended bed material discharge

$U_1^* = (gY_1S_{F1})^{1/2}$ is the shear velocity in the upstream section (m/s)
g = acceleration due to gravity (9.81 m/s^2)
Y_1 = average depth in the upstream main channel (m)
S_{F1} = slope of energy grade line in main approach channel (dimensionless). Usually assumed $S_F = S_O$.
w = the median fall velocity (m/s) of the bed material based on D_{50} (see Fig. 8.18)

After Richardson *et al.* (1993)

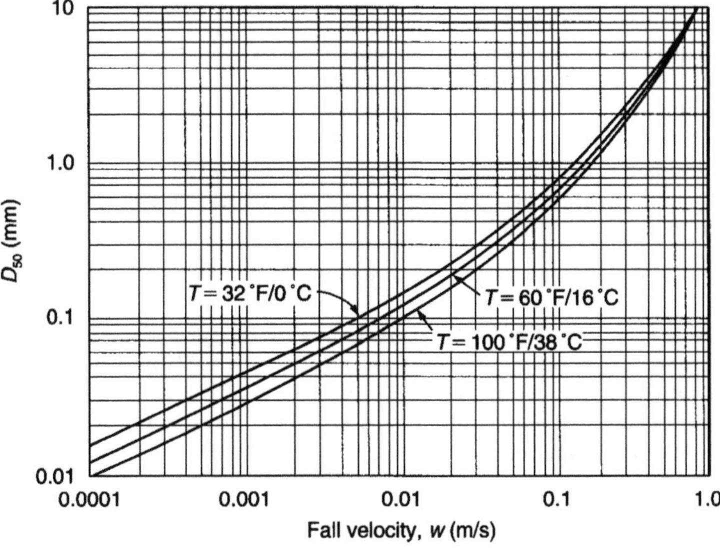

Fig. 8.18 Variation of fall velocity (w m/s) with median particle size (D_{50} mm) and temperature for use with Laursen live-bed contraction scour equation (8.13). (After Richardson *et al.*, 1993)

Example 8.5. The openings of multiple opening contractions have to be considered individually if the flow is not uniform, with part of the total discharge being allocated appropriately to each opening.

Richardson *et al.* (1993) recommended that the ratio of Manning's n be dropped from the equation. They also pointed out that this equation tends to overestimate scour depth if the bridge is located at the upstream end of a natural contraction or if the contraction is the result of bridge abutments and piers. However, they concluded that it was the best equation currently available. As before, the average depth of contraction scour, d_{SC}, can be obtained from equation 8.11. An approximate general solution of the equation is shown in Fig. 8.19. This can be useful during the early stages of bridge design, but should be confirmed by subsequent calculations.

Example 8.2

The upstream approach channel to a bridge is rectangular in section, 20 m wide, and carries a discharge of 25.84 m^3/s when the depth of flow is 1.70 m. The median particle size D_{50} = 0.5 mm (0.0005 m) while the energy gradient (S_{F1}) can be assumed equal to the bed slope at 1 in 3000. The bridge has two 5 m spans separated by a central pier. All of the approach flow passes through the bridge openings. The water temperature is 16 °C/60 °F. Determine the average scour depth in the waterway.

First, check that the live-bed condition exists in the upstream channel.
From equation 8.3 the scour-critical mean velocity is $V_S = 6.36\ Y^{1/6}\ D_{50}^{1/3} = 6.36 \times (1.70)^{1/6} \times (0.0005)^{1/3} = 0.55\,\text{m/s}$.
The actual mean velocity in the upstream channel $V = Q/A = 25.84/(20 \times 1.70) = 0.76\,\text{m/s}$.
Thus live-bed conditions exist ($V > V_S$) so equation 8.13 is appropriate.
To obtain the coefficient, k_1, calculate the shear velocity in the upstream section as shown in Table 8.5:

$$U_1^* = (gY_1S_{F1})^{1/2} = (9.81 \times 1.70 \times 0.00033)^{1/2} = 0.075\,\text{m/s}.$$

From Fig. 8.18, with $T = 16\,°\text{C}$ the median fall velocity (w) for a D_{50} particle with a diameter of 0.5 mm is 0.08 m/s. Thus $U_1^*/w = 0.075/0.08 = 0.94$. From Table 8.5 this corresponds to a combination of bed contact load with some suspended bed material for which $k_1 = 0.64$. Ignoring the Manning roughness ratio, equation 8.13 becomes

$$Y_2 = Y_1\left(\frac{Q_2}{Q_1}\right)^{6/7}\left(\frac{B}{b}\right)^{0.64}$$

Since all of the sediment carrying live-bed approach flow passes through the opening, $Q_1 = Q_2 = 25.84\,\text{m}^3/\text{s}$. The opening width is $b = 2 \times 5 = 10\,\text{m}$ (the pier is ignored), so:

$$Y_2 = 1.70\left(\frac{25.84}{25.84}\right)^{6/7}\left(\frac{20}{10}\right)^{0.64} = 2.65\,\text{m}$$

Consequently from equation 8.11 the average depth of contraction scour, d_{SC} is

$$d_{SC} = Y_2 - Y_1 = 2.65 - 1.70 = 0.95\,\text{m}$$

Note that the local scour caused by the pier and abutments has to be calculated separately, using the procedures described below, and then added to the 0.95 m to obtain the total maximum scour depth. As a quick check of the arithmetic, $B/b = 20/10 = 2.0$ and from above $(gY_1S_{F1})^{1/2}/w = 0.94$, so for a channel constriction Fig. 8.19 gives $d_{SC}/Y_1 = 0.55$ and hence $d_{SC} = 0.55 \times 1.70 = 0.94\,\text{m}$ (the chart is less accurate with more complex geometries).

8.4.2 Local pier scour depth (d_{SP})

Local pier scour is obviously influenced by pier shape and alignment in addition to the characteristics of the channel and the approach flow. As a general guide the ratio of pier scour depth to width (d_{SP}/b_P) for round-nosed piers aligned with the flow does not exceed about 2.3 or 2.4 when $F < 0.8$, rising to about 3.0 for larger F values. Significantly larger values may be

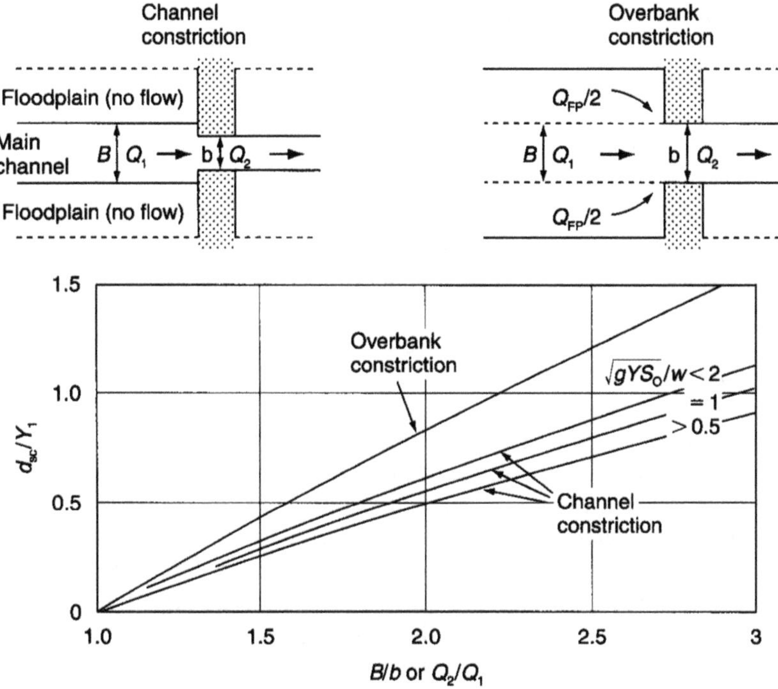

Fig. 8.19 Variation of the depth of live-bed scour in a long contraction (d_{SC}), expressed as a proportion of the average depth of flow in the upstream main channel (Y_1), with the severity of the channel constriction (Q_2/Q_1 or B/b) and $(gYS_O)^{1/2}/w$ as in Table 8.5. This is live-bed scour so the flow (Q_{FP}) on the (grassed) floodplains is ignored. Unless some flow bypasses the bridge, for a single opening Q_2 is the total flow. The solution was formulated for a constriction that narrowed both the main channel and the floodplain. If the abutments protrude into the main channel this is a channel constriction defined by B/b; if the main channel is unobstructed but the floodplain is constricted by the highway embankments then this is an overbank constriction defined by Q_2/Q_1. (After Laursen, 1962, Scour at Bridge Crossings, *Transactions of the ASCE*. Reproduced by permission of ASCE)

obtained for blunt-nosed piers, especially when at an angle to the flow (Fig. 1.15).

Under 'normal' conditions the foundations of the piers will be below bed level, but under some circumstances the footing or pile group may be exposed, and this too has to be allowed for when estimating scour depths. These scenarios are considered in turn below.

Clear-water and live-bed scour for conventional piers

For both clear-water and live-bed conditions Richardson *et al.* (1993) recommended the Colorado State University (CSU) equation for the estimation of equilibrium pier scour depth (d_{SP} m). Some alternative equations for pier scour depth are given in Appendix B, and these may be used cautiously to obtain additional estimates. However, the preferred CSU equation is

$$d_{SP} = 2.0\, Y_2\, K_{1P}\, K_{2P}\, K_{3P} \left(\frac{b_P}{Y_2}\right)^{0.65} F_2^{0.43} \tag{8.14}$$

where Y_2 is the flow depth (m) at the bridge section directly upstream of the pier (m), K_{1P} is an adjustment factor for pier nose shape obtained from Table 8.6 for $\phi < 5°$, K_{2P} is an adjustment factor for the angle of attack ($\phi > 5°$) obtained from Fig. 8.20, K_{3P} is an adjustment factor for bed configuration (explained below) obtained from Table 8.7, b_P is the pier width (m), V_2 is the mean velocity of flow (m/s) at the bridge directly upstream of the pier, and F_2 is the Froude number = $V_2/(gY_2)^{1/2}$. Note that if $\phi > 5°$ the factor K_{2P} dominates so K_{1P} can be taken as 1.0.

With piers that comprise trestle or pile bents (Fig. 5.6) or groups of cylinders as in Table 8.6, the scour depth depends upon the column spacing. If the spacing is ≥ 5 diameters, d_{SP} can be limited to about 1.2 times the scour depth at a single cylinder. With columns less than 5 diameters apart the spaces between the cylinders are ignored when calculating the equivalent pier dimensions perpendicular to the flow. For example, with three cylinders (as below) of 0.6 m diameter spaced at 2.0 m intervals, if $\phi = 0°$ then $b_P = 0.6$ m and $L = 1.8$ m. For other angles of attack the projected width

Table 8.6 K_{1P} – adjustment factor for pier nose shape applicable when $\phi < 5°$

Shape of pier		K_{1P}
Square nose	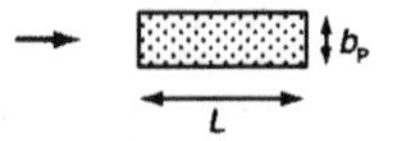	1.1
Round nose	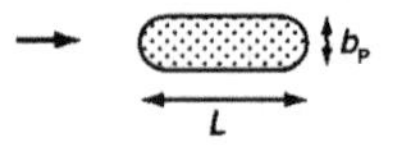	1.0
Circular cylinder		1.0
Sharp nose	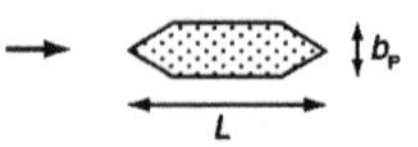	0.9
Group of cylinders	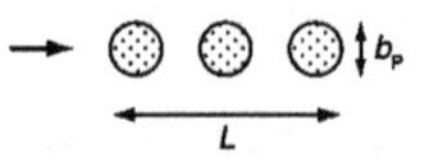	1.0

Note that these values are the ones most compatible with the CSU pier scour equation (8.14) recommended by Richardson *et al.*, 1993. They are applicable up to an angle of attack of 5° after which K_{2P} dominates and regardless of shape K_{1P} = 1.0. Pier length is not considered important, but for cylinder groups see the text above.

After Richardson *et al.* (1993)

Table 8.7 K_{3P}: increase in equilibrium pier scour depths for various bed conditions

Bed condition	*Dune height, H (m)*	K_{3P}
Clear-water scour	Not applicable	1.1
Plane bed and antidunes	Not applicable	1.1
Small dunes	0.6–3.0 m	1.1
Medium dunes	3.0–9 m	1.1–1.2
Large dunes	> 9 m	1.3

Note that these values are to be used with the CSU pier scour equation (8.14). See also Figs 8.9 and 8.10.

After Richardson *et al.* (1993)

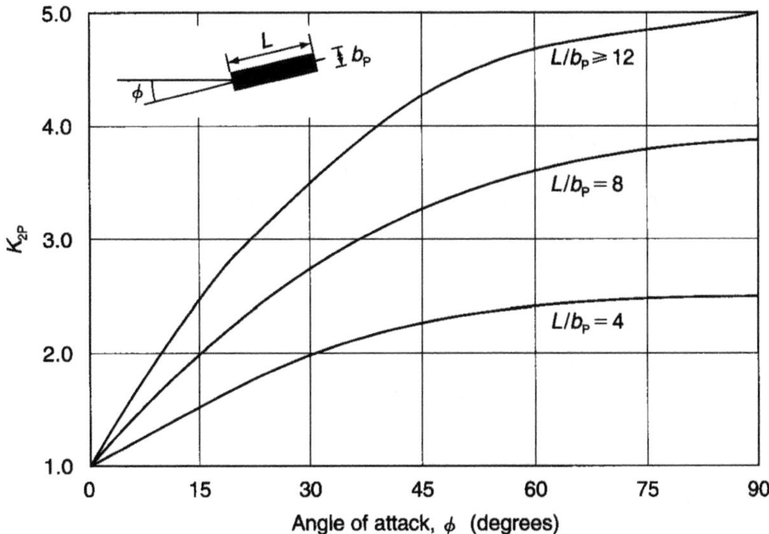

Fig. 8.20 Pier skew correction factor (K_{2P}) for use with the CSU equation (8.14). If $\phi < 5°$ the correction for skew can be ignored. If $\phi > 5°$ the skew dominates, so use the value of K_{2P} obtained from the diagram with $K_{1P} = 1.0$ regardless of shape. (After Richardson *et al.*, 1993)

will be between 0.6 m and 1.8 m. Because the angle of attack is allowed for in the projected width, K_{2P} is always taken as 1.0. The shape factor K_{1P} is also taken as 1.0 regardless of the actual column shape. The exception to this may be if debris becomes trapped between the columns, effectively turning them into a single solid pier of width (b_P) and length (L), which can be analysed as any normal pier.

The factor K_{3P} converts the equilibrium scour depth calculated from the CSU equation to the maximum scour depth according to the prevailing bed configuration (Table 8.7). For the plane-bed condition commonly encountered at bridge sites during a typical design flood, the maximum scour depth may be 10% larger than indicated by the CSU equation, so $K_{3P} = 1.1$. For various bed configurations and dune heights, the table shows the factor that should be used to increase the calculated pier scour depths (d_{SP}). For the unusual situation where the bed exhibits large dunes during flood, for very big rivers such as the Mississippi $K_{3P} = 1.3$, falling to 1.1–1.2 for smaller rivers. For the antidune bed condition $K_{3P} = 1.1$. Note that several different bed forms may occur during one flood as the velocities rise and fall, and that ripple-forming sediments usually have a grain size < 0.7 mm, larger diameters being non-ripple forming (Figs 8.9 and 8.10).

Sometimes K_σ and K_{FS} may be added to the other correction factors in equation 8.14, where K_σ is the factor that allows for sediment grading (Fig. 8.14) and K_{FS} is a factor of safety (Copp and Johnson, 1987). Richardson *et al.* (1993) felt that the former was still not proved satisfactorily by field data.

Example 8.3

Just upstream of a 0.9 m wide round-nosed pier the depth of flow is 1.3 m with a velocity of 1.6 m/s. The pier is skewed to the approach flow with ϕ = 15°. The length of the waterway in the direction of flow (L) is 14.4 m. Assume that the channel bed is plane. Calculate the local pier scour depth.

From Table 8.6, regardless of pier shape $K_{1P} = 1.0$ since $\phi > 5°$.
$L/b_P = 14.4/0.9 = 16$ so Fig. 8.20 gives $K_{2P} = 2.5$ (using the line labelled $L/b_P \geq 12$ and $\phi = 15°$).
$K_{3P} = 1.1$ (the plane bed condition in Table 8.7).

$$Y_2 = 1.3\,\text{m so } b_P/Y_2 = 0.9/1.3 = 0.69$$

$$F_2 = 1.6/(9.81 \times 1.3)^{1/2} = 0.45$$

$$d_{SP} = 2.0\, Y_2\, K_{1P}\, K_{2P}\, K_{3P}\, (b_P/Y_2)^{0.65}\, F_2^{0.43} \tag{8.14}$$

$$= 2.0 \times 1.3 \times 1.0 \times 2.5 \times 1.1 \times (0.69)^{0.65} \times (0.45)^{0.43} = 3.99\,\text{m}$$

As a quick check Fig. 1.15 gives an approximate $d_{SP} = 1.5b_P \times 2.5 = 3.38$ m (where 2.5 is the correction factor for 15° skew).

Pier scour for exposed footings

If the foundations of a bridge have been well designed with an appropriate allowance for scour this situation should not arise. Unfortunately it does, often as a result of long-term degradation, channel shifting, exceptionally large flows or poor design, so it is necessary to have some means of calculating likely scour depths. It should also be pointed out that having the foundations undermined, even if they are piled, is not desirable because lateral support may be lost and the piles may have to behave as columns subject to bending. Often the exposure of footings and the undermining of abutments is a prelude to failure.

The effect of an exposed footing or pile cap on scour is not always easy to predict. There is some evidence that if the top of the foundation is flush with the bed the scour may be reduced (because part of the downflow in Fig. 8.6 is intercepted by the concrete footing). However, as the scour holes become deeper, larger and more of the foundation becomes exposed and projects into the flow then Richardson *et al.* (1993) made the following recommendation for estimating pier scour: calculate the depth as described in case 1 and case 2 below, and then adopt the larger of the two values.

CASE 1: TOP OF THE FOOTING OR PILE CAP IS AT OR BELOW THE RIVER BED

After allowing for degradation and contraction scour, use the actual pier width (b_P) in the pier scour equations (as normal).

CASE 2: THE PIER FOOTING EXTENDS ABOVE THE RIVER BED

Use the width of the footing as b_P in the pier scour equation, in conjunction with the average depth and average velocity in the flow zone obstructed by the footing (Y_F and V_F respectively) instead of the depth and velocity of the approach flow just upstream (Y_2 and V_2). The value of V_F can be obtained from

$$V_F = V_2 \times \frac{\ln[1+10.93(Y_F/k_S)]}{\ln[1+10.93(Y_2/k_S)]} \tag{8.15}$$

where V_F is the average velocity (m/s) in the flow zone below the top of the footing, Y_F is the distance (m) from the bed to the top of the footing, and k_S is the grain roughness of the bed material, which is normally taken as the D_{84} value (m).

Pier scour for exposed pile groups and pile caps

The piled supports of piers and abutments may become exposed. The piles may be spaced across the flow in addition to along it. A particular problem with this situation is that the piles make excellent trash racks and tend to collect debris, which increases the effective size of the group and increases scour.

If local scour has resulted in the pile group being exposed to the flow, it is not necessary to consider the piles so the pier scour depth can be calculated using equation 8.14.

If bed degradation or contraction scour has resulted in the pile group being exposed to the flow, a conservative analysis can be undertaken by considering them as a single width equal to the projected width of the piles normal to the flow (the clear space between the piles is ignored). For example, three 0.4 m piles at 2 m centres at right angles to the flow would have a width $b_P = 3 \times 0.4 = 1.2$ m. Some allowance for debris may be needed. If a large amount of debris accumulates then the pile group may again be considered as a single pier using equation 8.14 (Jones, 1989; Richardson *et al.*, 1993). However, ignoring debris, for pile groups exposed by degradation or contraction scour the variables for use in equation 8.14 are as follows.

- b_P — use the total width of the piles only, as in the example in the text above
- L — is defined as for a group of cylinders in Table 8.6
- K_{1P} — use a factor of 1.0 regardless of shape
- K_{2P} — if the pile group is square (e.g. 3 piles by 3 piles) then $K_{2P} = 1.0$, but if the group is rectangular (e.g. 3 piles wide by 6 piles long) to determine K_{2P} use the dimensions for a single pier of appropriate L/b_P with L and b_P as defined immediately above and in Fig 8.20
- K_P — assume a value of 1.1

If degradation or scour results in the piles, pile cap and part of the pier all being exposed to the flow then the scour caused by each element should be calculated separately and the largest scour depth adopted. When calculating the scour due to the pile cap assume that the cap is resting on the bed and use the values of V_F and Y_F as described in connection with equation 8.15.

Width of pier scour holes, W_{SP}

The size of a scour hole can be important with respect to pier spacing and the stability of banks and abutments. In cohesionless bed material the top width, W_{SP} (m), of the hole measured on one side only of a pier or footing can be estimated as

$$W_{SP} = d_{SP}\,(K_W + \cot\theta) \tag{8.16}$$

where d_{SP} is the pier scour depth (m), K_W is a coefficient representing the bottom width of the scour hole (on one side of the pier) as a fraction of the scour depth, and θ is the angle of repose of the bed material, which is typically between 30° and 44°. For example, if the bottom width on one side of the scour hole equals the depth of scour then $K_W = 1$ and if the angle of repose is 30° then $W_{SP} = 2.73d_{SP}$. As a general guideline, the bottom width decreases as d_{SP} increases, but an assumed top width of $2.8d_{SP}$ has been recommended (Richardson *et al.*, 1993). Thus scour from a single pier may extend over a total width of $(5.6d_{SP} + b_P)$ measured across the waterway opening.

8.4.3 Local scour depth at abutments (d_{SA})

The evaluation of abutment scour is more problematical than for pier scour: piers are usually located in the centre of the channel where the flow is reasonably uniform, whereas abutments and the associated approach embankments may cross the entire width of a floodplain from the outside edge where the depth and velocity may be zero to a point in the main channel where the velocity is relatively large. Thus the bridge approaches cross a compound channel of variable topography, where the flow conditions vary greatly and the mean values are difficult to determine accurately. This contrasts sharply with laboratory investigations that generally have a uniform approach flow across the entire width of a rectangular channel so that the full length of the approach embankments experiences practically the same flow. Consequently it is not surprising that many laboratory-based equations that assume a relationship between embankment–abutment length and scour depth overestimate significantly.

Some of the factors that influence abutment scour include the topography of the site, the abutment shape, and the hydraulic and sediment characteristics. Scour is worst where conditions result in the overbank flow returning suddenly to the main channel, perhaps as a consequence of flow along the upstream face of the embankment being relatively easy, as shown in Fig. 7.11. Vertical wall abutments have approximately twice the scour depth of spillthrough types (see Fig. 4.21 for an illustration of the abutment types). Channel migration may result in a larger angle of attack and scour depth than envisaged during design, while hydraulic changes above or below the bridge may lead to channel degradation. The conditions in the approach channel combined with the type of bed material can result in either clear-water or live-bed scour.

Clear-water and live-bed abutment scour

Richardson *et al.* (1993) recommended the use of the live-bed Froehlich equation for the calculation of both clear-water and live-bed abutment scour depths, d_{SA}. The equation yields large scour depths as a result of the assumption that most of the overbank flow returns to the main channel at the end of the abutment, which is entirely feasible, particularly if the abutments protrude into the main channel. However, if the floodplain, channel and abutments are all covered in vegetation then smaller depths may be experienced. The predicted abutment scour depth, d_{SA} (m), is

$$d_{SA} = Y_{M1} + 2.27\, Y_{M1}\, K_{1A}\, K_{2A}\, (L_A/Y_{M1})^{0.43}\, F_{M1}^{0.61} \tag{8.17}$$

where Y_{M1} is the mean depth of flow (m) on the upstream floodplain, K_{1A} is the coefficient for abutment shape as in Table 8.8, $K_{2A} = (\Phi/90)^{0.13}$ is the coefficient for the angle of the embankment-abutment relative to

Table 8.8 K_{1A}: coefficients for abutment type

Abutment type	K_{1A}
Vertical-wall abutment	1.00
Vertical-wall abutment with wingwalls	0.82
Spillthrough abutment	0.55

Note that these values are to be used only with the abutment scour equations (8.17 or 8.18).

After Richardson *et al.* (1993)

the approach flow (see below), L_A is the length (m) of the embankment-abutment projected normal to the flow, and F_{M1} is the Froude number of the approach flow upstream of the abutment. In this case $F_{M1} = V_{M1}/(gY_{M1})^{1/2}$ where V_{M1} is the mean velocity (m/s) on the floodplain, which is calculated as $V_{M1} = Q_A/A_A$ where Q_A is the approach flow (m^3/s) obstructed by the embankment-abutment and A_A ($= L_A Y_{M1}$) is the flow area (m^2) of the approach cross-section obstructed by the embankment-abutment.

Note that with respect to the angle of the approach flow, $\Phi < 90°$ if the embankment-abutment points downstream and $\Phi > 90°$ if it points upstream. For example, with a crossing built diagonally at 45° over a floodplain then one abutment will point downstream so $\Phi = 45°$ and $K_{2A} = (45/90)^{0.13} = 0.91$, while the other abutment points upstream so $\Phi = 135°$ and $K_{2A} = (135/90)^{0.13} = 1.05$. Thus scour is reduced if the embankment-abutment points downstream and increased if it points upstream, so often only the latter needs to be calculated, being the worst case.

One difficulty with equation 8.17 is estimating accurately the values of the variables: depth and velocity will vary across a floodplain and may not have been measured, so there is scope for error. Often the conveyance or discharge over the entire floodplain will have to be calculated as in Examples 3.2 and 3.3, or by breaking the floodplain down into subsections.

Some alternative equations for abutment scour are listed in Appendix B, and they may be applied cautiously to obtain additional estimates of scour depth. The equation below arises from the study of scour at the end of spur dykes, which, apart from the difference in orientation, behave similarly to bridge approach embankments provided that $(L_A/Y_{M1}) > 25$ where L_A and Y_{M1} are defined above (Simons and Senturk, 1976; Melville, 1988; Richardson *et al.*, 1993). Assuming this similarity exists at the site under investigation then a rough estimate of the abutment scour depth (d_{SA} m) is

$$d_{SA} = 4\, Y_{A2}\, (K_{1A}/0.55)\, K_{A\Phi}\, F_{A2}^{0.33} \tag{8.18}$$

where Y_{A2} and V_{A2} are the depth of flow (m) and velocity (m/s) passing through the bridge opening at the end of the abutment, and F_{A2} is the corresponding Froude number. The original study employed spillthrough

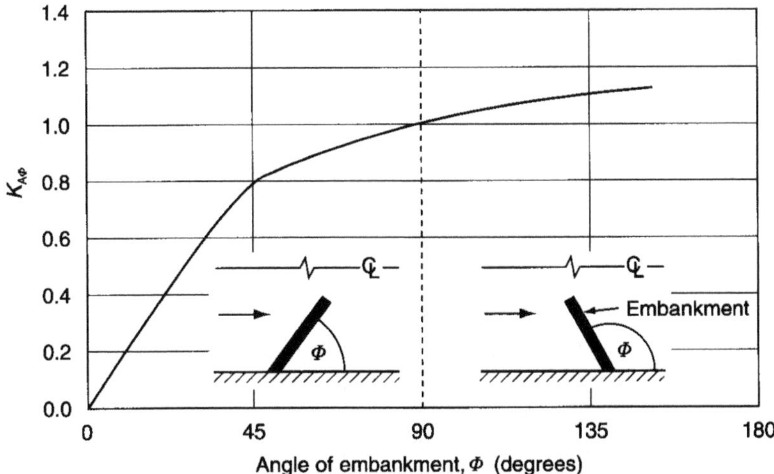

Fig. 8.21 Abutment-embankment skew correction factor ($K_{A\Phi}$) for use with equation 8.18. Embankments pointing downstream have $\Phi < 90°$; those pointing upstream have $\Phi > 90°$. $K_{A\Phi}$ represents the ratio of the depth of scour at the skewed embankment to that at a perpenicular crossing. (After Richardson *et al.*, 1993)

abutments for which $K_{1A} = 1.0$ so the term $K_{1A}/0.55$ converts the values in Table 8.8 to this scale. The adjustment factor for skew ($K_{A\Phi}$) can be obtained from Figure 8.21, embankments-abutments pointing upstream again having values of $\Phi > 90°$. Under appropriate circumstances the equation can be used to check the results from equation 8.17, as shown later in Example 8.5.

Example 8.4

The longitudinal centreline of an embankment leading to a bridge abutment is skewed at an angle of 30° compared with a perpendicular crossing. The length of the embankment/abutment is 33 m measured along the centreline. The abutments are of the vertical-wall type. It is estimated that the mean depth on the upstream floodplain is 1.2 m with a mean velocity of 0.7 m/s. Calculate the maximum abutment scour depth.

The scour depth will be calculated using the recommended equation 8.17.
$Y_{M1} = 1.2\,\text{m}$, $V_{M1} = 0.7\,\text{m/s}$ so $F_{M1} = 0.7/(9.81 \times 1.2)^{1/2} = 0.20$.
$K_{1A} = 1.00$ (from Table 8.8, vertical-wall abutments).
Scour will be greatest at the abutment pointing upstream when $\Phi = 30° + 90° = 120°$ (the definition of Φ is shown in Fig. 8.21, but not the value of K_{2A}) so $K_{2A} = (120/90)^{0.13} = 1.04$.

$L_A = 33 \cos 30° = 28.58\,\text{m}$ (i.e. the length normal to the approach flow)

$$d_{SA} = Y_{M1} + 2.27\, Y_{M1}\, K_{1A}\, K_{2A}\, (L_A/Y_{M1})^{0.43}\, F_{M1}^{0.61}$$

$$= 1.20 + 2.27 \times 1.20 \times 1.00 \times 1.04 \times (28.58/1.20)^{0.43} \times (0.20)^{0.61}$$

$$= 5.35\,\text{m} \qquad (8.17)$$

8.5 Designing for scour

8.5.1 General design philosophy

The return interval of the flood used to design scour prevention measures needs to be considered carefully. For instance, suppose that both the size of the waterway opening and the foundations of a bridge are designed for the 1 in 100 year flood (Q_{100}). If a discharge larger than Q_{100} occurred this would not be important hydraulically if the excess flow passed safely around or over the bridge, but it would be disastrous if the structure failed by scouring of the foundations. Thus the consequences of the design flood being exceeded may be more severe structurally than hydraulically.

The likelihood of the design flood being exceeded may be much higher than imagined from a cursory inspection of the problem. This is easily calculated from the equation

$$J = 1 - \left[1 - \left(\frac{1}{T}\right)\right]^n \qquad (8.19)$$

where J (fraction) is the probability that at least one event that equals or exceeds the T year return interval event will occur in any n year period. For example, if a bridge has a design life of $n = 120$ years and is designed for a $T = 1$ in 100 year flood then

$$J = 1 - \left[1 - \left(\frac{1}{100}\right)\right]^{120} = 0.70$$

so there is a 70% chance that the structure will encounter a larger flood during its existence. A 70% chance that the foundations will fail is not an acceptable risk: consequently the return interval assumed for the purpose of designing the foundations and scour prevention measures should be larger than this, and the design should have a reasonable factor of safety. Failure to adopt this principle may be one reason why so many bridges fail as a result of scour, and a good reason why bridge design needs to involve engineers with expertise in hydrology, hydraulics, structures and geotechnics.

Richardson *et al.* (1993) suggested that the geotechnical design of the foundations of important bridges should be conducted using a factor of safety of 1.5–2.0 with respect to a 1 in 100 year flood (Q_{100}), but the factor of safety should still be greater than 1.0 even with a super-flood. The super-flood should be the 1 in 500 year event, which can be estimated as $1.7Q_{100}$ if more accurate data are not available. If the super-flood results in water spilling over the bridge or its approaches, as described in Chapter 6, then it is possible that the worst-case scenario will be during a smaller event when all of the flow is forced to pass through the waterway opening. Obviously, for less important bridges the size of the design and super-floods may be reduced.

A failed bridge can disrupt commerce and cause loss of life (see Section 1.3), while the cost of replacing a collapsed bridge is normally many times larger than the original construction cost, so it is usually better value to design and construct a bridge to a standard that will withstand scour and the super-flood rather than to have to conduct repairs and fit scour protection measures at a later date.

When initially designing the bridge there are a number of elements or features that can be incorporated, often at relatively little expense, that can improve the hydraulic performance and/or reduce scour and improve structural integrity. Some of these are considered briefly below.

8.5.2 Hydraulic considerations

The list below summarises some of the things that may be considered to help reduce scour problems. They may not all be practical under all circumstances, nor is every possibility listed.

- Is the control of the longitudinal profile of the water surface constant (e.g. controlled by rock outcrops) or is it likely to change (e.g. removal of old weirs) leading to degradation or aggradation?
- Shifting or meandering channels create uncertainty in scour prediction. Can the bridge be located to avoid problems of this nature? Or can river training works be used effectively to eliminate such concerns?
- Skewed crossings are not as efficient as those perpendicular to the approach flow. Scour is increased at a skewed abutment pointing upstream. Can the bridge location or alignment be chosen to avoid skew?
- Can contraction scour be reduced by increasing the width of the bridge opening?
- Abutment scour can be reduced by using spillthrough abutments rather than vertical-wall types (Table 8.8).
- Abutment scour protection, such as riprap, may avoid the need to design for the full scour depth.
- Problems may be reduced if the abutments are set back from the edge of the main channel by 3–5 times the depth of flow, so that bank failure will not initiate abutment failure.
- Pier scour is reduced if the piers point directly into the flow (no skew).

Is this possible? Will channel shifting alter this in the future? If it is likely that the approach flow will change direction, can circular columns be used so the angle is unimportant?

- Pier scour increases with increasing pier thickness. How thin is it practicable to make the piers?
- Pier scour is reduced if rectangular shapes are avoided, sharp noses being best (Table 8.6). Can a more efficient shape be adopted economically?
- Can overlapping scour holes be avoided, by increasing pier spacing if necessary? When scour holes overlap the total depth of scour increases, although by how much is not clear.
- Generally a small number of large openings is better than a large number of small openings.
- Scour problems increase where overbank flow on wide floodplains returns suddenly to the main channel in order to pass through the bridge opening. Can guidewalls or spur dykes be used either to return the flow more gradually or to prevent transverse flow along the upstream face of the embankment? Can relief openings be used effectively on the floodplains?
- Have existing bridges in the area been assessed to provide some indication of the likely performance of the structure under consideration?
- When estimating scour depths, has adequate allowance been made for the accumulation of debris or ice on the piers and abutments?
- Has at least 0.6 m freeboard been allowed for the passage of debris between the design flood level and the underside of the bridge deck?
- Can the underside of the bridge deck be made as smooth and streamlined as possible to reduce the obstacle to flow if the deck becomes submerged, to reduce the vertical contraction, and to avoid snagging debris?
- Can the danger to the bridge during a super-flood be reduced by allowing flow over the approach embankments?
- Scour is additive, so where they overlap contraction scour (d_{SC}), pier scour (d_{SP}) and abutment scour (d_{SA}) must be added to any anticipated degradation (Δd) of the channel to get the total depth of scour (d_S), which is $d_S = d_{SC} + d_{SP} + d_{SA} + \Delta d$.
- Has the proposed design significant flaws? Does it involve some uncertainty? Is there some alternative that can be considered, perhaps using a different approach?

8.5.3 *Geotechnical and structural considerations*

Again, this is a brief and far from complete reminder of a few of the design factors that should be borne in mind.

- The foundations should withstand the design flood with a factor of safety of 1.5–2.0 and the super-flood with a factor of safety not less

Table 8.9 Some suggestions for foundations to resist scour

Foundation type	*Depth and other considerations*
Spread footing on soil	The bottom of the footing should be below the total scour and degradation depth calculated using equation 8.5.
Spread footings on erodable rock	Footings should be located on competent rock below the weathered zone and below the maximum scour and degradation depth. Excavation and blasting need to be conducted carefully, the excavation cleaned and completely filled with concrete.
Spread footing on resistant rock (e.g. granite)	The bottom of the footing should bear directly on the cleaned rock surface. Footings may be keyed to the rock by using dowels, but avoid operations such as blasting that may damage the rock structure and aid scour.
Deep foundations and piles with footings or caps	The top of the footing or pile cap should be below the total scour and degradation depth to minimise any possibility of being exposed, obstructing the flow, or suffering damage.
Stub abutments on piles	Stub abutments in the embankment should be piled to below lowest bed level in the bridge waterway to ensure structural integrity if the thalweg or channel shifts course.

After Richardson *et al.* (1993)

than 1.0. Factors of safety should reflect the importance of the crossing and the degree of uncertainty in the design.

- The design should assume that the material down to the total scour depth has been removed and is not available for bearing or lateral support. As a general rule, foundations should always be at least 2 m below the level of the streambed after allowing for the total depth of scour indicated by equation 8.5. Spread footings on soil or weathered rock should be at or below this level (see Table 8.9). If used, the piling should be designed for additional lateral restraint and column action.
- With piled foundations and a high scour potential, can the bearing load be carried by a small number of long piles instead of a large number of relatively short ones? The obstacle to flow is then reduced should the piles become exposed.
- When appropriate, the foundations of all piers (whether on the floodplain or main channel) should be taken to the same depth to allow for channel shifting over the life of the bridge.
- If likely to become submerged, or if significant debris or ice forces are expected, the superstructure should be securely anchored to the substructure. Buoyancy forces acting on the soffit of slab or arch bridges increase the risk of structural damage (see Appendix A).
- Forces arising from scour and foundation movement are resisted better by continuous span structures than by simple span bridges.

8.5.4 A general scour design procedure

The procedure for designing a bridge to resist scour can be broken down into the 10 steps shown below. These are illustrated in Example 8.5.

Step 1 *Assemble the data.* This should include. hydrological, climatic, hydraulic and topographic data relating to the site, with an indication of future catchment changes; field data including samples of the material forming the river channel and floodplains, a visual assessment of channel stability, evidence of the behaviour of other bridges in the area and any notable features; geotechnical data regarding depth to competent loadbearing material; the design brief, including an indication of the importance of the crossing, the level of expenditure that can be justified, and details of the proposed design such as the span between abutments, the number of piers and their spacing. Determine the return interval to be adopted for the hydraulic and geotechnical design.

Step 2 *Analyse long-term bed and channel changes.* On the basis of field observation and evidence, local development plans, industrial developments and possibly regime theory (Section 8.6) assess whether or not the channel bed is likely to remain at its current level, degrade, aggrade or shift course. Calculate the depth of degradation (Δd) or aggradation.

Step 3 *Determine whether clear-water or live-bed scour is involved.* Equation 8.3 can be used for this purpose while Table 8.2 summarises the equations that can be subsequently employed.

Step 4 *Calculate the depth of contraction scour,* d_{SC}. Either equation 8.10 or equation 8.13 can be used depending upon the answer from step 3.

Step 5 *Calculate the depth of local pier scour,* d_{SP}. Equation 8.14 is recommended. See also Appendix B.

Step 6 *Calculate the width of the pier scour holes.* Assume that the top width of the scour hole $= (5.6d_{SP} + b_P)$ and the angle of repose is 30° to the horizontal.

Step 7 *Calculate the depth of scour at the abutments,* d_{SA}. Equation 8.17 is recommended. Equation 8.18 and Appendix B offer alternatives.

Step 8 *Calculate the total scour depth,* d_S. This should include the estimated value for bed degradation (if any). On a cross-section of the bridge site, plot the bridge details, the existing bed level, and the scoured bed level and profile using the results of steps 2 and 4–7. If the bed elevation has been significantly reduced by scour then the enlarged channel will have lower velocities, so equilibrium bed level may be somewhat higher than indicated; if necessary repeat the calculations starting with an intermediate level. Concurrent with the above calculations, the scour depths during a super-flood may be estimated to ensure integrity of the structure.

Step 9 *Review the analysis and evaluate the design.* Have the correct equations been applied, within range, and are the results within normal guidelines? Have all relevant factors been considered? Glance through items 1–12 in Section 8.3. Has appropriate engineering judgement been applied? Is the design adequate? Glance down the checklists in Sections 8.5.2 and 8.5.3. Is it necessary to adjust the bridge span, pier or abutment type, width, spacing or orientation, or to provide river training works or riprap protection (Section 8.9)? Could it be done better or more cheaply some other way?

Step 10 *Design the foundations.* The design flood should be used with a suitable factor of safety, and foundations set below the maximum scour depth (see Table 8.9). The calculations should be repeated for a super-flood.

Example 8.5

This example is intended to illustrate the hydraulic aspects of the above procedure. For simplicity and brevity it is concerned mainly with the underlying principles and consequently lacks the scale and complexity that would be encountered in a real investigation.

The brief

A river in the lower stages of its course passes through a wide shallow valley (Fig. 8.22). The main channel is 60 m wide and has low banks about 1.0 m high that are rather prone to erosion, making the flow direction rather variable. The approach bed slope is 1 in 1000. The floodplains are grassed and used for grazing. The riverbed and floodplains comprise fine material with D_{50} = 1.0 mm. Rock can be found approximately 25 m below bed level. It is proposed that a road will cross the valley requiring a bridge, basic details of which are shown in Fig. 8.23. The proposed bridge waterway consists of a 60 m wide main channel section, with a raised overbank section 20 m wide. The latter is at current bank level and is provided partly to enable the landowner to move livestock from one side of the highway to the other during normal flow conditions. The bridge deck will be supported on three 1.2 m thick round-nosed piers at 20 m centres. The proposed abutments are of the vertical type with wingwalls. The intention is that both the piers and abutments will be perpendicular to the flow, their length in the direction of flow being 36 m. The bridge is still at the design stage, and modifications are permissible.

In this region the most severe floods occur in winter, and it is estimated that the 1 in 100 year flood is 490 m^3/s. The discharge through the main subsections of the approach channel and bridge waterway is indicated on the two diagrams. It is anticipated that parts of the catchment will become more urbanised in future (hence the road).

Step 1 Assemble the data

The data available are indicated above and in Figs 8.22 and 8.23. It would be advantageous at a later date to split the channel into a larger number of subsections (perhaps 10–20 streamtubes with an equal discharge) to get a

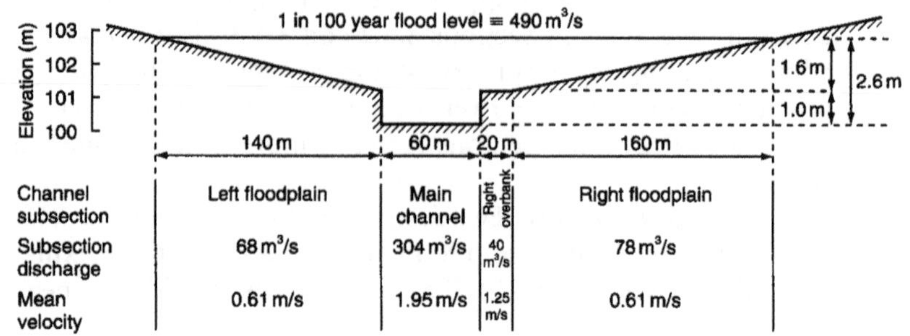

Fig. 8.22 Cross-section of the approach channel for Example 8.5.

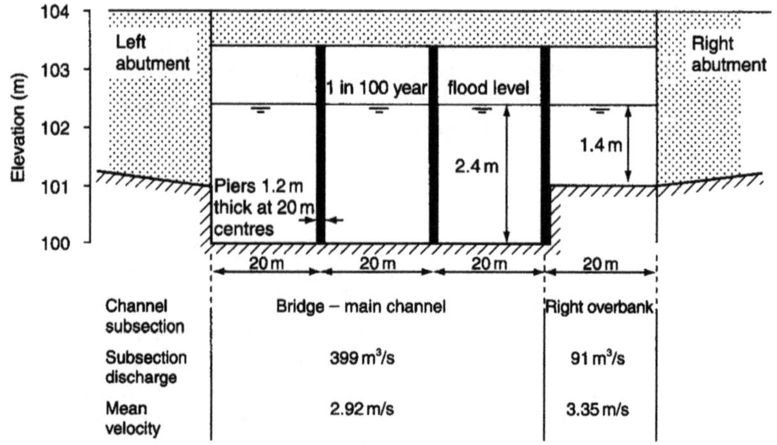

Fig. 8.23 Cross-section through the bridge waterway for Example 8.5.

more detailed picture of the flow patterns, and to have more data regarding the longitudinal profile.

Step 2 Analyse long-term bed and channel changes

The fact that urbanisation is expected locally may result in channel improvements, increased runoff, peakier hydrographs than at present, and an enhanced sediment transport capability. Consequently channel degradation may occur, so 0.5 m will be assigned to Δd. There is evidence of channel meandering, so although the intention is to construct the crossing at right angles to the river, a skew of 15° will be allowed initially in case this condition is not met exactly.

Step 3 Determine whether clear-water or live-bed scour is involved

From equation 8.3, the scour-critical mean velocity is $V_S = 6.36\ Y^{1/6} D_{50}^{1/3}$. For $D_{50} = 0.001$ m, the scour-critical and actual velocities in the subsections of the approach channel are shown below.

Approach channel	$Y = Y_M = A/B_T$ *(m)*	V_S *(m/s)*	*Actual* $V = Q/A$ *(m/s)*
Left floodplain	0.80	0.61	0.61
Main channel	2.60	0.75	1.95
Right overbank	1.60	0.69	1.25
Right floodplain	0.80	0.61	0.61

Thus the floodplains are borderline between the two conditions, but since they are grassed, clear-water scour is most likely. The main channel is live-bed scour ($V > V_S$). Although the right overbank area is also grassed this relatively narrow strip will probably be quite heavily trafficked and worn near the bridge; it will contain sediment that can be transported because of its proximity to the main channel, the possibility of bank caving and its relatively high velocity, so it will be considered as live bed.

Step 4 Calculate the depth of contraction scour, d_{SC}

Two areas need to be assessed: the main channel and the right overbank. This is the live-bed condition so equation 8.13 is applicable without the roughness ratio.

MAIN CHANNEL

Y_1 = depth in main approach channel = 2.6 m.
Q_1 = the flow carrying sediment in the main part of the approach channel = $304\,m^3/s$ (Fig. 8.22).
Q_2 = the flow passing through the main part of the bridge opening = $399\,m^3/s$ (Fig. 8.23).
B = width of main approach channel = 60 m.
b = net width of main bridge opening = 57 m (i.e. 60 m less 2.5 × 1.2 m pier widths).
From Table 8.5, with $S_F = S_O$ the shear velocity $U_1^* = (gY_1S_{F1})^{1/2} = (9.81 \times 2.6 \times 0.001)^{1/2} = 0.16\,m/s$.
Assume a winter flood with cold water (32 °F/0 °C) and $D_{50} = 1$ mm, so the median fall velocity $w = 0.14$ m/s from Fig. 8.18, hence $U_1^*/w = 0.16/0.14 = 1.14$.
From Table 8.5, $k_1 = 0.64$, which corresponds to a combination of bed contact and some suspended transport of sediment.

$$Y_2 = Y_1(Q_2/Q_1)^{6/7}(B/b)^{0.64} = 2.6 \times (399/304)^{6/7} \times (60/57)^{0.64} = 3.39\,\text{m}. \tag{8.13}$$

$$d_{SC} = Y_2 - Y_1 = 3.39 - 2.60 = 0.79\,\text{m} \text{ (below main channel level at bridge).} \tag{8.11}$$

Note that with $(Q_2/Q_1) = 399/304 = 1.31$ the overbank constriction line of Fig. 8.19 gives $d_{SC}/Y_1 = 0.3$ so $d_{SC} = 0.3 \times 2.60 = 0.78$ m.

RIGHT OVERBANK

Y_1 = depth on overbank at upstream section = 1.6 m (Fig. 8.22).
Q_1 = the flow in the overbank part of the approach channel = 40 m^3/s.
Q_2 = the flow passing through the overbank part of the bridge opening = 91 m^3/s (Fig. 8.23).
B = width of the overbank section in approach channel = 20 m.
b = net width of the overbank opening = 19.4 m (i.e. 20 m less 0.5 × 1.2 m pier width).
From Table 8.5, shear velocity $U_1^* = (gY_1S_{F1})^{1/2} = (9.81 \times 1.6 \times 0.001)^{1/2}$ = 0.13 m/s.
Assume T = 32 °F/0 °C and D_{50} = 1 mm so the median fall velocity w = 0.14 m/s from Fig. 8.18, hence $U_1^*/w = 0.13/0.14 = 0.92$.
Thus from Table 8.5, $k_1 = 0.64$, as above.

$$Y_2 = Y_1(Q_2/Q_1)^{6/7}(B/b)^{0.64} = 1.6 \times (91/40)^{6/7} \times (20/19.4)^{0.64} = 3.30\,\text{m} \quad (8.13)$$

$$d_{SC} = Y_2 - Y_1 = 3.30 - 1.60 = 1.70\,\text{m} \quad \text{(below the overbank level at the bridge)} \quad (8.11)$$

Note that with $(Q_2/Q_1) = 91/40 = 2.28$ the overbank constriction line of Fig. 8.19 gives $d_{SC}/Y_1 = 1.06$ so $d_{SC} = 1.06 \times 1.60 = 1.70$ m. The use of equation 8.10 for the clear-water scour condition on the overbank (with $D_M = 1.25D_{50}$ = 0.00125 m) gives: $Y_2 = [Q_2^2/(36b^2D_M^{2/3})]^{3/7} = [91^2/(36 \times 19.4^2 \times 0.00125^{2/3})]^{3/7}$ = 5.48 m.

Thus $d_{SC} = Y_2 - Y_1 = 5.48 - 1.60 = 3.88$ m below the overbank level. This appears excessive compared with the values above, and it would be difficult to believe that a scour hole this deep would not have sediment transported into it from the main channel (which is basically what clear-water scour assumes). Scour to this depth would alter the flow pattern and velocities considerably, requiring a recalculation of the problem. Unfortunately the application of scour equations frequently requires the use of judgement and experience, otherwise they can become random number generators!

Step 5 Calculate the depth of local pier scour, d_{SP}

Again two areas need to be assessed: the main channel and the right overbank. This is the live-bed condition so equation 8.14 is applicable:

$$d_{SP} = 2.0\,Y_2\,K_{1P}\,K_{2P}\,K_{3P}\,(b_P/Y_2)^{0.65}\,F_2^{0.43}$$

MAIN CHANNEL

Y_2 = 2.4 m upstream of the pier at the bridge cross-section in Fig. 8.23.
$V_2 = Q/A = 399/(57 \times 2.4) = 2.92$ m/s.
$F_2 = 2.92/(9.81 \times 2.4)^{1/2} = 0.60$.
The brief assumed a 15° angle of approach as a safety margin, so K_{1P} = 1.0 regardless of shape ($\phi > 5°$) and K_{2P} predominates (Fig. 8.20).

In this case $L/b_P = 36/1.2 = 30$ so use the maximum ratio of 12 giving $K_{2P} = 2.5$.

$K_{3P} = 1.1$ for the plane bed condition (Table 8.7)

$$d_{SP} = 2.0 \times 2.4 \times 1.0 \times 2.5 \times 1.1 \times (1.2/2.4)^{0.65} \times (0.60)^{0.43} = 6.75\,\text{m} \quad (8.14)$$

This scour depth is unacceptably large. Inspection of the above equation shows that this is mainly due to $K_{2P} = 2.5$, which was used to allow for some meandering of the approach channel. An alternative is to use river training works in the approach to the bridge to ensure that $\phi = 0°$ (as stated in the brief) so that $K_{1P} = 1.0$ (Table 8.6, round-nosed piers) and $K_{2P} = 1.0$ (Fig 8.20). Now: $d_{SP} = 2.0 \times 2.4 \times 1.0 \times 1.0 \times 1.1 \times (1.2/2.4)^{0.65} \times (0.60)^{0.43} = 2.70\,\text{m}$ (below the main channel).

This depth of scour will be assumed for all piers in the main channel; even if there is an identifiable thalweg it may shift, so design for the maximum depth of scour over the whole width. See also Example B1 in Appendix B, where some alternative pier scour equations are applied to this problem.

RIGHT OVERBANK

$Y_2 = 1.4\,\text{m}$ upstream of the pier at the bridge cross-section (Fig. 8.23).
$V_2 = Q/A = 91/(19.4 \times 1.4) = 3.35\,\text{m/s}$.
$F_2 = 3.35/(9.81 \times 1.4)^{1/2} = 0.90$.
As above, assume a 0° ($= \phi$) angle of approach, so $K_{1P} = 1.0$ for round-nosed piers, $K_{2P} = 1.0$ and $K_{3P} = 1.1$ for the plane bed condition.

$$d_{SP} = 2.0 \times 1.4 \times 1.0 \times 1.0 \times 1.1 \times (1.2/1.4)^{0.65} \times (0.90)^{0.43} = 2.66\,\text{m} \quad (8.14)$$

Thus the depth of scour is 2.66 m below overbank level, or 1.66 m below main channel level. It would be more sensible to adopt the pier scour depth (2.70 m) and level from the adjacent main channel. Note that $d_{SP}/b_P = 2.70/1.2 = 2.25$, which is consistent with the values normally expected for round-nosed piers aligned to the flow. The high velocity and Froude number in the overbank waterway may suggest that an additional overbank span should be considered.

Step 6 Calculate the width of the pier scour holes, W_{SP}

For both the main channel and the overbank the width of the scour holes on one side of a pier can be obtained from $W_{SP} = 2.8d_{SP} = 2.8 \times 2.70 = 7.56\,\text{m}$.
Thus the total width of the top of the hole $= 2W_{SP} + b_P = 2 \times 7.56 + 1.20 = 16.32\,\text{m}$.
The scour holes do not overlap since the piers are at 20 m centres. When drawing the hole this corresponds to an angle of repose of about 30° to the horizontal.

Step 7 Calculate the depth of scour at the abutments, d_{SA}

This must be done for the left and right abutments located at the left edge of the main channel and right overbank respectively. Equation 8.17 will be adopted initially. This essentially considers the flow on the upstream floodplain that is cut off by the abutment, and utilises the mean depth and Froude number.

$$d_{SA} = Y_{M1} + 2.27\, Y_{M1}\, K_{1A}\, K_{2A}\, (L_A/Y_{M1})^{0.43}\, F_{M1}^{0.61} \qquad (8.17)$$

LEFT ABUTMENT

The mean depth on the upstream floodplain, $Y_{M1} = A_1/B_{T1}$ where A_1 is the cross-sectional area of flow on the upstream floodplain and B_{T1} is the corresponding top width of the water surface, so

$A_1 = {}^1/_2 \times 1.6 \times 140 = 112\,\text{m}^2$ and $B_{T1} = 140\,\text{m}$ giving $Y_{M1} = 112/140 = 0.8\,\text{m}$.

From Fig. 8.22 the flow cut off by the abutment is $68\,\text{m}^3/\text{s}$, so $V_{M1} = 68/112 = 0.61\,\text{m/s}$.

$F_{M1} = V_{M1}/(gY_{M1})^{1/2} = 0.61/(9.81 \times 0.8)^{1/2} = 0.22$.

$K_{1A} = 0.82$ from Table 8.8 with vertical abutments and wingwalls.

$K_{2A} = 1.0$ (i.e. assuming that the abutments are now at right angles to the flow).

$$d_{SA} = 0.8 + 2.27 \times 0.8 \times 0.82 \times 1.0 \times (140/0.8)^{0.43} \times (0.22)^{0.61}$$
$$= 6.25\,\text{m}$$

This is very large. Modify the bridge design to use spillthrough abutments with $K_{1A} = 0.55$ so:

$$d_{SA} = 0.8 + 2.27 \times 0.8 \times 0.55 \times 1.0 \times (140/0.8)^{0.43} \times (0.22)^{0.61}$$
$$= 4.45\,\text{m}$$

This is still quite deep, but the Froehlich equation is known to give large values as it assumes that all of the flow passes the end of the abutment. If the flow returns to the main channel more gradually depths may be smaller. The presence of vegetation or the use of riprap can be sufficient to reduce these values further. Suggest the use of riprap or some other means of protection to the abutments (see Example 8.8).

The bridge abutment is not set 3–5 times the depth of flow back from the bank (Section 8.3, point 6), which combined with the scour depths suggests that the bank in the bridge approach may be destroyed without river training or bank stabilisation measures. The simplicity of the cross-section in Fig. 8.22 also raises questions about the application of the equations: the velocity and depth on each side of the dividing vertical at the left abutment are very different, whereas in reality the depth and velocity are likely to vary more gradually.

RIGHT ABUTMENT

The average depth on the upstream floodplain $Y_{M1} = A_1/B_{T1}$ where $A_1 = {}^1/_2 \times 1.6 \times 160 = 128\,m^2$ and $B_{T1} = 160\,m$ giving $Y_{M1} = 128/160 = 0.8\,m$.
From Fig. 8.22 the flow cut off by the abutment is $78\,m^3/s$, so $V_{M1} = 78/128 = 0.61\,m/s$.
$F_{M1} = V_{M1}/(gY_{M1})^{1/2} = 0.61/(9.81 \times 0.8)^{1/2} = 0.22$.
$K_{1A} = 0.55$ from Table 8.8 with a spillthrough abutment.
$K_{2A} = 1.0$ (i.e. assuming that the abutments are now at right angles to the flow).

$$d_{SA} = 0.8 + 2.27 \times 0.8 \times 0.55 \times 1.0 \times (160/0.8)^{0.43} \times (0.22)^{0.61}$$
$$= 4.67\,m \text{ (below the overbank)}$$

Note that the scour depth would be 6.57 m with vertical abutments and wingwalls instead of the spillthrough type now assumed.

A second estimate of scour depth can be obtained from equation 8.18, since $L_A/Y_{M1} = 160\,m/0.8\,m > 25$. At the end of the abutment in the main channel $Y_{A2} = 1.40\,m$, $V_{A2} = 3.35\,m/s$, $F_{A2} = 3.35/(9.81 \times 1.40)^{1/2} = 0.90$, so for spillthrough abutments:

$$d_{SA} = 4Y_{A2}(K_{1A}/0.55)K_{A\phi}F_{A2}^{0.33} = 4 \times 1.40 \times 1.0 \times 1.0 \times 0.90^{0.33}$$
$$= 5.41\,m \qquad (8.18)$$

The depth of 4.67 m will be adopted, but the abutments will be given scour protection so that these values become overly conservative (see Example 8.8).

Step 8 Calculate the total scour depth, d_S

According to the brief, rock occurs 25 m below bed level, so there appears to be nothing to stop the full calculated scour depths occurring. A summary of the scour depths relative to the bottom of the main channel or the right overbank is:

Scour type	*Left abutment (m)*	*Central piers (m)*	*Right overbank abutment (m)*
Degradation	0.50	0.50	Not applicable
Live-bed contraction	0.79	0.79	1.70
Live-bed pier	Not applicable	2.70	Not applicable
Live-bed abutment	4.45	Not applicable	4.67

These scour depths cannot be added numerically because it is not clear where the scour holes overlap. Instead the depths must be drawn on a cross-section of the bridge and channel as shown in Fig. 8.24.

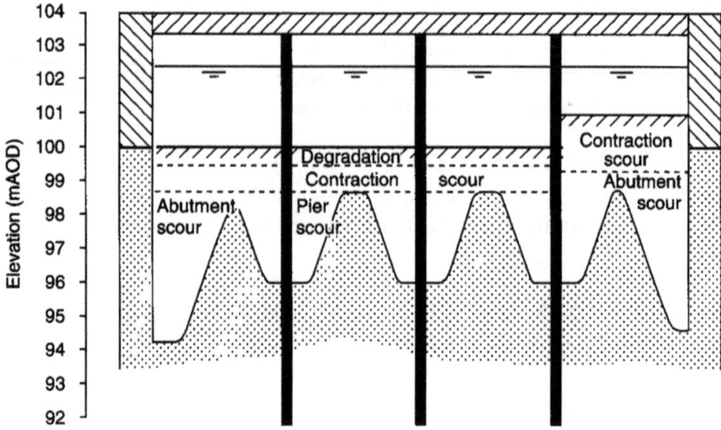

Fig. 8.24 Plot of the scoured prism using the results from Example 8.5.

Step 9 Review the analysis and evaluate the design

The scour depths are large, causing an increase in bridge waterway area that may result in the velocity of flow being less than assumed, and thus reduced scour. As a quick check, Fig. 1.14 suggests that with $D = 1\,\text{mm}$ and an average depth of flow of 5 m the competent velocity is about 1.4 m/s, so if the discharge through the main opening is $Q = 399\,\text{m}^3/\text{s}$ then $a = 399/1.4 = 285\,\text{m}^2$ giving $Y = 285/57 = 5.0\,\text{m}$ and $d_S = 5.0 - 2.4 = 2.6\,\text{m}$. This is similar to the scour level shown in Fig. 8.24, but the calculations should be repeated starting with an intermediate bed level to check this. Otherwise the appropriate equations have been used within range, but it would be preferable to have more detailed information regarding the longitudinal water profile and the conveyance of the various parts of the main channel and floodplains. The current lack of detail leads to some uncertainty. The repeat calculations should also take into consideration the revised opening geometry following the decision to use spillthrough abutments. An alternative method of analysis based on regime theory can be found in Section 8.6.

A brief review of the design is:

- abutments – spillthrough type required with riprap protection;
- piers – round nosed;
- angle of attack – assumed to be zero so river training works are needed;
- river training works – possibly riprap revetments or gabions, to be designed;
- scour – possible down to 94.3 m elevation;
- foundations – set below 92 m elevation.

It would be better if the left abutment was set back from the main channel so there was less risk of bank failure initiating failure of the abutment. This is probably desirable even if river training works are used to ensure that the approach flow is perpendicular to the crossing. Velocities and Froude numbers are relatively high in the right overbank area. Possibly a relief opening on the right floodplain should be investigated (see also Example 1.1).

Step 10 Design the foundations

In general terms it looks as though all foundations should be set below a level of 92 m following the general guidelines in Table 8.9.

8.6 Regime theory

Regime theory concerns the concept that the long-term width, depth and gradient of an alluvial channel carrying a certain amount of silt are dictated by nature and can be approximated by equations. The regime equations were developed mainly on the Indian subcontinent in the first half of the twentieth century from observation and experience arising from the design of irrigation systems (Kennedy, 1895; Lacey, 1929–30, 1933–34, 1939). It took many years for empirical equations of this form to gain acceptance and for a theoretical basis to be developed. Utilising data from a large number of canals Lacey summarised the relationships as

$$V = 0.625(f_S R)^{1/2} \tag{8.20}$$

$$P = 4.84Q^{1/2} \tag{8.21}$$

$$S_O = 0.000304 f_S^{5/3} Q^{-1/6} \tag{8.22}$$

$$\text{where } f_S = (2500D)^{1/2} \tag{8.23}$$

and V is the mean flow velocity (Q/A m/s), Q is the discharge (m^3/s), A is the cross-sectional area of flow (m^2), R is the hydraulic radius (A/P m), P is the wetted perimeter (m), S_O is the dimensionless channel slope, and f_S is a silt factor for a sediment of diameter D(m). In wide rectangular channels the width B (m) can be approximated by P and obtained from equation 8.21.

Numerous other equations have been proposed, notably by Blench (1939), who introduced separate factors for the channel bed and sides, but care must be taken when using them because they apply only to a particular type of material (e.g. sand bed, gravel bed, cohesive bed) or to a restricted range of discharge, sediment size, median bed material diameter (D_{50}), bank vegetation and channel slope. The equations can be applied successfully to channels with similar characteristics, but when applied to channels with significantly different features the results can be disastrous (Brandon, 1987). Another complication is that canals, which formed the basis of regime theory, essentially carry a relatively steady discharge and sediment load whereas rivers do not. This leads to the concept of a single 'dominant discharge' which would result in the same channel dimensions as a range of flows. The dominant discharge usually coincides with flow at, or about, the bankfull stage.

Because of the possibility of using the regime equations outside their recommended range, they will not be listed here. A useful guide to regime equations and their use to design stable channels was provided by Brandon (1987). Ackers (1992) gave the 1992 Lacey Memorial Lecture, which provided a good summary of the historical development of canal and river regime theory.

Although the equations in Section 8.4 normally provide a better approach, regime theory can be used to obtain an indication of the scour depth at a bridge. This involves calculating the stable regime width and depth of an alluvial channel (B_R and R_S) for a particular discharge (Q), and then adjusting the depth to take into account the effect of the bridge contraction. The depth of contraction scour (d_{SC}) is obtained from the difference in the depths of flow (Novak *et al.*, 1996).

The minimum stable regime width of an alluvial channel is

$$B_R = 4.75Q^{1/2} \tag{8.24}$$

where B_R is the channel regime width (m) measured along the water surface at 90° to the banks, and Q is the maximum flood discharge (m^3/s). If f is taken as the average of the Lacey silt factor $= 1.59D_{50}^{1/2}$ and Blench's bed factor $= 1.9D_{50}^{1/2}$, then

$$f = 1.75D_{50}^{1/2} \tag{8.25}$$

where D_{50} is the median diameter of the bed material (mm). The depth of flow during the flood measured from the high flood level (HFL) gives an indication of the scour at a cross-section having a width equal to the regime width (B_R), and so is called the regime scoured flow depth, R_S (m). In fact this is the regime hydraulic radius, which approximates the stable flow depth Y(m) in a wide rectangular channel. For a river channel that has a width (B) greater or equal to B_R then:

$$R_S = 0.475\left(\frac{Q}{f}\right)^{1/3} \tag{8.26}$$

If the actual width of the river channel (B) is less than B_R then this may be regarded as a natural contraction when the scoured depth of flow increases to Y_S (m), where

$$Y_S = 1.35\left(\frac{Q^2}{fB^2}\right)^{1/3} \tag{8.27}$$

In both cases R_S or Y_S is the flow depth measured below the HFL.

To obtain the scoured depth of flow in a bridge opening the value of R_S must be adjusted for the severity of the constriction (B_R/b), where b is the width of the bridge opening. The expression used is very similar to equation 8.13 for live-bed contraction scour with $Q_1 = Q_2$ and the Manning roughness ratio omitted. Thus the normal scoured depth of flow Y_{SN} (m) in the bridge opening is

$$Y_{SN} = R_S \left(\frac{B_R}{b}\right)^{0.61} \tag{8.28}$$

This value should be increased by about: 25% for a single-span bridge with a straight approach; 50% for a moderately curved approach; 75% for a severely curved approach; and 100% for a multispan bridge with a curved approach. If the constriction is predominant then either the modified value from equation 8.28 or the maximum scoured depth of flow, Y_{SMAX} (m), should be adopted depending which is the largest, where

$$Y_{SMAX} = R_S \left(\frac{B_R}{b}\right)^{1.56} \tag{8.29}$$

In either case the scour depth, d_{SC}, may be taken as the difference between Y_{SN} or Y_{SMAX} and R_S, in a similar manner to equation 8.11. It must be appreciated that these equations for the average scour depth are approximate, as are the various factors used to increase Y_{SN}, and do not indicate local scour depths. Normally the scour depth equations of previous sections will provide a much better means of evaluating both contraction and local scour.

Example 8.6

Use the regime equations to calculate the scour depth in the main bridge opening of Example 8.5.

Considering only the discharge (399 m^3/s) passing through the main bridge opening (i.e. excluding the overbank) then $B_R = 4.75 \times 399^{1/2} = 94.9$ m.
$D_{50} = 1$ mm so from equation 8.25, $f = 1.75\ (1.0)^{1/2} = 1.75$.
From equation 8.26 the regime scoured flow depth $R_S = 0.475(399/1.75)^{1/3} = 2.90$ m.
From equation 8.28, $Y_{SN} = 2.90(94.9/57)^{0.61} = 3.96$ m.
Since this is a multispan bridge with a straight approach increase the value by 30% so $Y_{SN} = 5.15$ m.
Assume the scour depth $d_{SC} = Y_{SN} - R_S = (5.15 - 2.90) = 2.25$ m.
Alternatively, from equation 8.29, $Y_{SMAX} = 2.90(94.9/57)^{1.56} = 6.42$ m.
The scour depth $d_{SC} = Y_{SMAX} - R_S = (6.42 - 2.90) = 3.52$ m.

Therefore the scour depth is between 2.3 m and 3.5 m depending upon whether the normal or maximum scour depth (constriction predominant) is adopted. In Example 8.5 (step 9) the competent velocity method indicated an average scour depth of 2.6 m. From Fig. 8.23 it appears that these figures are acceptable as average depths, but of course they do not indicate the local scour depths.

8.7 Scour in cohesive materials

Relatively little research has been conducted relating to scour depths when the bed material is cohesive. However, the fact that a cohesive material is likely to be more scour resistant (than a cohesionless one) does not necessarily mean that the depth will be less, only that the scour hole will take longer to reach its full depth. The collapse of the Schoharie Creek bridge in the USA was the result of local scour accumulated over a 10 year period, the bed material being glacial till (Lagasse *et al.*, 1995). Sometimes a few large floods, not 10 years, may be sufficient to cause significant scour.

Some theoretical approaches to sediment transport and erosion involve considering the forces on individual particles exposed to the flow as incipient motion occurs. This concept does not apply to a cohesive material such as glacial till. Saturated cohesive materials may tend to break away in unpredictable lumps, rather than a particle at a time. Nevertheless, in the absence of any special equations for scour depth in cohesive materials the only options are either to use the equations above for cohesionless materials, or to use the competent mean velocity as a guide to the size of waterway required for the bed material not to be eroded (see Table 1.3b, Example 1.1 and Section 8.4.1).

8.8 Scour in tidal waterways

The tidal cycle with its associated flow reversal can increase degradation, contraction and local scour compared with the equivalent non-tidal crossing depending upon the amount of sediment transported in the flow and where and when scour occurs. The introduction of salinity, saline wedges, storm tides, reversible flow and unsteady conditions makes the tidal situation more complex than its riverine counterpart, and as yet there are no proven scour depth equations specifically for tidal waterways (Fig. 8.25). Additionally there are very few measurements of scour at tidal bridges to prove or disprove the various equations. Consequently the riverine equations described above are usually considered to be the best available option for evaluating tidal scour (Richardson *et al.*, 1993). It has been speculated that future research may show these equations to be too conservative or safe (Sheppard, 1993), but this is usually regarded as the lesser of two evils.

In all tidal or non-tidal streams the basic concept of continuity of sediment transport applies so that for the reach under consideration during a given period of time,

$$V_{SE} - V_{SL} = \Delta V_S \tag{8.30}$$

where V_{SE} is the volume of sediment entering (m^3), V_{SL} is the volume of sediment leaving (m^3), and ΔV_S is the change in the volume of sediment stored in the bed (m^3). In the normal riverine environment the sediment enters and

Fig. 8.25 Laira rail bridge over the River Plym, Plymouth: an example of a tidal crossing.

leaves in the downstream direction borne by the river flow, whereas with a tidal situation the river is carrying sediment downstream while the flood tide is carrying sediment upstream, some of which may be derived from littoral drift along the coastline. Thus the tidal situation is more complex. However, the scour may again be classified as clear-water ($V < V_S$) or live-bed ($V > V_S$) depending upon whether or not bed material is being transported into the bridge reach. If live-bed material is carried into the scour hole then the depth will be less than with the equivalent clear-water condition.

One of the first steps when planning a tidal crossing is to assess how much the tidal process affects the conditions at the site and whether the river is tidally affected or tidally controlled. The magnitude of the tidal influence depends upon the distance from the sea, bed slope, channel roughness, cross-sectional area and tidal volume. For example, at the tidal limit, river flow conditions predominate. Just downstream of the tidal limit (A in Fig. 8.26a) the bridge is tidally affected, but the tidal influence will be small. As the distance from the tidal limit increases and the sea gets nearer (B) both river flow and tidal fluctuations become important: the tides provide a cyclic variation in the controlling downstream water level and thus influence the depth, velocity, discharge and direction of flow at the bridge. Eventually a point (C) is reached at which the tidal influence will be sufficient at high tide to reduce the discharge through the bridge to zero, with river flow being stored upstream. This is the limiting case, and between this point and the sea bridges are classed as tidally controlled. They will experience flow in two directions (freshwater downstream and the tide in both directions), with the discharge and velocity being higher on the ebb as the stored river water is discharged.

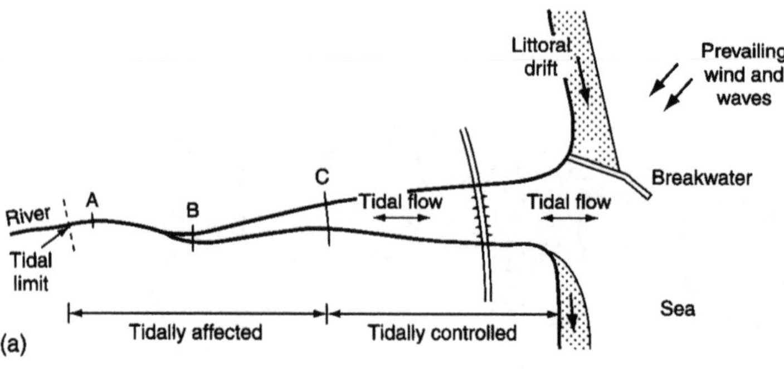

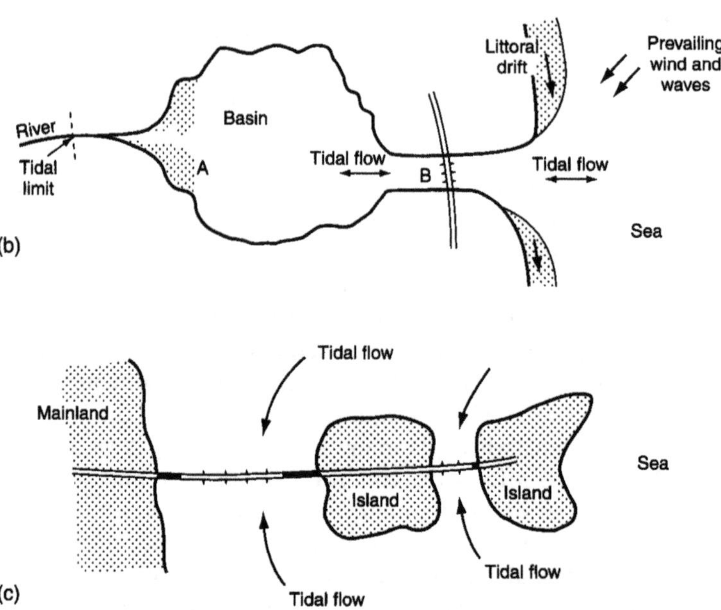

Fig. 8.26 The three types of tidal inlet defined by Neill (1973): (a) an unconstricted inlet; (b) a constricted inlet consisting of a narrow inlet or 'neck' leading to a larger inland tidal basin; (c) offshore islands separated by tidal passages and linked to the mainland by causeways, which experience tidal flow in two directions simultaneously. Often tidal passages can be completely closed by a causeway since there is no absolute necessity for water to flow through them.

The maximum tidal discharge (Q_{MT}) usually occurs near the midpoint of the tidal cycle, the maximum and minimum tidal levels being associated with slack (still) water at the turn of the tide (Fig. 8.27). Note that the bed level in the estuary may be above or below the low water level shown in the diagram, so the depth measured from the mean water (sea) level has to be calculated accordingly. Needless to say, the design of a tidal crossing is greatly facilitated if measurements of depth and velocity at the location have been recorded over several tidal cycles, especially spring tides when the range is largest. Storm tides, which arise from the combination of normal astronomical tides with storm surges resulting from wind and barometric pressure effects, may be much higher than normal tides and may have a much different tidal period.

If the tidal influence is small or insignificant then tidally affected bridges can be designed using the riverine procedures described previously, typically using the 1 in 100 and 1 in 500 year flood (Q_{100} and Q_{500}). If the tidal influence is larger, but the greatest risk to the bridge is still river floods, then again the riverine procedures can be used. If the greatest risk to the bridge is from a combination of river flooding and tide and storm surges, or from the latter alone, then the bridge must be designed using the procedures for a tidal crossing, possibly using the 1 in 100 and 1 in 500 year storm surge with a full consideration of all the coastal engineering processes (e.g. resonance, wind and wave effects). When the river flood discharge is relatively small and arises from the same storm that created the tidal surge then the river and tidal discharges can be added, or the runoff hydrograph volume can be added to the tidal volume. If the river flood and tidal surge might result from different storms a joint probability approach may be used to determine the Q_{100} and Q_{500} flows. If it is not clear what presents the greatest risk, both the riverine and tidal approach should be used and the largest scour depth adopted. Obviously the return

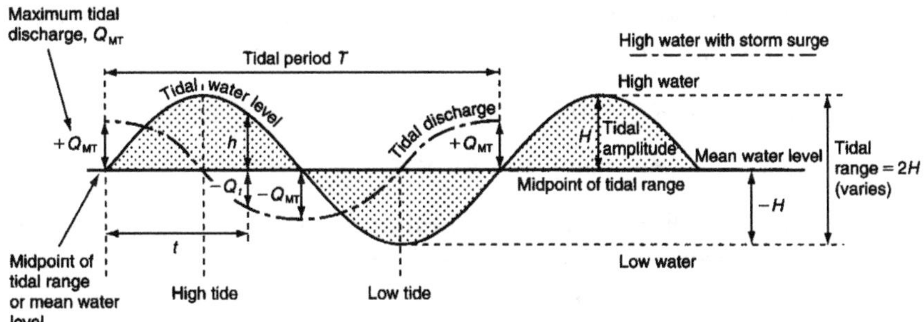

Fig. 8.27 The tidal cycle. Note that the tidal amplitude (H) varies significantly depending upon whether the tides are neap, high water springs, etc. Similarly the normal semidiurnal astronomical tidal period, T, is 12 hours 24 minutes but this may be significantly different in some parts of the world or during a severe storm surge. Starting at the mean water level the tidal height (h) above mean water at any time (t) is $h = H \sin \theta$ where $\theta = 360(t/T)$. This represents one tidal cycle (T) as 360°. Values below the axis are negative. The discharge at any time (t) is $Q_t = Q_{MT} \cos \theta$ where $\theta = 360(t/T)$: the maximum tidal discharge (Q_{MT}) occurs at the midpoint of the tidal range; the mean discharge $Q_{MEAN} = 2Q_{MT}/\pi$.

interval employed for design purposes varies with the importance of the crossing.

The main differences between scour in rivers (as considered previously) and scour in tidal waterways are described briefly below.

- The flow in a tidal stream can be in two directions: that is, freshwater flow downriver with the tidal flow changing direction periodically.
- The continuity equation $Q = AV$ does not apply in the normal way because Q is not constant: as the bed level is reduced by scour and the cross-sectional area of flow increases it is possible that both the tidal discharge and velocity will also increase, resulting in long-term degradation.
- The values of the variables (e.g. Q, A, V, Y) are harder to estimate in the tidal situation. It is possible that the construction of a bridge will reduce the tidal discharge and level on the landward side of the structure.
- The normal semidiurnal period between successive high or low astronomical tides is about 12.4 hours, but the tidal period during a design event may be much different as a result of storm surge, wind, river floods or the geometry of the estuary and coastline. Semidiurnal tides generally occur at low earth latitudes with diurnal variations ($T \approx 25$ hours) becoming increasingly dominant in higher latitudes.
- The mean maximum velocity, $V_{MEANMAX}$, on the ebb (corresponding to the maximum discharge at the midpoint of the tidal cycle, Q_{MT}) at a cross-section of a tidal inlet is usually around 1.00 m/s (Bruun, 1990). Taking the hydraulic radius (R m) as being equal to the mean depth of flow, for spring tides the values of the mean maximum velocities in tidal inlets fall within a relatively narrow band, approximately as follows:

  $$V_{MEANMAX} = (R^{1/8} - 0.2) \text{ m/s for } R \geq 5\,\text{m}$$

  $$V_{MEANMAX} = (R^{1/8} - 0.1) \text{ m/s for } R < 5\,\text{m}$$

  Considering all flood, ebb and littoral drift conditions, $V_{MEANMAX}$ may be between 0.89 m/s and 1.08 m/s, with local values in the centre of the inlet being around 33% greater (Neill, 1973). Over a full tidal cycle the average velocity in an inlet is typically around 0.77 m/s, being slightly lower (0.70 m/s) for semidiurnal tides and higher (0.87 m/s) for diurnal tides (Bruun, 1990).
- The relatively small velocities in tidal inlets (compared with rivers in flood) result in smaller Froude numbers, typically ranging up to a maximum of about 0.2.
- In addition to the usual data, for tidal scour assessments it is necessary to have a knowledge of sediment transport on the ebb and flood tides, littoral drift, the magnitude of tidal flows and storm surges, and the interaction between tides and river flow.

- Because of material originating as littoral drift, tidal inlets usually have a clean sandy bottom with an average particle diameter of around 0.1–0.5 mm, whereas rivers carrying fine sediments often have clayey or silty bottoms (Bruun, 1990).
- Storm surges (e.g. from hurricanes) may have very high peaks and short durations (less than 24 hours) compared with riverine events of the same magnitude. Consequently, 'steady' conditions never really exist, so scour holes may never fully develop and scour depth equations may not be valid.
- The unsteady storm condition just described can result in large changes in bottom roughness as the bedform changes, ranging from sand ripples to dunes to a flat bed.
- The deposition, consolidation and susceptibility to erosion of fine sediments are influenced by the ions present in saline water, and will be different from those in fresh water (Sheppard, 1993).

Guidelines for assessing scour in tidal locations are given below. They should be used in conjunction with common sense and engineering judgement, as appropriate.

8.8.1 Degradation and aggradation

The tidal flow arising from the flood and ebb of normal astronomical tides will usually have the most influence on long-term degradation (Δd) or aggradation patterns. If there is transport of sediment on both the ebb and flood tide then there may be no net loss of bed material in the vicinity of the bridge and no degradation. If there is a net loss of sediment in one or both directions then there is degradation, while a net gain would be aggradation. Sometimes this may be difficult to judge. For example, if a river flows into a wide basin (Fig. 8.26b) then riverine sediments may be mostly deposited as sandbanks in the tranquil areas of the basin (A) and not reach the crossing at B. This will result in clear-water scour and degradation at B unless sea-derived sediment is transported on both the flood and ebb tides. If in doubt the general principle outlined in equation 8.30 should be remembered. To obtain a worst-case estimate of the degradation depth apply equation 8.10 for clear-water contraction scour to the inlet assuming the maximum tidal discharge (Q_{MT}). Obviously the presence of solid rock would restrict degradation and scour, so knowledge of the local geology is required.

The assessment of long-term changes to bed level is complex and must include a consideration of current and future dredging practice, the construction of new beach groynes or entrance breakwaters, the removal of old groynes or breakwaters, land reclamation, and any coastal defence measures that may affect littoral drift. For example, if in Fig. 8.26b the construction of beach groynes or a breakwater at the estuary mouth (as in diagram a) cut off the littoral drift, the flood tide would erode the mouth and deposit sediment in the basin, only some of which may be carried out to sea on the ebb tide. Thus there may be aggradation in the basin and degradation in the estuary mouth. As degradation occurs, the enlarged cross-sectional area of flow will increase the tidal discharge and velocity and cause further degradation.

As with any bridge, the stability of the tidal channel and its tendency to shift laterally must be fully assessed during the early stages of design. The comparison of old and new charts, maps and aerial photographs can be useful in this respect.

8.8.2 Contraction scour and local scour

Tidal flow also affects the depths of contraction and local scour, but in this case the largest depths may result from storm surges and tsunamis rather than from regular astronomical tides. Local scour may occur at either end of the piers and abutments, and may be most severe on the landward side on the ebb and the seaward side on the flood tide.

The live-bed contraction scour formula (equation 8.13) can be applied to estuaries or inlets that are long enough to develop fully live-bed conditions, but not to short inlets. For example, the equation can be applied to situations such that shown in Fig. 8.26a, where B/b is the relative width of the estuary to the total length of the bridge openings and Q_2/Q_1 is the ratio of the discharge passing through the bridge to that in the upstream channel. The equation cannot be applied successfully to short inlets similar to that in Fig. 8.26b. The clear-water equation (8.10) could be applied, but would overestimate because there will generally be a supply of sediment from the upstream river and/or the tidal flow. With either equation the assumption of a long constriction experiencing steady uniform flow is unlikely to be met, so engineering judgement and experience are needed when interpreting the results to avoid overestimating scour depth.

The CSU equation (8.14) for local pier scour can be applied using the appropriate tidal depths and velocities. The calculation of abutment scour is problematical: unless the bridge opening is very narrow, tidal velocities are relatively low compared with those in rivers during flood, and even lower at the banks of the estuary where the abutments are likely to be situated. Thus it may not be necessary to calculate abutment scour, but if it is the depths obtained may be unrealistically large since the riverine equations tend to overestimate anyway. As in the riverine situation, riprap or some other measure can be easily adopted to limit scour at the abutments.

8.8.3 Analysis of tidal crossings

The procedure used depends upon whether the tidal inlet is unconstricted or constricted. The constriction may be due to either the geometry of the inlet or the bridge itself. The essential distinction is that the tidal level in an unconstricted inlet is assumed to be constant so the water surface is horizontal, whereas at any time in a constricted inlet there is a significant difference in surface water level so the surface is not horizontal. The existence of large mudflats, thick vegetation (rushes, reeds) in the estuary or heavily vegetated overbank areas will result in the attenuation of tidal levels and a sloping water surface, and so should be classed as constricted for analysis purposes. Figure 8.26a shows a typical unconstricted river mouth, and Fig. 8.26b shows a typical constricted inlet. It may not always be obvious whether an estuary is unconstricted or constricted, and it is possible that an estuary that is unconstricted at normal tides may be considered constricted during a 1 in 100 year or 1 in 500 year design storm surge.

Unconstricted inlets can be analysed using a procedure outlined by Neill (1973), which assumes a horizontal water surface and well-defined vertical sides to the channel. Constricted inlets can be analysed using an 'orifice' equation. Both cases are considered below. If it is not clear which approach to employ, use both and adopt the largest scour depth.

Richardson *et al.* (1993) recommended a three-stage analysis for tidal waterways. Stage 1 includes: a qualitative assessment of the stability of the tidal inlet; an investigation of future changes and development plans for the estuary and adjacent coastline; an evaluation of whether scour is predominantly controlled by river flow, tidal flow or both; classification of the estuary as constricted or unconstricted; and an assessment of whether clear-water or live-bed scour is most likely. All of these judgements are facilitated by site visits, sonic sounding to obtain sediment data, and the installation of equipment to record flow depths and velocities simultaneously at different points in the estuary. Stage 2 is the quantitative evaluation of the variables (notably discharge, velocity and water depth) that are used in the riverine scour depth equations presented earlier. This will probably result in an overestimation of scour depths because the assumption of steady-state equilibrium is not valid with unsteady tidal flows. For complex problems, stage 3 involves using physical or computer models to verify the stage 1 and 2 analysis.

The sections below are principally concerned with the stage 2 analysis.

Scour depth in an unconstricted estuary

The steps required to evaluate the bridge scour in an unconstricted estuary with a horizontal water surface are summarised below.

1. Decide upon the location of the bridge and its preliminary design.
2. Determine the type and size of the bed material (D_{50}) and locate any rock outcrops that will limit scour. Many coastal inlets have sandy beds in the 0.1–0.5 mm range.
3. Assess the pattern of sediment transport, potential degradation depth and the lateral stability of the channel.

4. Plot a graph of the elevation of the water surface against net bridge opening area (a). This is the total cross-sectional area of the bridge openings perpendicular to the direction of flow, or the cross-sectional area of the estuary channel minus the area of the piers and abutments (e.g. see Fig. 8.28).
5. For the site, determine the normal tidal range, which is the height between the mean high and low tidal water levels (or the estuary bottom, if this is higher) as shown in Fig. 8.27. Note that the mean sea level in Britain is often assumed to be 0.000 m relative to Ordnance Datum (m OD) at Newlyn, but it varies with location.
6. For the 1 in 100 year and 1 in 500 year design storm tide, or the return interval appropriate to the importance of the crossing, determine the tidal range ($2H$ m) and tidal period (T s). These data are unlikely to have been recorded and so may have to be estimated.
7. For the the return interval adopted, use the storm tidal amplitude (H m) to plot the storm tidal level (elevation) as a function of time from the information in Fig. 8.27 using normal mean sea level as the datum for elevation (e.g. see Fig. 8.29). Determine the midpoint of the tidal range, which is assumed to coincide with the maximum tidal discharge, Q_{MT}.
8. For various water surface elevations calculate the volume of water stored in the estuary or basin upstream of the bridge. This can be obtained from the average area enclosed by successive contours and the contour interval. Plot a graph of water surface elevation against storage (e.g. Fig. 8.30). From this determine the tidal volume (VOL m^3) between the high and low storm tide levels (or the estuary bottom, if this is higher). The tidal volume is the amount of water (m^3) that must flow past the bridge between low and high tide to fill the estuary upstream.
9. Determine the maximum design discharge, Q_{MAX} (m^3/s). Assuming a sinusoidal variation of discharge with time as in Fig. 8.27, then according to Neill (1973) for an ideal estuary the maximum tidal discharge (Q_{MT} m^3/s) occurs at the midpoint of the tidal range when the tidal energy gradient is steepest. With the storm tidal period (T s) defined by the plot in step 7 and with π = 3.14, Q_{MT} can be estimated from the area between the sine curve and $Q = 0$ axis as

$$Q_{MT} = \frac{3.14VOL}{T} \tag{8.31}$$

$$\text{giving } Q_{MAX} = Q_{MT} + Q \tag{8.32}$$

where Q is the significant riverine contribution to flow which (as discussed earlier) may be zero if the river flow is insignificant, the peak flood discharge if it is relatively large, or the routed flow through a basin (Fig. 8.26b) as appropriate. Alternatively, the volume of water under the river hydrograph can be added to VOL with $Q = 0$ if the timing of the events coincide.

10. Determine the corresponding average midtide velocity (V_{MT} m/s) in the bridge openings from

$$V_{MT} = Q_{MAX}/a_{MT} \tag{8.33}$$

where a_{MT} is the net area (m^2) of the bridge openings between the bed and the midpoint of the storm tidal range obtained from the graph in step 4. Note that V_{MT} is the average maximum velocity, which may be exceeded in the thalweg or centre of the estuary, with lower values near the banks. Thus when applying the scour depth equations, V_{MT} should be adjusted for individual piers and abutments. According to Neill (1973) the maximum velocity in estuaries (V_{MAX}) is about 33% higher than V_{MT}. The horizontal velocity distribution can be estimated by splitting the estuary into subsections and calculating their relative conveyance and discharge, as for any other river channel (Section 3.12). Compare the calculated velocities with any field measurements to verify the analysis.
11. The average depth of flow in the estuary can be approximated by $Y_{MT} = a_{MT}/B$ where a_{MT} is the value in step 10 and B is the total width of the estuary. In an unconstricted analysis the bridge cannot be a significant constriction so $a_{MT} \approx A_{MT}$, which is often a reasonable assumption (e.g. Fig. 8.25).
12. Use the value of D_{50} and Y_{MT} in equation 8.2 or 8.3 to calculate the scour-critical velocity (V_S) and hence determine whether the scour is clear-water ($V_{MT} < V_S$) or live-bed ($V_{MT} > V_S$).
13. Calculate contraction scour (d_{SC}) by applying the appropriate riverine scour depth equation (8.10–8.13) using the values of Q_{MAX}, V_{MT} and Y_{MT}.
14. Calculate the local pier scour depth (d_{SP}) from equation 8.14 using V_{MAX}, as appropriate.
15. Carefully consider the values obtained, the quality of the input data and the inherent limitations of the analysis. This is difficult without experience, and should not be interpreted as an invitation to adjust answers to more acceptable figures without good reason. If there are existing bridges in the vicinity of the proposed crossing, then they should be carefully monitored to provide a comparison.
16. Where the scour holes overlap the total scour depth = $d_{SC} + d_{SP} + \Delta d$. Abutment scour has not been included, on the assumption that the abutments are located at the banks of the estuary where the depth of flow and velocity are likely to be low and/or scour protection can be used easily (Fig. 8.25). This contrasts with a riverine crossing where all of the overbank flow is forced to return to the main channel by flowing around the upstream corners of the abutments, which may be located in the main channel. However, where the abutments are near the centre of a tidal estuary and there is considerable flow around them then abutment scour should be included, as in the riverine situation.

17. For important crossings, physical or computer modelling should be undertaken to verify the analysis, particularly if the hydraulics or geometry of the site are complex or unusual.

Scour depth in a constricted estuary

Constricted estuaries are characterised by having a sloping water surface as a result of significant tidal flow resistance, Fig. 8.26b being a possible example. It is assumed that the maximum difference in water level (Δh m) between the sea and the inland part of the estuary or basin occurs at the midtide level when the tidal discharge is greatest. When the constriction is caused by the channel (not by the bridge) the velocity and discharge in the inlet channel can be obtained from the following 'orifice' and continuity equations (Kreeke, 1967; Bruun, 1990; Richardson *et al*, 1993):

$$V = C_d\,(2g\Delta h)^{1/2} \qquad (8.34)$$

$$\text{and } Q_{MT} = A_{MT}\,V \qquad (8.35)$$

where V is the average maximum velocity in the inlet neck (m/s), C_d is a dimensionless coefficient of discharge (see below), g is 9.81 m/s^2, Q_{MT} is the maximum tidal discharge, and A_{MT} is the average cross-sectional area (m^2) of the estuary perpendicular to the direction of flow at the midpoint of the tidal range. When the channel causes the constriction the value of C_d is less than 1.0 and may be calculated from

$$C_d = (1/k)^{1/2} \qquad (8.36)$$

where k is explained below. In fact the 'orifice' may be easier to visualise as the familiar problem of flow between two large reservoirs (the sea and the basin) via a conduit (inlet channel). Thus the head difference (Δh m) between the sea and basin equals the total head loss ($kV^2/2g$) in the channel, which is the sum of the entrance, exit and friction losses, all of which are expressed as a multiple of the velocity head in the neck of the inlet channel. Solving for V and using the continuity equation results in the expressions above, while the value of k is obtained from

$$k = k_S + k_B + \frac{2gn^2L_C}{h_C^{1.33}} \qquad (8.37)$$

If the flow is from sea to basin, k_S is the entry velocity head loss coefficient on the seaward side where $k_S = 1.0$ if the head loss equals the velocity head or $k_S = 0.0$ if head loss is negligible, k_B is the exit velocity head loss coefficient on the basin side where $k_B = 1.0$ if the velocity reduces to zero, n is Manning's roughness coefficient (s/m$^{1/3}$), L_C is the length (m) of the inlet 'neck' or narrow part commensurate with Δh, and h_C is the average depth of flow (m) in the inlet neck at the midtide level. Since Δh is often taken as the average value on the flood and ebb tides the actual direction of flow may not be important.

If the measurement of Δh does not include riverine storm flow (Q) then, when the circumstances warrant it, equation 8.32 can be used to obtain the maximum design discharge, $Q_{MAX} = Q_{MT} + Q$.

Equation 8.33 gives the average midtide velocity in the bridge openings as $V_{MT} = Q_{MAX}/a_{MT}$. Again it may be prudent to increase the average value of V_{MT} by 33% to obtain V_{MAX}, although in a constricted 'orifice' inlet there may be a smaller variation of transverse velocity. Apart from the depth, h_C, and the method of calculating Q_{MAX} and V_{MAX}, most of the steps listed for the investigation of unconstricted estuaries can be applied to constricted inlets as well. The depth of local scour during the design storm surge can be obtained from the appropriate riverine clear-water or live-bed equations using the values of Q_{MAX}, V_{MAX} and h_C. The estimation of contraction scour using V_{MT} in the riverine equations results in excessively large values if the inlet is too short to be a long contraction and the storm surge is too short for scour depths to be fully achieved. The worst-case degradation depth can be obtained from the clear-water contraction scour equation (8.10), as described in Section 8.8.1, using the value of Δh for normal ebb and flood tides (not a storm surge) in equation 8.34 to obtain the corresponding values of Q_{MT} and V_{MT}.

Example 8.7

A small tidal basin at the head of a river estuary has a similar geometry to that shown in Fig. 8.26b. The basin lies within an urban area that is to be opened up for redevelopment by having a new river crossing constructed at the narrowest point of the channel. It is not clear whether the estuary is unconstricted or constricted, or whether the crossing will be tidally affected or tidally controlled. The proposed crossing point is 132 m wide. The bridge will be perpendicular to the flow with abutments on the same alignment as the existing quay walls (i.e. they do not protrude into the channel). Thus the contraction will result from three round-nosed piers each 4.0 m wide (including some protection from impact by small craft). At the centre of the channel the depth is about −3.9 m OD. A preliminary investigation of the site has yielded the data below.

Bed material: fine to medium sand, D_{50} = 0.2 mm.
Tide heights for various return intervals: 1 in 1 year: ±2.85 m OD
1 in 50 years: ±3.28 m OD
1 in 100 years: ±3.36 m OD
Tidal period corresponding to the above levels = 12.41 hours.
Riverflow into the basin during an observed 1 in 100 year storm = 30.4 m^3/s.
Length of inlet neck L_C = 800 m.
Manning's n for the inlet bed = 0.03 $s/m^{1/3}$.
Estimated midtide difference in water level between the estuary and the inland basin during a 1 in 100 year tide (but with normal river flow), Δh = 0.31 m.
Corresponding average midtide depth in the 800 m length of inlet = 2.4 m.

Initial assessment

An unconstricted and constricted analysis will be undertaken since it is not clear which procedure is the most appropriate. A return interval of 1 in 100 years will be adopted. Because the river catchment is quite small it is possible that both the 1 in 100 year river flood and the storm tide will be the result of the same event, so the same return interval will be used for both. The basin is developed and its edges are bounded by well-defined quay walls, so the lateral stability of the channel is not a problem. However, further development of the catchment may increase river floods and cause degradation in future. An allowance of 1.0 m (Δd) is made for this. The D_{50} = 0.2 mm and it is suspected that there is sediment transport in both directions on the tide, so live-bed scour is likely.

Unconstricted estuary

The net cross-sectional area of the bridge openings (i.e. with the area of the piers subtracted) is plotted in Fig. 8.28 as a function of water level.

Mean sea level is 0.000 m OD. The normal tidal range is ± 2.85 m OD but during the 1 in 100 year storm this increases to ±3.36 m OD. Thus the design storm amplitude, H, is 3.36 m. The equivalent tidal period is 12.41 hours. Assuming a sinusoidal variation the equations in Fig. 8.27 can be used to plot the design tidal cycle as shown in Fig. 8.29. The maximum discharge occurs at the midpoint of the tidal cycle, which is at normal mean sea level in this instance.

The volume of water stored in the estuary is shown as a function of water level in Fig. 8.30. This gives the tidal volume between low and high tides (±3.36 m) as $VOL = 3.4 \times 10^6\ \text{m}^3$. Using equation 8.31:

$$Q_{MT} = \frac{3.14VOL}{T} = \frac{3.14 \times 3.4 \times 10^6}{12.41 \times 60 \times 60} = 239.0\ \text{m}^3/\text{s} \qquad (8.31)$$

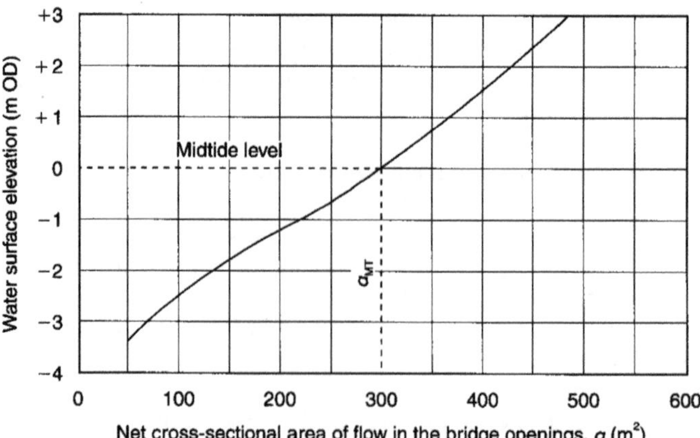

Fig. 8.28 Graph of water level against the net cross-sectional area of flow in the bridge openings perpendicular to the direction of flow (a) for Example 8.7.

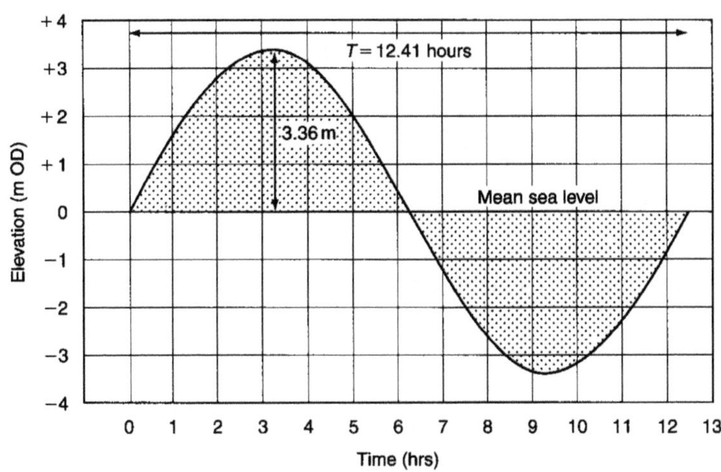

Fig. 8.29 The storm tidal cycle calculated for Example 8.7.

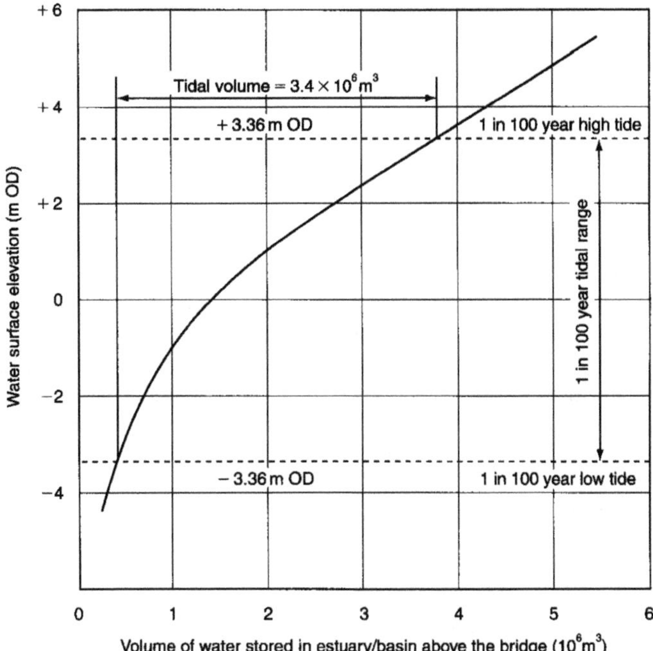

Fig. 8.30 Graph of sea level against the volume of water in the estuary. The difference between the storage at high and low tides is the tidal volume (VOL m^3) in Example 8.7.

The 1 in 100 year river flow is $Q = 30.4\,\mathrm{m^3/s}$ so from equation 8.32:

$$Q_{MAX} = Q_{MT} + Q = 239.0 + 30.4 = 269.4\,\mathrm{m^3/s} \qquad (8.32)$$

From Fig. 8.28 at midtide level the net area of flow in the bridge openings $a_{MT} = 295\,\mathrm{m^2}$. Thus the average midtide velocity in the openings is

$$V_{MT} = \frac{Q_{MAX}}{a_{MT}} = \frac{269.4}{295} = 0.91\,\mathrm{m/s}. \qquad (8.33)$$

Assume that the maximum velocity in the centre of the channel is 33% higher so:

$$V_{MAX} = 1.33 \times 0.91 = 1.21\,\mathrm{m/s}.$$

The average depth of flow at the bridge is approximately $Y_{MT} = a_{MT}/B = 295/132 = 2.23\,\mathrm{m}$.
From equation 8.3 the scour-critical velocity is:

$$V_S = 6.36\ Y^{0.167} D_{50}^{0.333} = 6.36 \times 2.23^{0.167} \times 0.0002^{0.333} = 0.42\,\mathrm{m/s}. \qquad (8.3)$$

This value is less than both 0.91 m/s and 1.21 m/s so live-bed scour can be assumed. Therefore equation 8.13 will be used to calculate the depth of contraction scour (d_{SC}) with $Q_1 = Q_2$, $B = 132\,\mathrm{m}$ and $b = (132 - 3 \times 4) = 120\,\mathrm{m}$. In this situation it is assumed that most of the material will be transported as suspended bed load, so from Table 8.5 the exponent $k_1 = 0.69$. Thus:

$$Y_2 = Y_1\ (Q_2/Q_1)^{6/7} (B/b)^{k_1} = 2.23 \times 1 \times (132/120)^{0.69} = 2.38\,\mathrm{m} \qquad (8.13)$$

$$d_{SC} = Y_2 - Y_1 = 2.38 - 2.23 = 0.15\,\mathrm{m} \qquad (8.11)$$

The local scour at the piers is given by equation 8.14, which uses the conditions immediately upstream of the pier. The known water depth at the pier in the centre of the channel at the maximum discharge midtide level is $Y_2 = 3.90\,\mathrm{m}$ and the velocity $V_2 = V_{MAX} = 1.21\,\mathrm{m/s}$. For round-nosed piers Table 8.6 gives $K_{1P} = 1.0$, $K_{2P} = 1.0$ since the crossing is perpendicular to the flow, for a plane bed $K_{3P} = 1.1$ from Table 8.7, $b_P = 4.0\,\mathrm{m}$, and $F_2 = V_2/(gY_2)^{1/2} = 1.21/(9.81 \times 3.90)^{1/2} = 0.20$.

$$d_{SP} = 2.0\ Y_2\ K_{1P}\ K_{2P}\ K_{3P}\ (b_P/Y_2)^{0.65}\ F_2^{0.43} \qquad (8.14)$$

$$= 2.0 \times 3.90 \times 1.0 \times 1.0 \times 1.1\ (4.0/3.90)^{0.65} \times 0.20^{0.43} = 4.37\,\mathrm{m}.$$

The abutments are inline with the quay walls and can be ignored, so the total scour depth is: 1.00 m allowance for degradation (Δd) plus 0.15 m contraction scour (d_{SC}) plus 4.37 m local pier scour (d_{SP}). Thus considering the estuary to be unconstricted gives the future maximum scour depth as 5.52 m, so scour is possible to −9.42 m OD.

Constricted estuary

Many of the values above can be used for the constricted analysis, which is based on the head loss as the sea floods into (or ebbs out of) the tidal basin. Equation 8.37 gives the velocity head loss coefficient as

$$k = k_S + k_B + 2gn^2L_C/h_C^{4/3} \quad (8.37)$$

On the seaward side assume that half of the velocity head is lost ($k_S = 0.5$) but in the basin the entire velocity head is dissipated and the velocity reduces to zero ($k_B = 1.0$), $g = 9.81\,\text{m/s}^2$, $n = 0.03\,\text{s/m}^{1/3}$, and the length of the inlet channel, $L_C = 800\,\text{m}$. Assuming the maximum velocity in the estuary occurs at midtide level (0 m OD) and that the corresponding average depth of flow in the channel at this level over the full 800 m length (not at the bridge) is $h_C = 2.40$ m, then:

$$k = 0.5 + 1.0 + (2 \times 9.81 \times 0.03^2 \times 800/2.40^{1.33}) = 5.90. \quad (8.37)$$

$$C_d = (1/k)^{1/2} = (1/5.90)^{1/2} = 0.41. \quad (8.36)$$

$$V = C_d\,(2g\Delta h)^{1/2} = 0.41\,(2 \times 9.81 \times 0.31)^{1/2} = 1.01\,\text{m/s}. \quad (8.34)$$

$$Q_{MT} = A_{MT}V = 132 \times 2.4 \times 1.01 = 320.0\,\text{m}^3/\text{s}. \quad (8.35)$$

$$Q_{MAX} = 320.0 + 30.4 = 350.4\,\text{m}^3/\text{s}. \quad (8.32)$$

If $a_{MT} = 295\,\text{m}^2$ as above, then $V_{MT} = 350.4/295 = 1.19\,\text{m/s}$. (8.33)

Assume a velocity 33% higher in the centre of the channel so V_{MAX} = 1.19 x 1.33 = 1.58 m/s. The depth of live-bed contraction scour (d_{SC}) can be calculated using $Y_1 = h_C$ (or Y_{MT} as above):

$$Y_2 = Y_1\,(Q_2/Q_1)^{6/7}(B/b)^{k_1} = 2.40 \times 1 \times (132/120)^{0.69} = 2.56\,\text{m}. \quad (8.13)$$

$$d_{SC} = Y_2 - Y_1 = 2.56 - 2.40 = 0.16\,\text{m}. \quad (8.11)$$

With $V_{MAX} = 1.58$ m/s and $Y_2 = 3.90$ m (the known depth at the pier, as before) then $F_2 = V_{MAX}/(gY_2)^{1/2} = 1.58/(9.81 \times 3.90)^{1/2} = 0.26$. For round-nosed piers, no skew and a flat bed the pier scour depth is

$$d_{SP} = 2.0\,Y_2\,K_{1P}\,K_{2P}\,K_{3P}\,(b_P/Y_2)^{0.65}\,F_2^{\,0.43} \quad (8.14)$$

$$= 2.0 \times 3.90 \times 1.0 \times 1.0 \times 1.1\,(4.0/3.90)^{0.65} \times 0.26^{0.43} = 4.88\,\text{m}.$$

With a degradation depth of 1.0 m, the total potential scour depth is 1.0 + 0.16 + 4.88 = 6.04 m (i.e. scour to −9.94 m OD).

The estimates of V_{MT} and F are lower than normally expected, probably because the basin is at the head of the estuary. Under different circumstances the average velocity for a 1 in 100 year storm tide may be expected to exceed significantly the value of $V_{MEANMAX}$, which is often around 1.00 m/s for normal spring tides. However, much depends upon the validity of the assumptions and data (see Kreeke, 1967; Bruun, 1990). This inlet may be too short to meet the conditions for a long contraction, while the value of Δh is not easy to determine accurately, particularly in situations where a significant and variable riverflow is also involved. The existence of mudflats above low tide level that submerge at different times can make it difficult to define precisely the 'midtide level' that corresponds to maximum discharge. Regardless of which method is used there is a considerable scope for error when analysing tidal crossings.

8.9 Combatting scour

The best and most cost-effective way to ensure that scour is not a threat to the structural integrity of a bridge is to design the structure so that its foundations are either well below the scour level or sufficiently well protected (by riprap or a suitable alternative) for scour not to be a problem. The former is the safest and most reliable method, assuming that the scour depth and channel stability have been correctly assessed over the lifetime of the bridge, whereas protective measures need regular monitoring and maintenance to be continually effective. Another measure that can be adopted at the design stage is to optimise the location and alignment of the bridge, particularly in relation to potentially destructive flood flows. A good hydraulic design is also required, with suitably sized waterways and efficient streamlined piers and abutments positioned to avoid skew and eccentricity. The provision of spur dykes, guidebanks, channel training works, channel improvements, relief openings and allowing the overtopping of embankments have all been discussed previously.

With existing bridges that are found to be suffering from scour, if it is not possible to alter the foundations some of the ancillary works just described can be added. Bridge piers can also be protected by placing in front of them sacrificial piles in a V formation that points directly into the flow (Anon, 1992). This approach has been adopted for the Over rail viaduct on the River Severn, England (Anon, 1994). However, one of the most common and time-honoured methods of alleviating scour is to dump large pieces of rock (riprap) in the affected area, usually at piers and abutments. When rock is available in a suitable size, riprap is generally regarded as the most economical protective material. A riprap blanket has the advantage of being flexible should settlement occur, easily repaired, easy to place, durable, and recoverable should the stones be moved.

Riprap can be an effective scour prevention measure, but if incorrectly placed it may also initiate scour. In either case, during a sequence of floods the large velocities and turbulence experienced in bridge openings can move riprap, so regular monitoring is required. The collapse of the bridge at

Schoharie Creek in the USA in 1987, which caused the death of 10 people, started with the removal of riprap around the bridge piers. Severe scour of the glacial till beneath the spread footings then occurred over a 10 year period (Richardson *et al.*, 1993; Lagasse *et al.*, 1995).

8.9.1 Riprap at abutments and piers

There are many design guides that advise on the use of riprap (Brown and Clyde, 1989; Li *et al.*, 1989; Richardson *et al.*, 1990). One recommendation is that riprap should not be used on slopes steeper than 1 in 1.5. This simple limitation, along with using a correctly sized and graded stone (Table 8.10), may eliminate many potential problems. When designing riprap at abutments common failure modes are:

- the movement and erosion of particles as a result of high velocities and powerful vortices, which are exacerbated if the stone size is not large enough, the gradation is too uniform or the embankment too steep relative to the angle of repose of the riprap;
- a translational slide of the riprap down an embankment, which may be initiated by channel scour eroding the toe of the riprap blanket, sliding along the plane of the filter blanket, excess pore water pressure, or bank slopes that are too steep. This type of failure may be indicated by cracks parallel to the channel in the upper part of the riprap blanket;
- a modified slump failure along an internal slip surface within the riprap, which may occur if the slope is near the riprap's angle of repose so that movement of individual stones results in a slide. The causes are similar to a translational slide;
- a slump failure involving a rotational-gravitational movement along a slip surface within the embankment material, causes being excess pore water pressure reducing friction, the use of non-homogeneous material that results in 'fault lines', and slopes that are too steep.

Chiew (1995) reviewed the mechanics of riprap failure at cylindrical bridge piers, and identified three failure modes:

- shear failure due to the stones not being large or heavy enough to withstand the downflow and horseshoe vortex;
- winnowing failure due to the erosion of the underlying bed material through the voids or interstices of the coarse riprap stones;
- edge failure, where the erosion of the natural bed material next to the riprap results in a small scour hole into which the outer stones of riprap roll or slide, leading to a progressive failure.

These problems can be avoided by making the riprap large and heavy enough to resist motion, ensuring the thickness of the riprap layer is adequate and at least greater than one stone thick (this facilitates rearmouring of any scour hole and helps to eliminate winnowing and edge failure), providing a stone or fabric filter layer to retain the bed material, and ensuring that the riprap extends uniformly over the areas prone to scour (see below). Thus it is important that the riprap layer is thick enough and composed of stones of the necessary diameter and gradation to withstand the prevailing conditions. There are many equations that can be used to calculate the required diameter, and these share some of the disadvantages of scour depth equations: they are either theoretical or based on small-scale laboratory experiments, while the field data available to verify them are scarce and sometimes of uncertain accuracy. The factors that should be considered include the velocity to be resisted, stone density, the shape and angularity of the rock, the depth of flow, the degree of turbulence or eddying, the curvature of the flow, and the slope angle (Neill, 1973). Velocity-based design procedures are thought to be as good as any, provided that an accurate estimate of the velocity can be obtained (Richardson *et al.*, 1990). Less scientific approaches to the problem are to use the size of stone that has been observed to work under similar conditions during similar floods, or to use a rule of thumb. Some of these are considered later.

Riprap at abutments

When deciding the stone diameter to use with a Froude number (F) ≤ 0.80 the following equation, based on the Isbash (1935) formula, was recommended by Pagán-Ortiz (1991, 1992) and Richardson *et al.* (1993):

$$D_{50} = \frac{KY}{(s_S - 1)}\left(\frac{V^2}{gY}\right) \tag{8.38}$$

where D_{50} is the median stone diameter (m), V (m/s) is the characteristic average velocity in the the contracted section as explained below, s_S is the dimensionless specific gravity (relative density) of the rock riprap (usually around 2.65), g is 9.81 m/s^2, Y is the depth of flow (m) in the bridge opening, and $K = 0.89$ for spillthrough abutments or 1.02 for vertical-wall abutments. The Ys cancel but have been included so that the equation has the same basic form as the one below. The term ($s_S - 1$) is the difference between the specific gravity of the stone ($s_S \approx 2.65$) and fresh water ($s = 1.0$; for sea water use 1.025).

For $F > 0.80$, using the same notation as above:

$$D_{50} = \frac{KY}{(s_S - 1)}\left(\frac{V^2}{gY}\right)^{0.14} \tag{8.39}$$

where $K = 0.61$ for spillthrough abutments or 0.69 for vertical-wall abutments.

In both equations the factor K allows for the local increase in velocity at the point of riprap failure. The recommended procedure for obtaining the characteristic average velocity in the above equations is as follows.

(a) Determine the distance from the toe of the abutment to the nearest edge of the main channel (Fig. 8.31), which is called the set-back distance, L_{SB}.

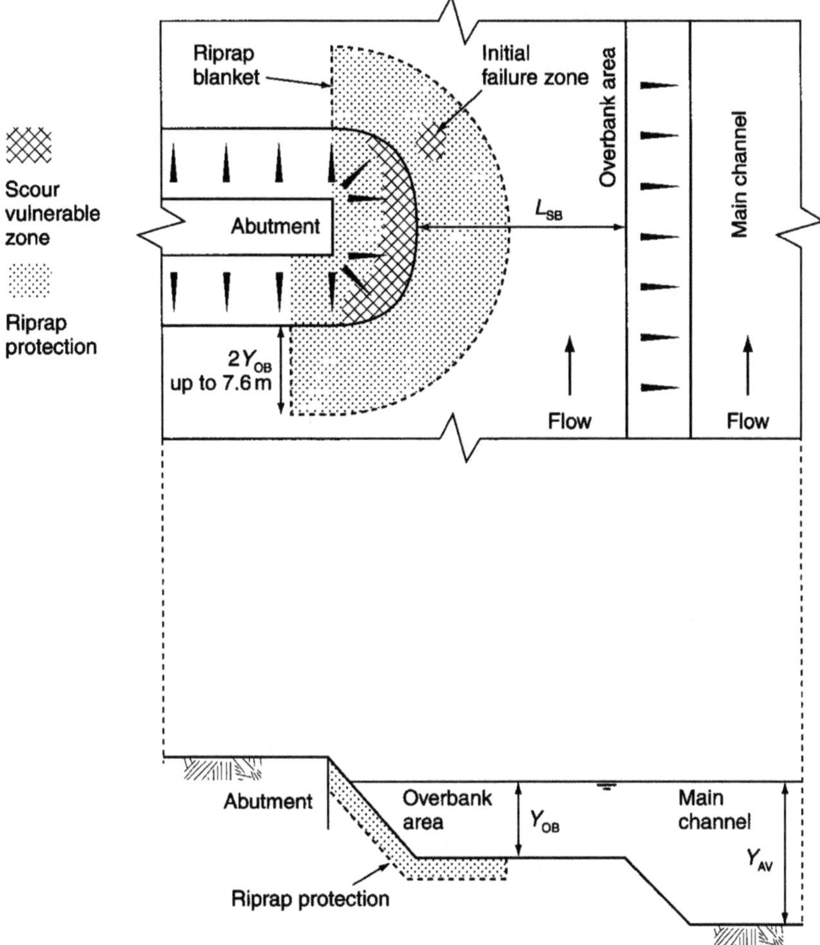

Fig. 8.31 Riprap protection and zones of failure at setback abutments. The cross-hatched area shows where the riprap is most likely to fail. Without the riprap protection the channel bed is most vulnerable to scour at the upstream corner of the abutment as shown in Fig. 8.7, the scoured zone extending into the waterway (perpendicular to the flow direction) by a distance of $2.75d_{SA}$ where d_{SA} is the abutment scour depth. See also Fig. 8.35. (After Richardson *et al.*, 1993)

(b) Calculate the average depth of flow in the main channel, Y_{AV}.
(c) Calculate the set-back ratio (SBR) = L_{SB}/Y_{AV}.
(d) If SBR < 5 for both abutments then $V = Q/a$ where Q is the total upstream flow but excluding any that overtops the roadway, and a is the total area of the bridge openings.
(e) If SBR > 5 for an abutment then $V = Q/A$ for the overbank flow at that abutment. This effectively assumes an imaginary wall along the edge of the main channel so that all of the flow on that floodplain (Q) passes through the corresponding overbank area of the bridge opening (A).
(f) If SBR < 5 for one abutment but > 5 for the other then the velocity for the former obtained from point (d) may be too low so the average velocity for this abutment should be based on the area (A) between the abutment and an imaginary wall on the opposite bank. The corresponding discharge (Q) is all of the flow between the outer edge of the floodplain associated with the abutment and the imaginary wall on the opposite bank.

The area of the riprap protection required is illustrated in Fig. 8.31, and should extend outwards to a distance twice the depth of flow on the overbank area ($2Y_{OB}$) up to a maximum of 7.6 m. This area should include the abutment slopes. With spillthrough abutments the zone where initial failure of the riprap usually occurs is on the overbank near the downstream toe (shown cross-hatched in the diagram). It then expands outwards to include the toe of the abutment slope, which can endanger the abutment, as described above. Another vulnerable zone is at the upstream corner of the abutment. With vertical-wall abutments the initial failure zone usually occurs in the bed at the upstream corner, where the flow accelerates as it enters the waterway opening (Fig. 8.7). The zone then spreads into the opening (Richardson *et al.*, 1990; Pagán-Ortiz, 1991, 1992).

The minimum thickness of the riprap layer should be the larger of D_{100} or $1.5D_{50}$. This thickness should be increased by at least 50% if the riprap is placed underwater or in less than ideal conditions. Some guides (Li *et al.*, 1989) suggest that the thickness need not exceed $2D_{100}$. Thickness is

Table 8.10 Typical riprap grading and classification (all values in metres)

		D_{50} *size (m)*						
% passing	*Particle size*	*0.15*	*0.20*	*0.30*	*0.45*	*0.60*	*0.75*	*0.90*
100–90	2.0 D_{50}	0.300	0.400	0.600	0.900	1.200	1.500	1.800
85–70	1.5 D_{50}	0.225	0.300	0.450	0.675	0.900	1.125	1.350
50–30	1.0 D_{50}	0.150	0.200	0.300	0.450	0.600	0.750	0.900
15–5	0.67 D_{50}	0.100	0.125	0.200	0.300	0.400	0.500	0.600
5–0	0.33 D_{50}	0.050	0.075	0.100	0.150	0.200	0.250	0.300

After Li *et al.* (1989); Richardson *et al.* (1990)

measured perpendicular to a sloping surface. The need for a filter layer or fabric to support and reinforce the riprap and separate it from the bed material should also be considered (Section 8.9.2). There is a large body of trade literature that can help the designer in this regard.

Riprap at piers

Planning to use riprap at piers is to some extent tantamount to accepting that the basic design is inadequate. At best riprap is a temporary means to alleviate pier scour problems. The large velocities and strong vortices near piers make regular inspection of the riprap essential when it is used (for this reason the riprap should never be buried). The stone diameter required in fresh water is (Brown and Clyde, 1989; Pagán-Ortiz, 1991, 1992; Richardson *et al.*, 1993)

$$D_{50} = \frac{0.692(KV)^2}{(s_S - 1)2g} \tag{8.40}$$

where D_{50} is the median stone diameter (m); $K = 1.5$ for a round-nosed pier; or 1.7 for a rectangular pier; V is the velocity at the pier (m/s) as described below; s_S is the dimensionless specific gravity (relative density) of the riprap, which is normally around 2.65 with fresh water being 1.00; and $g = 9.81\,\text{m/s}^2$. The average velocity (V) is that in the channel in which the piers are located, which is

$$V = \frac{CQ}{A} \tag{8.41}$$

where Q and A are the discharge and area of the main channel, $C = 0.9$ for a pier near the bank in a straight uniform reach, while $C = 1.7$ for a pier in the fastest part of the flow around a bend. Intermediate values of C are used in between these two extremes.

The riprap protection should be flush with the river bed and extend horizontally over a distance of $2b_P$ to $3b_P$ from all faces of the pier, where b_P is the width of the pier normal to the flow. The greater the penetration of the riprap into the bed the less likely it is to be moved, but the minimum thickness should be $3D_{50}$. The largest rock size used should not exceed $2D_{50}$. In some situations a stone or fabric filter may be needed (see below).

Highway embankments and streambanks

If riprap is placed on a steeply sloping surface it is easier for flowing water to roll or slide the stones down the slope. Consequently a larger diameter (heavier) stone may be needed on sloping revetments, abutments, spurs and embankments. A theoretical treatment of riprap stability on slopes was given by Brown and Clyde (1989), Li *et al.*, (1989) and Richardson *et al.*,

(1990), while Figs 8.32 and 8.33 provide a simple and interesting comparison of the diameters required on flat and sloping surfaces which are parallel to the flow. The charts may be used to design riprap protection to highway embankments or eroding stream banks. The trial and error procedure is as follows.

1. Estimate the mean flow velocity (V m/s) and guess the 50% stone size, D_{50} (m).
2. If the depth of flow $Y < 3$ m calculate D_{50}/Y or if $Y > 3$ m calculate $D_{50}/(0.4 \times Y)$.
3. Enter Fig. 8.32 with the value from step 2 and obtain the corresponding value of V_{IMP}/V, which is the ratio of the velocity impacting on the stone (V_{IMP} m/s) to the mean flow velocity (V m/s).
4. Calculate $V_{IMP} = (V_{IMP}/V) \times V$. If the flow hits the riprap perpendicularly the velocity, V_{IMP}, should be increased by a factor between 1.0 and 2.0. A factor of 2.0 may be prudent to allow for future changes in stream alignment.
5. Enter Fig. 8.33 with the adjusted value of V_{IMP} and read off the stone size/weight needed for the embankment slope in question. This chart assumes a spherical stone of density 2650 kg/m^3 (i.e. specific gravity = 2.65).

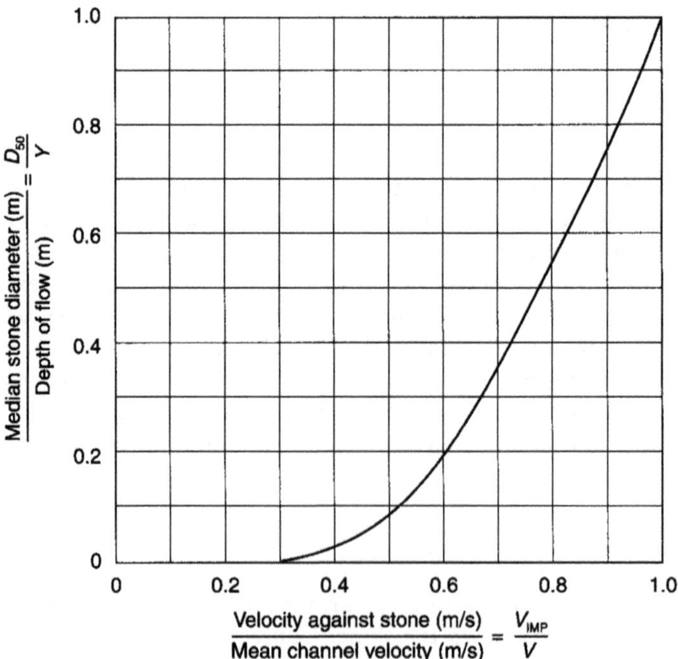

Fig. 8.32 Velocity against riprap stone (V_{IMP}) on the channel bottom, for use with Fig. 8.33. (After HEC 11, Brown and Clyde, 1967)

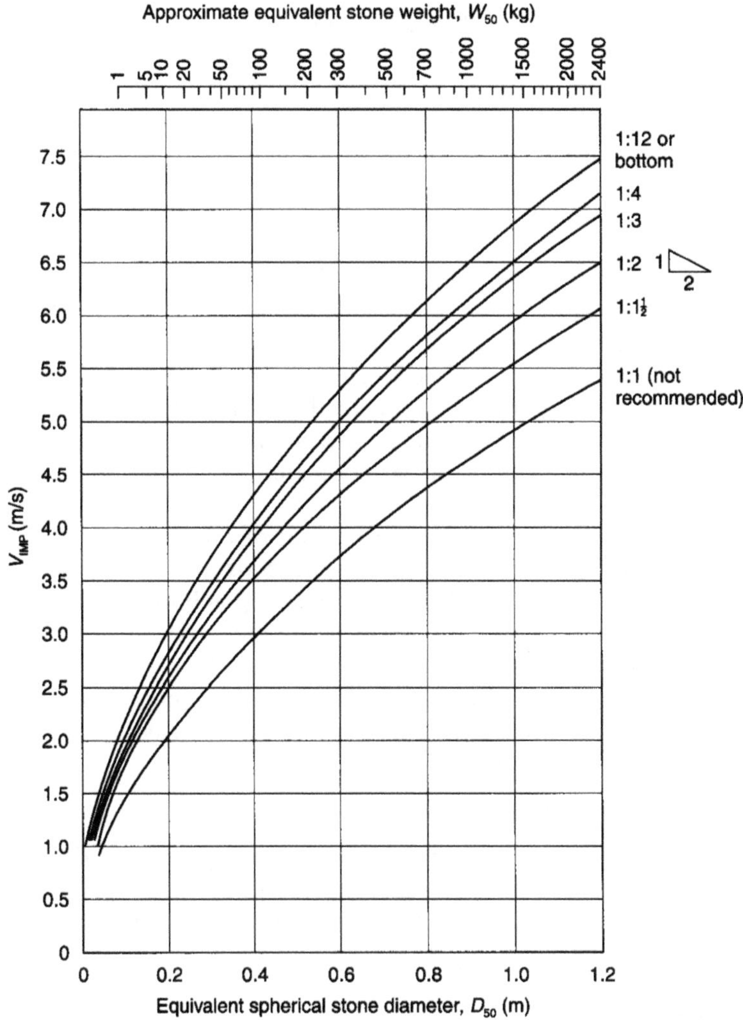

Fig. 8.33 Approximate size of riprap (D_{50} or W_{50}) required to resist displacement as a function of impact velocity against the stone (V_{IMP} m/s) and the embankment slope. (After HEC 11, Brown and Clyde, 1967)

6. If the diameter of stone obtained from step 5 is not the same as assumed in step 1 then repeat the procedure until it is.

Remember that slopes steeper than 1 in 1.5 are not recommended and that if significant drawdown of the water surface occurs or if there is a change in flow direction around the corner of an abutment the results must be used cautiously and/or the stone size increased.

Alternative approaches

Different equations will give different diameters of riprap, assuming that the velocity and depth are known – which may not be the case, so it is important that common sense and experience are also employed. Existing bridges in the region may provide a guide as to what works in practice. Rules of thumb can serve as a means of learning from the experience of others, provided individual judgement is also applied. Paraphrasing Neill (1973) quoting Blench (1969), a rough guide is:

- large sand bed rivers without a very large bed load – normally need stone about 68 kg;
- small sand bed rivers – stone as small as 23 kg may suffice;
- gravel bed river with a small bed load and moderate attack – twice the diameter of the largest material that rolls on the bed;
- gravel bed river with a very violent attack (as at the nose of a spur) – three times the diameter of the largest material that rolls on the bed.

Frequently, trade literature can provide a means of obtaining 'second opinions' or useful design suggestions. For channel beds (not specifically in the vicinity of piers or abutments) Fig. 8.34 shows a graph for the determination of stable riprap diameter or weight. Although very simplistic it can serve as a handy guide; note that it gives the D_{40} or W_{40} size, which will be smaller than the D_{50} or W_{50} (for a typical riprap, very approximately $D_{40} \approx 0.8D_{50}$). Assuming spherical particles of density 2650 kg/m^3 the conversion from diameter (D m) to weight (W kg) is

$$W = 0.524D^3 \times 2650 = 1388D^3 \qquad (8.42)$$

When using Fig. 8.34, to ensure that adequate protection is afforded and to avoid winnowing, the velocity of water in the voids (V_V m/s) of the riprap layer should be calculated. For a uniformly graded material of diameter, D, this is given by (Stephenson, 1979)

$$V_V = 0.092V\left(\frac{D}{Y}\right)^{2/3} \qquad (8.43)$$

which, when combined with equation 8.42, gives (Netlon Ltd)

$$V_V = \frac{0.019V(W_{40})^{0.22}}{Y^{0.66}} \qquad (8.44)$$

where V is the flow velocity (m/s) and Y is the corresponding depth (m). If V_V exceeds the threshold values in Table 1.3a then the bed material underneath the riprap may be scoured, so the design should be reviewed.

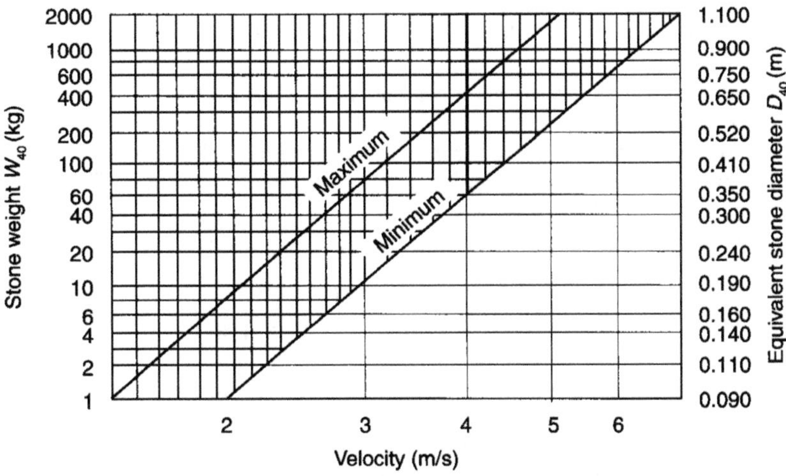

Fig. 8.34 Approximate riprap stone weight and diameter required for various velocities. Use the maximum line for very turbulent flows and the minimum line for normal flows. (Reproduced by permission of US Army Engineer Waterways Experiment Station)

Example 8.8

For the crossing in Example 8.5, determine the riprap requirements at the abutments.

The design in Example 8.5 was left incomplete in that the calculations were not repeated after the decision was made to change to spillthrough abutments, but as a preliminary estimate of the riprap requirements assume that the conditions in the contracted section are as shown in Fig. 8.23.

Left abutment/main channel

In the main channel $Y_2 = 2.4$ m, $V_2 = 2.92$ m/s, $F_2 = 0.60$.

Note that in this case the SBR = 0 and 8.3 for the left and right abutments respectively so either from case (d) $V_2 = 399/(2.4 \times 57) = 2.92$ m/s or from (f) and Fig 8.22 $V_2 = (68 + 304)/(2.4 \times 57) = 2.72$ m/s. The higher value will be adopted.

Using equation 8.38 for $F < 0.80$, with spillthrough abutments $K = 0.89$ and $s_S = 2.65$, then

$$D_{50} = \frac{KY}{(s_S-1)}\left(\frac{V^2}{gY}\right) = \frac{0.89}{(2.65-1)}\left(\frac{2.92^2}{9.81}\right) = 0.47\,\text{m} \qquad (8.38)$$

Use a 0.45 m class riprap as in Table 8.10. The area covered should extend 4.8 m (2Y) from the abutment as in Fig. 8.31. The minimum thickness should be D_{100} to $2D_{100}$, say 1.8 m (the same as below).

Right overbank area

On the overbank $Y_2 = 1.40$ m, $V_2 = 3.35$ m/s, $F_2 = 0.90$ so use equation 8.39 for $F > 0.80$.

Say $L_{SB} = 20$ m and in the main channel $Y_{AV} = 2.40$ m so SBR = 20/2.40 = 8.3 > 5.

Thus V should be calculated from all of the overbank flow at that abutment, which (from Fig. 8.22) is $Q = (40 + 78) = 118\,\text{m}^3/\text{s}$. Thus $V = Q/A = 118/(1.40 \times 19.4) = 4.35$ m/s.

With spillthrough abutments $K = 0.61$ and $s_S = 2.65$ then

$$D_{50} = \frac{KY}{(s_S - 1)}\left(\frac{V^2}{gY}\right)^{0.14} = \frac{0.61 \times 1.40}{(2.65-1)}\left(\frac{4.35^2}{9.81 \times 1.40}\right)^{0.14} = 0.54\text{ m} \quad (8.39)$$

Use a 0.6 m class riprap (Table 8.10). The area of protection required is as shown in Fig. 8.31 and should extend outwards from the abutment to a distance of twice the depth on the overbank area, that is 2 × 1.4 = 2.80 m (it may be sensible to make it 4.8 m, the same as at the other abutment). The minimum thickness of the layer should be about D_{100} to $2D_{100}$, say 1.8 m.

Piers

Riprap should not be needed at the piers of an appropriately designed new bridge but, as an illustration, for the main channel $Y_2 = 2.4$ m and $V_2 = 2.92$ m/s (from Fig. 8.23). Assume $s_S = 2.65$, for a round-nosed pier $K = 1.5$, and that the piers are in a straight where the flow is fastest so $C = 1.2$. From equation 8.41 $V = 1.2 \times 2.92 = 3.50$ m/s.

$$D_{50} = \frac{0.692(KV)^2}{(s_S - 1)2g} = \frac{0.692(1.5 \times 3.50)^2}{(2.65-1) \times 2 \times 9.81} = 0.59\text{ m} \quad (8.40)$$

Use a 0.60 m class riprap as in Table 8.10. The riprap should extend a distance of $2b_p = 2 \times 1.2 = 2.4$ m from the pier. The minimum thickness is $3D_{50} = 3 \times 0.60 = 1.80$ m with the surface of the riprap at bed level.

As a quick check, using the same velocity as above, $V = 3.50$ m/s and assuming very turbulent flow around the pier so that the maximum curve applies, the chart in Fig. 8.34 gives $W_{40} = 200$ kg and $D_{40} = 0.52$ m (so very approximately if $D_{40} = 0.8D_{50}$ then $D_{50} = 1.25 \times 0.52 = 0.65$ m, similar to above). Check the void velocity using equation 8.44:

$$V_V = \frac{0.019V(W_{40})^{0.22}}{Y^{0.66}} = \frac{0.019 \times 3.50 \times (200)^{0.22}}{2.4^{0.66}} = 0.12\text{ m/s}$$

Since 0.12 m/s is less than 0.6 m/s (the value for sand with $D = 1$ mm in Table 1.3a), this is acceptable.

8.9.2 *Riprap specification*

The required properties of riprap are that it is hard, durable, dense and angular. The rock must be able to withstand abrasion, freeze–thaw action, weathering, and dissolution. Typically, igneous and metamorphic rock is used, although some sedimentary rocks may be suitable. A relatively high density is desirable because it increases the weight of a rock of any given size, so the current is less likely to move it. Angular riprap is preferred to rounded stones because it interlocks better, but it also has a higher angle of repose, which increases stability. The angle of repose of very angular material is fairly constant at around 40° regardless of size, whereas with very rounded material large diameters may have an angle of repose of 40° but it falls off to about 33° below 0.15 m diameter. As explained in Section 8.9.1, on sloping surfaces a low angle of repose can initiate failure of the riprap blanket. However, not more than 25% of the stones should have a length 2.5 times their breadth. Very flat, slab-like stones are easily dislodged by the flow. The use of broken concrete riprap was not recommended by Richardson *et al.* (1990).

Riprap should be a graded material that comprises a range of sizes rather than being entirely composed of stones with the same (D_{50}) diameter. The different-sized stones interlock and prevent flow through the void spaces. Typical grading classifications are shown in Table 8.10. Generally the grading is not critical: usually a tolerance is specified and any material that falls in this band is acceptable. The availability and cost of stone in a particular location must always be borne in mind.

Riprap can be placed by hand, or by backhoe, skip, bucket or dragline. Some sources say that hand placing is better than machine placing, but others say it is worse than tamped, machine-placed riprap. In most cases hand placing is too expensive anyway. If a backhoe, skip or bucket is used the objective should be to place the stone as near as possible to its final position and rework it as little as possible. Excessive re-handling or dragging tends to break the stone and cause segregation. For this reason it should not be tipped down an embankment or pushed down by bulldozer.

The riprap at the toe of a bank is particularly vulnerable to scour, and any loss of material can initiate sliding or failure of the entire blanket. Generally a toe trench, taken well below normal bed level along the whole length of the riprap blanket, is filled with riprap to anchor the toe and support the upslope material (Fig. 8.35a). If a trench is not possible the toe may be anchored by placing a deep, narrow strip of riprap on the bed (shown dashed), but this has its drawbacks: if badly done it can channel flow along the base of the slope and cause scour. With both of these designs the objective is to form a reserve of riprap in the most vulnerable area, so when scour occurs the riprap falls into the scour hole and armours it. The quantity of riprap required in the reserve is 1.5 times the volume calculated assuming the design blanket thickness at a slope of 1 in 2 over the anticipated scour depth (Brown and Clyde, 1989). Care must also be taken at the interface between the start/end of the riprap protected area and the natural channel.

Riprap is not usually placed directly on the base (the bed or side-slope of the channel) but on a filter layer, which is designed to prevent base material leaching through the riprap. Thus the filter material must be relatively fine

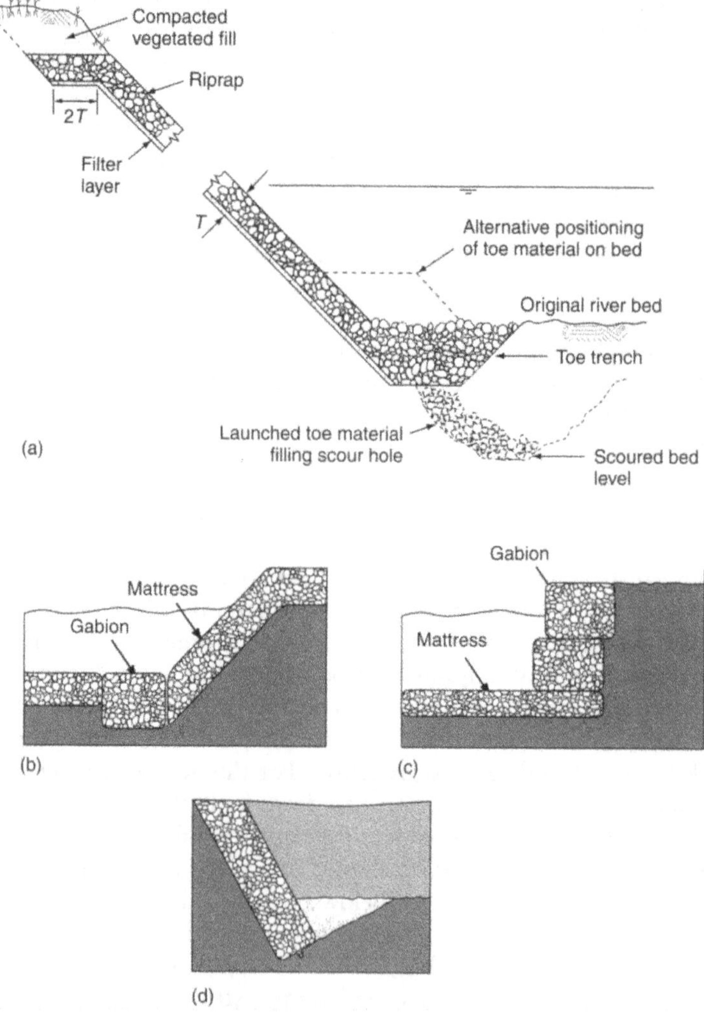

Fig. 8.35 Some typical bank protection measures: (a) bank protection only using riprap, with extra material at the toe, which can be launched into the toe scour hole when erosion occurs (after Brown and Clyde, 1989); (b) gabion and mattress protection with a moderate embankment slope, and (c) a steep slope; (d) bank protection using a tubular geotextile gabion taken to below bed scour level. (b–d after Netlon Ltd, reproduced with permission)

to prevent winnowing of the bed but more permeable than the bed material, the choice being between a gravel or synthetic fabric filter.

Very fine base material may require a gravel filter which consists of several layers, each around 0.1–0.2 m thick and meeting the stability and permeability requirements of the layer underneath, otherwise one layer will usually suffice. A filter ratio of 5 or less between layers usually results in a stable condition, where the filter ratio is defined as the ratio of the 15% particle size (D_{15}) of the coarser layer to the 85% particle size (D_{85}) of the finer layer (Brown and Clyde, 1989; Richardson *et al.*, 1990). An additional requirement is that the ratio of the D_{15} of the coarser layer to the D_{15} of the finer layer should exceed 5 but be less than 40. Thus:

$$\frac{D_{15}\,\text{(coarser layer)}}{D_{85}\,\text{(finer layer)}} < 5 < \frac{D_{15}\,\text{(coarser layer)}}{D_{15}\,\text{(finer layer)}} < 40 \tag{8.45}$$

$$\text{also}\ \frac{D_{50}\,\text{(coarser layer)}}{D_{50}\,\text{(finer layer)}} < 40 \tag{8.46}$$

If the riprap itself meets this requirement then a filter is not needed; if it does not then the above relationships can be used to establish the D_{15}, D_{50} and D_{85} size of the filter material, and hence its grading. Gravel filter blankets generally consist of material ranging in diameter from about 5 mm to 90 mm depending upon the riprap size. A filter thickness one-half that of the riprap layer is satisfactory, but should not be less than 0.225 m. With fine cohesive materials these relationships do not apply, and the filter requirement is that its $D_{15} < 0.4$ mm.

As an illustration of the above relationships, say that the bed material is sand and that the proposed filter and riprap materials have the specifications shown below. Is a filter needed and, if so, would the proposed filter material be suitable?

	Sand bed (mm)	*Proposed filter (mm)*	*Riprap (mm)*
D_{15}	0.8	30	300
D_{50}	1.5	45	450
D_{85}	7	68	675

The answer is that for the riprap (coarser layer) and sand bed (finer layer) the D_{50}/D_{50} in equation 8.46 is $450/1.5 = 300 > 40$, so a filter is needed. Checking the filter specification starting with the filter as the coarse layer and the bed as the fine layer, $D_{15}/D_{85} = 30/7 = 4.3 < 5$, $D_{15}/D_{15} = 30/0.8 = 38 < 40$ and $D_{50}/D_{50} = 45/1.5 = 30 < 40$, as required. Now considering the riprap and filter as the coarse and fine layer respectively, $D_{15}/D_{85} = 300/68 = 4.4 < 5$, $D_{15}/D_{15} = 300/30 = 10 < 40$ and $D_{50}/D_{50} = 450/45 = 10 < 40$, so the filter is satisfactory.

An alternative to the traditional gravel filter is a filter cloth, which is quick and easy to install, economical, and not reliant upon the availability of suitable stone. However, placement underwater is not easy, the in-situ performance of such filters over the lifetime of a bridge is unproven, and they may initiate sliding or slump failure of the riprap on steep slopes. Often fabrics with an open area of about 25–30% are selected, their function being to provide both filtration and drainage. If not well designed and specified with respect to piping resistance, clogging resistance, strength and installation procedure the performance may be unsatisfactory. A 100–150 mm protective layer of sand or gravel may be spread on the cloth to prevent puncturing and local stretching when using large riprap, but heavy rock must still be placed carefully. The sides and toe of the cloth must be sealed or embedded in a trench to prevent leaching of the base material at the edges. Similarly, care must be taken at the joins, which are usually sewn or overlapped. Folds should be included to allow stretching under settlement.

8.9.3 Summary of some protective measures for bridges and riverbanks

Below is a brief reminder of some of the options available when trying to decide how to combat scour. Obviously which is adopted depends upon a combination of many factors. Generally rock riprap, rock or timber groynes, jacks and vegetation are at the cheaper end of the spectrum, while cellular block revetments, concrete mattresses and concrete walls are more expensive (Petersen, 1986; Richardson *et al.*, 1990).

- Riprap, as described above.
- A paved invert through the bridge opening (Highways Agency, 1994). This must extend a sufficient distance up and downstream of the structure, and must be toed in to the river bed (by about 1.5 m in the UK) to prevent undermining by scour. Note that a smooth surface to the paving may increase velocities and cause scour downstream.
- Soil–cement mixtures using in situ soil might be an alternative to riprap if rock is scarce, but they have many disadvantages.
- Gabions: stone in wire baskets. Hand filling can be expensive.
- Tubular gabions: a geotextile tube mechanically filled with stone.
- Grass or woody vegetation (possibly in combination with a geotextile) can be used to reinforce embankments naturally (see Hemphill and Bramley, 1989; Coppin and Richards, 1990). This may include the use of fascines (typically living willow poles that grow, providing bank stabilisation) and buried thorn or willow faggots (bundles of branches), which are used as reinforcement to weak soils.
- Sand–cement-filled sacks; commonly used for emergency repairs.
- Precast cellular concrete blocks; used as a facing material.

- Concrete (reinforced) if frequent impact by ice flows or logs or other floating debris is expected.
- Geotextile mattresses for covering larger areas: a stone blanket wrapped in a geotextile.
- Articulated concrete mattresses; difficult to place and expensive.
- Timber or concrete cribs: large open 'cubes' formed from timber or concrete members, which are filled with rock or earth. Used for bulkheads or retaining walls, perhaps to hold a highway embankment. An alternative to a reinforced concrete retaining wall.
- Drop structures: concrete or gabion structures that are used to reduce and stabilise the slope of a channel. They can take the form of either a weir or a straight drop (step) without the structure projecting above the upstream bed level. Both usually have a protective downstream apron to prevent erosion near the drop.
- Groynes (or spurs); used to stabilise a channel, reposition the flow or encourage siltation (Section 7.6.1).
- Jetties (USA) consisting of triangular steel frames (jacks) tied together into units. By collecting debris they increase the roughness of parts of the channel or floodplain and help train the stream (like groynes). They also reduce velocities and erosion near the banks.
- Spur dykes (guide banks); used where large overbank flows have to return to the main channel (Section 7.4).

For bank protection Hemphill and Bramley (1989) quoted a typical cost per metre (1978–85 prices) of £1 for seeded grass, £9–18 for geotextiles, £13 for thorn faggots, £19 for geotextile-reinforced vegetation, £25 for ungrouted riprap, £85 for concrete-filled bags, £87 for gabions, £116 for interlocking concrete blocks (ungrouted), £130 for steel sheet piling, £291 for concrete-filled mattress, and £347 for in-situ concrete wall.

References

Ackers, P. (1992) Gerald Lacey Memorial Lecture. Canal and river regime in theory and practice: 1929–92. *Proceedings of the Institution of Civil Engineers, Water Maritime and Energy*, **96**, September, 167–178.

Ahmad, M. (1953) Experiments on design and behaviour of spur dikes, in *Proceedings of the International Association of Hydraulic Research*, American Society of Civil Engineers Joint Meeting, University of Minnesota, August 1953.

Anon (1992) Sacrificial action may counter pier pressure. *Construction Weekly*, 26 February, 14–15.

Anon (1994) Rail rethink leaves research in limbo. *New Civil Engineer*, 24 February, 8.

Apt, S.R., Richardson, J.R., Hogan, S.A., Van Zanten, B.L. and Siller, T.J. (1992) Laboratory testing of scour-monitoring devices, in *Transportation Research Record 1350, Hydrology and Bridge Scour*, Transportation Research Board/ National Research Council, Washington DC, pp. 19–27.

Blench, T. (1939) A new theory of turbulent flow of liquids of small viscosity. *Journal of the Institution of Civil Engineers*, Paper 5185.

Blench, T. (1969) *Mobile-Bed Fluviology*, University of Alberta Press, Edmonton.

Blodgett, J.C. (1984) Effect of bridge piers on streamflow and channel geometry, in *Transportation Research Record 950*, Second Bridge Engineering Conference, Vol. 2, Transportation Research Board/National Research Council, Washington DC, pp. 172–183.

Brandon, T.W. (ed.) (1987) *River Engineering – Part 1, Design Principles*, Institution of Water and Environmental Management, London.

Brice, J.C. (1984) Assessment of channel stability at bridge sites, in *Transportation Research Record 950*, Second Bridge Engineering Conference, Vol. 2, Transportation Research Board/National Research Council, Washington DC, pp. 163–171.

Brice, J.C. and Blodgett, J.C. (1978) *Countermeasures for Hydraulic Problems at Bridges*. Vol. 1, *Analysis and Measurement*, report FHWA-RD-78-162; Vol. 2, *Case Histories for Sites 1–283*, report FHWA-RD-78-163. Federal Highways Administration, US Dept of Transportation.

Brown, S.A. and Clyde, E.S. (1989) Design of riprap revetment. Report no FHWA-IP-39-016, Hydraulic Engineering Circular 11 (HEC 11). Federal Highways Administration, McLean, VA.

Bruun, P. (1990) *Port Engineering*, 4th edn, Vol. 2, *Harbor Transportation, Fishing Ports, Sediment Transport, Geomorphology, Inlets and Dredging*; Chapter 9, Tidal inlets on alluvial shores, Gulf Publishing Company, Houston, TX.

Bryan, B.A., Simon, A., Outlaw, G.S. and Thomas, R. (1995) Methods for assessing the channel conditions related to scour-critical conditions at bridges in Tennessee. Final Report Project No. TN-RES1012, United States Geological Survey, Nashville.

Chadwick, A. and Morfett, J. (1993) *Hydraulics in Civil and Environmental Engineering*, 2nd edn, E & FN Spon, London.

Chang, F.F.M. (1973) A statistical summary of the cause and cost of bridge failures. Federal Highway Administration, US Department of Transportation, Washington DC.

Chiew, Y.-M. (1995) Mechanics of riprap failure at bridge piers. *American Society of Civil Engineers, Journal of the Hydraulics Division*, **121**(9), September, 635–643.

Chin, C.O., Melville, B.W. and Raudkivi, A.J. (1994). Streambed armoring. *American Society of Civil Engineers, Journal of Hydraulic Engineering*, **120**(8), August, 899–918.

Copp, H.D. and Johnson, J.P. (1987) Riverbed scour at bridge piers. Final Report WA-RD 118.1, Washington State Department of Transportation, Olympia, WA.

Coppin, N.J. and Richards, I.G. (1990) *Use of Vegetation in Engineering*, CIRIA/Butterworths, London.

Ettema, R. (1976) Influence of bed gradation on local scour. Report no. 124, University of Auckland School of Engineering.

Ettema, R. (1980) Scour at bridge piers. Report no. 216, University of Auckland School of Engineering.

Farraday, R.V. and Charlton, F.G. (1983) *Hydraulic Factors in Bridge Design*, Hydraulics Research Station, Wallingford, England.

Hemphill, R.W. and Bramley, M.E. (1989) *Protection of River and Canal Banks*, CIRIA/Butterworths. London.

Henderson, F.M. (1966) *Open Channel Flow*, Macmillan, New York.

Highways Agency (1994) *Design Manual for Road Bridges*, Vol. 1, Section 3, Part 6, BA 59/94, *The Design of Highway Bridges for Hydraulic Action.* HMSO, London.

Hjulström, F. (1935) The morphological activity of rivers as illustrated by river Fyris. *Bulletin of the Geological Institute of Uppsala*, Vol. 25, Chap III.

Isbash, S.V. (1935) Construction of dams and other structures by dumping stones into flowing water. *Trans. Res. Inst. Hydrot. Leningrad*, **17**, 12–66.

Johnson, P.A. (1995) Comparison of pier scour equations using field data. *American Society of Civil Engineers, Journal of Hydraulic Engineering*, **121**(8), August, 626–629.

Johnson, P.A. and Ayyub, B.M. (1996) Modeling uncertainty in predictions of pier scour. *American Society of Civil Engineers, Journal of Hydraulic Engineering*, **122**(2), February, 66–72.

Johnson, P.A. and Simon, A. (1997) Effect of channel adjustment processes on reliability of bridge foundations. *American Society of Civil Engineers, Journal of Hydraulic Engineering*, **123**(7), July, 648–651.

Jones, J.S. (1984) Comparison of prediction equations for bridge pier and abutment scour, in *Transportation Research Record 950*, Second Bridge Engineering Conference, Vol. 2, Transportation Research Board/National Research Council, Washington DC, pp. 202–209.

Jones, J.S. (1989) Laboratory studies of the effects of footings and pile groups on bridge pier scour. US Interagency Sedimentation Committee Bridge Scour Symposium, US Department of Transportation, Washington DC.

Kennedy, R.G. (1895) The prevention of silting in irrigation canals. *Proceedings of the Institution of Civil Engineers*, **119**, 281–290.

Kreeke, J. van de. (1967) Water-level fluctuations and flow in tidal inlets. *Proceedings of the American Society of Civil Engineers, Journal of the Waterways and Harbors Division*, **93**(WW4), November, 97–106.

Lacey, G. (1929–30) Stable channels in alluvium. *Proceedings of the Institution of Civil Engineers*, **229**, 258–384.

Lacey, G. (1933–34) Uniform flow in alluvial rivers and canals. *Proceeding of the Institution of Civil Engineers*, **237**, 421–453.

Lacey, G. (1939) Regime flow in incoherent alluvium. Central Board of Irrigation and Power, India, Publication 20, July.

Lagasse, P.F., Thompson, P.L. and Sabol, S.A. (1995) Guarding against scour. *Civil Engineering*, **65**(6), June, 56–59.

Lane, E.W. (1955) The importance of fluvial morphology in hydraulic engineering. *American Society of Civil Engineers Proceedings*, **81**(795), 1–17.

Laursen, E.M. (1962) Scour at bridge crossings. *Transactions of the American Society of Civil Engineers*, **127**, Part 1, 166–180.

Laursen, E.M. (1963) An analysis of relief bridge scour. *American Society of Civil Engineers, Journal of the Hydraulics Division*, **89**(HY3), 93–118.

Laursen, E.M. (1984) Assessing vulnerability of bridges to floods, in *Transportation Research Record 950*, Second Bridge Engineering Conference, Vol. 2, Transportation Research Board/National Research Council, Washington DC, pp. 222–229.

Li, R.M., MacArthur, R. and Cotton, G. (1989) Sizing riprap for the protection of approach embankments and spur dikes and limiting the depth of scour at bridge piers and abutments, Vol. 2, *Design Procedure*. Report no. FHWA-AZ89-260, Arizona Department of Transportation, Phoenix.

Melville, B.W. (1988) Scour at bridge sites, in *Civil Enginering Practice, 2/Hydraulics/Mechanics* (eds P.N. Cheremisinoff, N.P. Cheremisinoff and S.L. Cheng), Technomic Publishing Company, Lancaster, PA, pp. 327–362.

Melville, B.W. and Dongol, D.M. (1992) Bridge pier scour with debris accumulation. *American Society of Civil Engineers, Journal of Hydraulic Engineering*, **118**(9), September, 1306–1310.

Miller, A.C., Johnson, D. and Steinhart, R. (1992) Predictive equations for bridge scour applicable to streams in Pennsylvania, Field Manual, Research Project 89-03. Environmental Resources Research Institute, The Pennsylvania State University.

Neill, C.R. (1968) Note on initial movement of coarse uniform bed material. *Journal of Hydraulics Research*, **17**(2), 247–249.

Neill, C.R. (ed.) (1973) *Guide to Bridge Hydraulics*, Roads and Transportation Association of Canada/University of Toronto Press, Toronto.

Netlon Ltd. *Maritime and Waterway Engineering with Tensar and Netlon Geogrids*, Blackburn, England.

Novak, P., Moffat, A.I.B., Nalluri, C, and Narayanan, R. (1996) *Hydraulic Structures*, 2nd edn, Unwin Hyman, London.

Pagán-Ortiz, J.E. (1991) Stability of rock riprap for protection at the toe of abutments located at the floodplain. Federal Highways Administration, McLean, VA.

Pagán-Ortiz, J.E. (1992) Stability of rock riprap for protection at toe of abutments and floodplain. *Transportation Research Record 1350*, Transportation Research Board, Washington DC, 34–46.

Petersen, M.S. (1986) *River Engineering*, Prentice-Hall, Englewood Cliffs, NJ.

Raudkivi, A.J. (1997) Ripples on stream beds. *American Society of Civil Engineers, Journal of Hydraulic Engineering*, **123**(1), January, 58–64.

Rice, R.J. (1977) *Fundamentals of Geomorphology*, Longman, London.

Richardson, E.V., Simons, D.B. and Julien, P.Y. (1990) Highways in the river environment. Report no. FHWA-HI-90-016, National Highways Institute/Federal Highways Administration, McLean, VA.

Richardson, E.V., Harrison, L.J., Richardson, J.R. and Davis, S.R. (1993) *Evaluating Scour at Bridges* 2nd edn. Publication no. FHWA-IP-90-017, Hydraulic Engineering Circular No. 18, National Highways Institute/Federal Highways Administration (FHWA), McLean, VA.

Sheppard, D.M. (1993) Bridge scour in tidal waters. *Transportation Research Record 1420*, Transportation Research Board, Washington DC, pp. 1–5.

Shields, A. (1936) Anwendung der ähnlichkeitsmechanic und turbulenzforschung auf die geschiebebewegung. Mitteil, PVWES, Berlin, no. 26.

Simons, D.B. and Senturk, F. (1976) *Sediment Transport Technology*, Water Resources Publication, Fort Collins, CO.

Stephenson, D. (1979) *Rockfill in Hydraulic Engineering*, Elsevier, Amsterdam.

Sundborg, A. (1956) The river Klaralven: a study of fluvial processes. *Geogr. Annlr.*, **38**, 125–316.

Trent, R.E. and Brown, S.A. (1984) An overview of factors affecting river stability, in *Transportation Research Record 950*, Second Bridge Engineering Conference,

Vol. 2, Transportation Research Board/National Research Council, Washington DC, pp. 156–163.

Tyagi, A.K. (1989) Scour around bridge piers in Oklahoma streams. Final Summary Report, No. 89-1. Oklahoma State University, Stillwater.

Appendix A
Hydrodynamic forces on bridges

A1 Hydrodynamic forces on piers

These are the drag force (F_D kN) in the direction of flow and the lift force (F_L kN) perpendicular to it. According to Apelt and Isaacs (1968):

$$\text{drag force, } F_D = \frac{C_D \rho V^2 YL}{2000} \text{ kN} \tag{A1}$$

$$\text{lift force, } F_L = \frac{C_L \rho V^2 YL}{2000} \text{ kN} \tag{A2}$$

where C_D and C_L are the dimensionless coefficients of drag and lift respectively, ρ is the density of the water (kg/m^3), V is the approach flow velocity (m/s), Y is the depth upstream of the pier (m), and L is either the length of the pier in the direction of flow or the diameter of a single cylindrical pier (m).

The hydrodynamic forces on piers are usually small (e.g. compared with ship impact), which is often convenient because C_D and C_L depend upon factors such as the shape and spacing of the piers, the angle of attack, and the Reynolds number of the flow (Apelt and Isaacs, 1968; Farraday and Charlton, 1983). Note that (unlike coefficients of discharge) C_D and C_L can have values above 1.0. Very approximately, C_D values around 0.2–0.5 may be typical for some pier shapes pointing into the flow, but rise with the angle of attack to somewhere around 1.0–2.0. Since equations A1 and A2 are the same apart from the coefficients, a blunt pier, which would be expected to experience a larger drag than lift force, would have a higher C_D value than C_L, and vice versa for aerofoil shapes.

A2 Hydrodynamic force on submerged superstructures

Hydraulic Engineering Circular 20 of the US Federal Highways Authority (FHWA, 1991) gave the drag force per metre length (F_D kN/m) of a submerged or partially submerged bridge deck as

$$F_D = \frac{C_d \rho H V^2}{2000} \text{ kN/m} \tag{A3}$$

where C_d is the dimensionless coefficient of drag, which has a suggested value of between 2.0 and 2.2, H is the depth of submergence (m), and the other variables are as above.

A3 Ice forces

Neill (1973) suggested that piers with semicircular noses in plan and slightly inclined inwards to the vertical (Fig. 1.15) are effective in discouraging ice accumulations. The worst-case scenario may be large sheets of hard ice hitting the piers. Unfortunately the forces generated depend upon the type and strength of the ice, and how the ice fails (e.g. crushing, splitting, shearing, bending). Farraday and Charlton (1983), the American Association of State and Transport Officials (AASHTO, 1993) and the UK's Highways Agency (1994) presented an equation for the horizontal force (F_H kN) on a pier having 'substantial mass and dimensions':

$$F_H = C_n\, s_i\, t_i\, b_P\, (C_P) \text{ kN} \tag{A4}$$

where C_n is a coefficient for the inward inclination of the nose (0–15° = 1.0; 15–30° = 0.75; 30–45° = 0.5), s_i is the strength of the ice (between 700 and 2800 kN/m^2: the low value represents disintegrating melting ice and the high value major ice flows with freezing temperatures), t_i is the thickness of the ice in contact (m), and b_P is the width of the pier (m). The value of C_n may be modified according to pier width or pile diameter and ice thickness by multiplying by another coefficient (C_P), which ranges in value from 1.8 to 0.8 for b_P/t_i = 0.5 and ≥ 4.0 respectively (Canadian Standards Association, 1978).

A4 Debris forces

Australian specifications recommend allowing for a 2 tonne log travelling at the normal stream velocity being arrested within 150 mm and 75 mm for column type and solid type concrete piers respectively. However, according to the UK's Highways Agency (1994), 3 tonne logs travelling at almost 4.5 m/s have been observed. The average collision force (F kN) on the pier is

$$F = \frac{MV^2}{2d} \text{ kN} \tag{A5}$$

where M is the mass of the moving body (tonnes), V is its velocity (m/s), and d is the distance before it comes to rest (m). According to Farraday and Charlton (1983) some UK engineers assume a 10 tonne mass being arrested in 75 mm.

Forces can also be generated on the pier as a result of the flow impacting on debris trapped against the pier or across the waterway opening. Australian guidelines suggest calculating the hydrodynamic force on a minimum depth of 1.2 m of debris over a width of half the sum of the spans adjacent to the pier up to a maximum of 21 m. The following equation gives the force (F kN) due to trapped debris of area A (m^2) being hit by a flow with an approach velocity, V m/s:

$$F = 0.517\ V^2 A \text{ kN} \qquad \text{(A6)}$$

References

AASHTO (1993) *Standard Specification for Highway Bridges,* American Association of State Highway and Transportation Officials.

Apelt, C.J. and Isaacs, L.T. (1968) Bridge piers – hydrodynamic force coefficients. *Proceedings of the American Society of Civil Engineers, Journal of the Hydraulics Division,* 94(HYI), January, 17–30.

Canadian Standards Association (1978) *Design of Highway Bridges.* National Standards of Canada, CAN 3-56-M78, Rexdale, Ontario.

Farraday, R.V. and Charlton, F.G. (1983) *Hydraulic Factors in Bridge Design,* Hydraulics Research Station, Wallingford, England.

Federal Highways Administration/US Department of Transportation (1991) *Stream Stability at Highway Structures.* Report number FHWA-IP-90-014, Hydraulic Engineering Circular 20.

Highways Agency (1994) *Design Manual for Road Bridges,* Vol. 1, Section 3, Part 6, BA 59/94, *The Design of Highway Bridges for Hydraulic Action,* HMSO, London.

Neill, C.R. (ed.) (1973) *Guide to Bridge Hydraulics,* Roads and Transportation Association of Canada/University of Toronto Press, Toronto.

Appendix B
Some alternative equations for local scour

See Jones (1984), Melville (1988) and Richardson *et al.* (1993) for further details.

B1 Pier scour

All equations have to be used with the appropriate adjustment factors to allow for pier shape and orientation to the flow. The equations below are for square-nosed piers.

The Laursen–Toch equation is

live-bed scour $$d_{SP} = 1.5\, b_P^{\;0.7}\, Y^{0.3} \tag{B1}$$

clear-water scour $$d_{SP} = 1.35\, b_P^{\;0.7}\, Y^{0.3} \tag{B2}$$

where d_{SP} is the pier scour depth (m), b_P is the width of the pier (m) measured across the channel, and Y is the depth of the approach flow (m). Figure 8.20 can be used to adjust for angle of attack.

The Shen II equation is

$$d_{SP} = 3.4\, b_P^{\;0.67}\, F^{0.67}\, Y^{0.33} \tag{B3}$$

where F is the Froude number of the approach flow.

The Laursen–Toch equation assumes that flow depth is the most important factor in determining the depth of scour, whereas the Shen II equation assumes that velocity is important by including the Froude number.

Example B1

Using the data for the main channel in Example 8.5, which involved live-bed scour with no skew, with square-nosed piers

$Y = 2.4\,\text{m}$, $b_P = 1.2\,\text{m}$, $V = 2.92\,\text{m/s}$, giving $F = 2.92/(9.81 \times 2.4)^{1/2} = 0.60$.

Laursen–Toch: $d_{SP} = 1.5\, b_P^{\;0.7}\, Y^{0.3} = 1.5 \times 1.2^{0.7} \times 2.4^{0.3} = 2.22\,\text{m}$

Shen II: $d_{SP} = 3.4\, b_P^{0.67}\, F^{0.67}\, Y^{0.33} = 3.4 \times 1.2^{0.67} \times 0.60^{0.67} \times 2.4^{0.33}$
$= 3.64\,\text{m}$

CSU: $d_{SP} = 2.0\, Y\, K_{1P}\, K_{2P}\, K_{3P}\, (b_P/Y)^{0.65}\, F^{0.43}$ (8.14)
$= 2.0 \times 2.4 \times 1.0 \times 1.1 \times 1.1 \times (1.2/2.4)^{0.65}$
$\times\, 0.60^{0.43} = 2.97\,\text{m}$

For the CSU equation it is assumed the bed is plane ($K_{3P} = 1.1$).

B2 Abutment scour

All equations have to be used with the appropriate adjustment factors to allow for abutment shape and orientation to the flow. These equations tend to assume that there is no overbank flow, so that abutment or embankment length is measured from the side of the main channel.

The following simplified Laursen equation will be very conservative for wingwall and spillthrough abutments:

live-bed scour $$d_{SA} = 1.57\,(YL)^{1/2} \qquad \text{(B4)}$$

clear-water scour $$d_{SA} = 1.89\,(YL)^{1/2} \qquad \text{(B5)}$$

where d_{SA} is the abutment scour depth (m), Y is the depth of the approach flow (m), and L is the length by which the abutment protrudes into the main channel measured perpendicularly to the bank.

For live-bed scour at spillthrough abutments the equation attributed to Liu *et al.* (1961) is

$$d_{SA} = 1.1\, Y^{0.6}\, L^{0.4}\, F^{0.33} \qquad \text{(B6)}$$

where the variables are as above and L is defined as the abutment and embankment length measured at the water surface normal to the side of the channel from where the design flood hits the bank to the outer edge of the abutment in the opening.

For live-bed scour at wingwall or other vertical-wall abutments:

$$d_{SA} = 2.15\, Y^{0.6}\, L^{0.4}\, F^{0.33} \qquad \text{(B7)}$$

Example B2

Using the data in Example 8.1 where $B = 7.5\,\text{m}$, $b = 4.0\,\text{m}$, $V = 0.92\,\text{m/s}$ and $Y = 1.0\,\text{m}$, assuming no skew, spillthrough abutments but assuming live-bed scour, then

$L = (7.5 - 4.0)/2 = 1.75\,\text{m}$.

$F = 0.92/(9.81 \times 1.0)^{1/2} = 0.29$.

Laursen: $d_{SA} = 1.57\,(YL)^{1/2} = 1.57 \times (1.0 \times 1.75)^{1/2} = 2.08\,\text{m}$.

Liu *et al.*: $d_{SA} = 1.1\ Y^{0.6}\ L^{0.4}\ F^{0.33} = 1.1 \times 1.0^{0.6} \times 1.75^{0.4} \times 0.29^{0.33}$
$= 0.91\,\text{m}.$

Froehlich: $d_{SA} = Y + 2.27\ Y\ K_{1A}\ K_{2A}\ (L/Y)^{0.43}\ F^{0.61}$ (8.17)
$= 1.0 + 2.27 \times 1.0 \times 0.55 \times 1.0 \times (1.75/1.0)^{0.43}$
$\times\ 0.29^{0.61} = 1.75\,\text{m}.$

References

Jones, S.J. (1984) Comparison of prediction equations for bridge pier and abutment scour, in *Transportation and Research Record 950*, Second Bridge Engineering Conference, Vol. 2, Transportation Research Board/National Research Council, Washington DC, pp 202–209.

Liu, H.K., Chang, F.M. and Skinner, M.M. (1961) Effect of bridge constriction on scour and backwater. Engineering Research Center, Colarado State University, CER 60 KHL22.

Melville, B.W. (1988) Scour at bridge sides, in *Civil Engineering Practice, 2/Hydraulics/Mechanics* (eds P.N. Cheremisinoff, N.P. Cheremisinoff, and S.L. Cheng), Technomic Publishing Company, Lancaster, PA, pp 327–362.

Richardson, E.V., Harrison, L.J., Richardson, J.R. and Davis, S.R. (1993) Evaluating scour at bridges (2nd edn), publication no. FHWA-IP-90-017, Hydraulic Engineering Circular No. 18, National Highways Institute/Federal Highways Administration (FHWA), McLean, VA.

Index

Milton Keynes UK
Ingram Content Group UK Ltd.
UKHW021819071024
449327UK00021B/1357